AF581914

Shoreline Dynamics and Beach Erosion

Shoreline Dynamics and Beach Erosion

Editors

Gianluigi Di Paola
Germán Rodríguez
Carmen M. Rosskopf

MDPI • Basel • Beijing • Wuhan • Barcelona • Belgrade • Manchester • Tokyo • Cluj • Tianjin

Editors
Gianluigi Di Paola
University of Molise
Italy

Germán Rodríguez
University of Las Palmas de Gran Canaria
Spain

Carmen M. Rosskopf
University of Molise
Italy

Editorial Office
MDPI
St. Alban-Anlage 66
4052 Basel, Switzerland

This is a reprint of articles from the Special Issue published online in the open access journal *Geosciences* (ISSN 2076-3263) (available at: https://www.mdpi.com/journal/geosciences/special_issues/Beach_Erosion).

For citation purposes, cite each article independently as indicated on the article page online and as indicated below:

LastName, A.A.; LastName, B.B.; LastName, C.C. Article Title. *Journal Name* **Year**, *Volume Number*, Page Range.

ISBN 978-3-0365-7438-7 (Hbk)
ISBN 978-3-0365-7439-4 (PDF)

Cover image courtesy of Carmen M. Rosskopf

Contents

MDPI

Editorial

Shoreline Dynamics and Beach Erosion

Gianluigi Di Paola [1,*], Germán Rodríguez [2] and Carmen M. Rosskopf [1]

1 Department of Biosciences and Territory, University of Molise, I-86090 Pesche, Italy
2 Departamento de Física, Institute of Environmental Studies and Natural Resources (iUNAT), Universidad de Las Palmas de Gran Canaria, E-35001 Las Palmas de Gran Canaria, Spain
* Correspondence: gianluigi.dipaola@unimol.it

Abstract: Coasts are highly dynamic and geomorphologic complex systems that evolve under the increasing pressure of climate change and anthropogenic activities, having direct or indirect impacts on the coastal environment. Among the major adverse effects, coastal erosion represents one of the most pressing global issues, especially in flat and low-lying coastal areas that appear to be particularly susceptible to beach erosion and related shoreline retreat. This Special Issue collects a set of twelve papers on "Shoreline Dynamics and Beach Erosion". Of course, this collection of papers does not cover all the broad number of topics concerning the dynamics and spatial-temporal evolution of shorelines and beach systems, but, in our opinion, they contribute to the growing body of knowledge. Coastal systems of variable complexity located in different geographic and climatic contexts are investigated from various points of view by using multi- and interdisciplinary approaches, as well as new experimental ones. The major topics covered concern the morphodynamics and hydrodynamics of coastal systems, the driving factors of coastal erosion, and the use of models/indexes to study coastal vulnerability and the mitigation of human/natural pressures affecting coastal ecosystems.

Keywords: morpho-sedimentological characterization; coastal system analysis and modeling; hydro-meteorological extremes; sea level rise; coastal hydrodynamics; beach erosion drivers; coastal susceptibility and risk; coastal dune and beach management; coastline defense and anthropization

Citation: Di Paola, G.; Rodríguez, G.; Rosskopf, C.M. Shoreline Dynamics and Beach Erosion. *Geosciences* **2023**, *13*, 74. https://doi.org/10.3390/geosciences13030074

Received: 19 January 2023
Accepted: 15 February 2023
Published: 7 March 2023

1. Editorial for the Special Issue Shoreline Dynamics and Beach Erosion

The coastal zone is a unique physical space including the transition between the land and sea, whose surface area totals approximately 5% of the earth's surface [1]. However, this percentage and the related features change based on the definitions used and the related spatial boundaries attributed to them, which can vary depending on the way they are studied, exploited, and/or managed (e.g., [2]).

Although the coastal zone represents only a narrow strip of the Earth's surface, there is no doubt about its great socio-economic and ecological importance, as it provides a wide range of services of fundamental importance to human well-being, health, and subsistence [3–5]. The high socio-economic and strategic value of coastal areas explain why around two-thirds of the world's population live within 100 km from the coastline [6] and about 17% in low-elevation coastal zones, i.e., in low-lying areas that are less than 10 m above sea level [7,8]. In particular, sandy coasts are largely preferred as places for living and leisure activities, thus playing a fundamental role in the lucrative and growing tourism market, providing easy access to the sea [9,10].

From a dynamic perspective, the coastal zone represents one of the most energetic environments on Earth. Although the physical processes governing coastal dynamics show moderate and reasonably predictable behaviors most of the time, forcing processes can undergo unexpected abrupt and rapid changes leading to extreme and dangerous events. In particular, coastal storms, during which the wave energy reaches values significantly higher than those observed under average conditions, are among the most damaging natural events, primarily due to the high number of victims, but also to huge monetary losses (e.g., [11]).

The occurrence of storm events is particularly important along open, low-lying coastlines where beach erosion is one of the most common impacts, often causing the loss of land of high socio-economic and natural value (e.g., [10,12–14] and references therein). To succeed in planning and developing resilient and sustainable coastal built environments [15], it is crucial that coastal planners, engineers, and decision makers are aware of the fact that coastal areas are highly dynamic and hazard-prone zones.

In the light of the above, the progressive expansion of our knowledge on coastal morphodynamics and shoreline evolution and the modelling of shoreline changes and beach erosion in response to natural and anthropogenic stressors are of enormous relevance. Progress in this direction requires advances in various branches of science and technology, including techniques used for data acquisition, processing, and analysis, as well as the development of efficient numerical procedures for the computational implementation of theoretical models based on detailed studies and monitoring the activities of investigated coastal systems.

2. Special Issue

This Special Issue collects a set of twelve contributions authored by forty researchers on the topic of "Shoreline Dynamics and Beach Erosion". Of course, this collection of papers does not cover the entire range of subjects concerning the dynamics and spatial-temporal evolution of shorelines and beach systems, but, in our opinion, it contributes to the growing body of knowledge. In particular, these papers address the study of coastal systems of variable complexity located in different geographic areas (Figure 1) and climatic contexts from various points of view by using multi- and interdisciplinary approaches and providing new experimental data and methodologies.

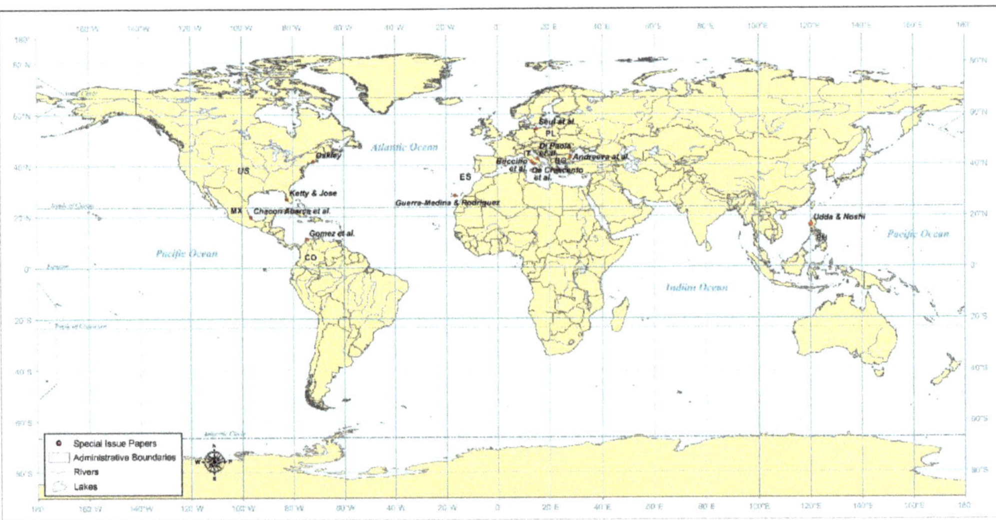

Figure 1. Geographical distribution of the contributions included in the Special Issue.

We hope that the papers of this issue serve to foster additional research, stimulate scientific discussion, and promote reflections on how to achieve a better understanding of the coastal systems morphodynamics, or even present research opportunities in this field, contributing, in the end, to the safeguarding of coastal zones and beaches. From an operational point of view, we expect that their results may contribute to the development of pragmatic approaches through which the competent institutions and administrations can implement intervention strategies to eliminate, reduce, and/or compensate coastal erosion, as well as support the resilience and sustainable development of coastal areas.

The contributions included in this issue refer to different continents, from the Americas to Africa, Europe, and South East Asia (Figure 1). The review paper written by Pranzini and William [16], which deals with theoretical concepts and, therefore, does not refer to a specific geographic area, represents the only exception.

The papers have been categorized into three groups according to the major research topics addressed, as presented in Table 1.

Table 1. Grouping of contributions according to major research topics covered.

	Main Topics	Papers
Group 1	Morphodynamics and hydrodynamics of coastal systems	Pranzini and William (2021) [16] Andreeva et al. (2021) [17] Uda and Noshi (2021) [18] Seul et al. (2020) [19] Buccino et al. (2021) [20]
Group 2	Driving factors of coastal erosion in different geomorphological contexts	Gomez et al. (2021) [21] Oakley (2021) [22] Kelly and Jose (2021) [23] Guerra-Medina and Rodríguez (2021) [24]
Group 3	Models/indexes for the assessment of vulnerability and risk aspects of coastal systems, and the mitigation of human/natural pressures on coastal ecosystems	Di Crescenzo et al. (2021) [25] Di Paola et al. (2022) [26] Chacón Abarca et al. (2021) [27]

A first group of papers (Group 1, Table 1) focuses on the morphodynamics and hydrodynamics of coastal systems. In detail, the papers written by Andreeva et al. [17] and Uda and Noshi [18] highlight that coastal morphodynamics assessed by analyzing the formation, location, and movement of coastal bars and sand spits strongly depends on the wave climate that controls the erosion and movement of sediments along the coast. Seul et al. [19], based on the petrographic analysis and characterization of beach sediments, have created an index that aids in understanding both the nature and origin of sediments, as well as the characteristics of wave climate responsible for their transport. The paper by Buccino et al. [20] illustrates how numerical modelling can be used for investigating the relationships between incident wave characteristics and beach shape. Finally, the review paper by Pranzini and William [16], which uses examples from the literature to analyze some of the major processes responsible for the evolution of a beach, highlights that the use of the term "beach equilibrium" is not entirely appropriate, since the sedimentary budget depends on numerous non-feedback-regulated factors.

A second consistent group of papers (Group 2, Table 1) analyzes some of the major driving factors of coastal erosion in different geographical and geomorphological contexts. In detail, these papers deal with the effects of subsidence [21], storm surges [22], and hurricanes [23] on the studied coastal systems, which adapt to new morphological conditions in different times. Furthermore, the paper by Guerra-Medina and Rodríguez [24] highlights how such control factors can decisively influence the economic and social decisions of a community and, especially, tourism activities.

Finally, the third group of papers (Group 3, Table 1) focuses on the development and application of models/indexes suitable for coastal erosion vulnerability assessments for developing mitigation strategies for coastal ecosystem in relation to human and natural pressures. In detail, the papers by Di Crescenzo et al. [25] and Di Paola et al. [26] illustrate the identification and application of indexes that allowed them to assess the susceptibility of the coastline to undergo erosion, respectively, in high rocky coast and sandy beach contexts. The paper by Chacón Abarca et al. [27] deals with the evaluation of a model to help the community reducing the negative effects on ecosystems caused by physical and human changes in lagoon coastal environments. It illustrates a developed tool that considers the inter-relationships between natural systems and the factors inducing alterations in the coastal lagoon environment, providing insights into human actions that can reduce these negative consequences on ecosystems.

Funding: This research received no external funding.

Data Availability Statement: No new data were created or analyzed in this study. Data sharing is not applicable to this article.

Acknowledgments: We thank our informants and supporting staff for their contribution to the Special Issue.

Conflicts of Interest: The authors declare no conflict of interest.

References

1. Hossain, M.S.; Gain, A.K.; Rogers, K.G. Sustainable coastal social-ecological systems: How do we define "coastal"? *Int. J. Sustain. Dev. World Ecol.* **2020**, *27*, 577–582. [CrossRef]
2. Masselink, G.; Hughes, M.G. *Introduction to Coastal Processes and Geomorphology*; Arnold: London, UK, 2004; 354p.
3. Costanza, R.; d'Arge, R.; de Groot, R.; Farber, S.; Grasso, M.; Hannon, B.; Limburg, K.; Naeem, S.; O'Neill, R.V.; Paruelo, J.; et al. The value of the world's ecosystem services and natural capital. *Nature* **1997**, *387*, 253–260. [CrossRef]
4. Martínez, M.L.; Intralawan, A.; Vázquez, G.; Pérez-Maqueo, O.; Sutton, P.; Landgrave, R. The coasts of our world: Ecological, economic and social importance. *Ecol. Econ.* **2007**, *63*, 254–272. [CrossRef]
5. Costanza, R.; Groot, R.; Sutton, P.; van der Ploeg, S.; Anderson, S.J.; Kubiszewski, I.; Farber, S.; Turner, R.K. Changes in the global value of ecosystem services. *Glob. Environ. Change* **2014**, *26*, 152–158. [CrossRef]
6. Maul, G.A.; Duedall, I.W. Demography of Coastal Populations. In *Encyclopedia of Coastal Science*; Encyclopedia of Earth Sciences Series; Finkl, C.W., Makowski, C., Eds.; Springer: Cham, Switzerland, 2019. [CrossRef]
7. Agardy, T.; Alder, J.; Dayton, P.; Curran, S.; Kitchingman, A.; Wilson, M.; Catenazzi, A.; Restrepo, J.; Birkeland, C.; Blaber, S.; et al. Coastal systems. In *Millennium Ecosystem Assessment: Ecosystems & Human Well-Being, Volume 1: Current State and Trends*; Reid, W., Ed.; Island Press: Washington, DC, USA, 2005; pp. 513–549.
8. McGranahan, G.; Balk, D.; Anderson, B. The rising tide: Assessing the risks of climate change and human settlements in low elevation coastal zones. *Environ. Urban.* **2007**, *19*, 17–37. [CrossRef]
9. Williams, A.; Micallef, A. *Beach Management: Principles and Practice*, 1st ed.; Earthscan: London, UK, 2009.
10. Houston, J.R. The economic value of beaches—A 2013 update. *Shore Beach* **2013**, *81*, 3–10.
11. Harley, M.D. Coastal Storm Definition. In *Coastal Storms: Processes and Impacts*; Ciavola, P., Coco, G., Eds.; Wiley-Blackwell: Hoboken, NJ, USA, 2017; pp. 1–21. [CrossRef]
12. Alexandrakis, G.; Manasakis, C.; Kampanis, N.A. Valuating the effects of beach erosion to tourism revenue. A management perspective. *Ocean. Coast. Manag.* **2015**, *11*, 1–11. [CrossRef]
13. Rosskopf, C.M.; Di Paola, G.; Atkinson, D.E.; Rodríguez, G.; Walker, I.J. Recent shoreline evolution and beach erosion along the central Adriatic coast of Italy: The case of Molise region. *J. Coast. Conserv.* **2018**, *22*, 879–895. [CrossRef]
14. Flayou, L.; Snoussi, M.; Raji, O. Evaluation of the economic costs of beach erosion due to the loss of the recreational services of sandy beaches—The case of Tetouan coast (Morocco). *J. Afr. Sci.* **2021**, *182*, 104257. [CrossRef]
15. Burningham, H.; Fernandez-Nunez, M. Shoreline change analysis. In *Sandy Beach Morphodynamics*; Jackson, D.W.T., Short, A.D., Eds.; Elsevier: Amsterdam, The Netherlands, 2020; pp. 439–460.
16. Pranzini, E.; Williams, A.T. The Equilibrium Concept, or … (Mis)concept in Beaches. *Geosciences* **2021**, *11*, 59. [CrossRef]
17. Andreeva, N.; Saprykina, Y.; Valchev, N.; Eftimova, P.; Kuznetsov, S. Influence of Wave Climate on Intra and Inter-Annual Nearshore Bar Dynamics for a Sandy Beach. *Geosciences* **2021**, *11*, 206. [CrossRef]
18. Uda, T.; Noshi, Y. Recent Shoreline Changes Due to High-Angle Wave Instability along the East Coast of Lingayen Gulf in the Philippines. *Geosciences* **2021**, *11*, 144. [CrossRef]
19. Seul, C.; Bednarek, R.; Kozłowski, T.; Maciąg, Ł. Beach Gravels as a Potential Lithostatistical Indicator of Marine Coastal Dynamics: The Pogorzelica–Dziwnów (Western Pomerania, Baltic Sea, Poland) Case Study. *Geosciences* **2020**, *10*, 367. [CrossRef]
20. Buccino, M.; Tuozzo, S.; Ciccaglione, M.C.; Calabrese, M. Predicting Crenulate Bay Profiles from Wave Fronts: Numerical Experiments and Empirical Formulae. *Geosciences* **2021**, *11*, 208. [CrossRef]
21. Gómez, J.F.; Kwoll, E.; Walker, I.J.; Shirzaei, M. Vertical Land Motion as a Driver of Coastline Changes on a Deltaic System in the Colombian Caribbean. *Geosciences* **2021**, *11*, 300. [CrossRef]
22. Oakley, B.A. Storm Driven Migration of the Napatree Barrier, Rhode Island, USA. *Geosciences* **2021**, *11*, 330. [CrossRef]
23. Kelly, E.W.; Jose, F. Geomorphologic Recovery of North Captiva Island from the Landfall of Hurricane Charley in 2004. *Geosciences* **2021**, *11*, 358. [CrossRef]
24. Guerra-Medina, D.; Rodríguez, G. Spatiotemporal Variability of Extreme Wave Storms in a Beach Tourism Destination Area. *Geosciences* **2021**, *11*, 237. [CrossRef]
25. Di Crescenzo, G.; Santangelo, N.; Santo, A.; Valente, E. Geomorphological Approach to Cliff Instability in Volcanic Slopes: A Case Study from the Gulf of Naples (Southern Italy). *Geosciences* **2021**, *11*, 289. [CrossRef]

26. Di Paola, G.; Minervino Amodio, A.; Dilauro, G.; Rodriguez, G.; Rosskopf, C.M. Shoreline Evolution and Erosion Vulnerability Assessment along the Central Adriatic Coast with the Contribution of UAV Beach Monitoring. *Geosciences* **2022**, *12*, 353. [CrossRef]
27. Chacón Abarca, S.; Chávez, V.; Silva, R.; Martínez, M.L.; Anfuso, G. Understanding the Dynamics of a Coastal Lagoon: Drivers, Exchanges, State of the Environment, Consequences and Responses. *Geosciences* **2021**, *11*, 301. [CrossRef]

Review

The Equilibrium Concept, or . . . (Mis)concept in Beaches

Enzo Pranzini [1,*] and Allan T. Williams [2]

1 Dipartimento di Scienze della Terra, Università degli Studi di Firenze, 50121 Firenze, Italy
2 Faculty of Architecture, Computing and Engineering, Trinity St. David, University of Wales, Swansea SA1 5HF, Wales, UK; allanwilliams512@outlook.com
* Correspondence: enzo.pranzini@unifi.it

Abstract: Beaches, as deposits of unconsolidated material at the land/water interface, are open systems where input and output items constitute the sediment budget. Beach evolution depends on the difference between the input/output to the system; if positive the beach advances, if negative the beach retreats. Is it possible that this difference is zero and the beach is stable? The various processes responsible for sediment input and output in any beach system are here considered by taking examples from the literature. Results show that this can involve movement of a volume of sediments ranging from few, to over a million cubic meters per year, with figures continuously changing so that the statistical possibility for the budget being equal can be considered zero. This can be attributed to the fact that very few processes are feedback-regulated, which is the only possibility for a natural system to be in equilibrium. Usage of the term "beach equilibrium" must be reconsidered and used with great caution.

Keywords: beach sediment budget; coastal dynamics; beach evolution; coastal morphology; feedback processes

Citation: Pranzini, E.; Williams, A.T The Equilibrium Concept, or . . . (Mis)concept in Beaches. *Geosciences* **2021**, *11*, 59. https://doi.org/10.3390/geosciences11020059

Academic Editors: Jesus Martinez-Frias and Gianluigi Di Paola

Received: 30 December 2020
Accepted: 25 January 2021
Published: 29 January 2021

1. Introduction

Equilibrium is a term having many different meanings. In physics, it describes the average condition of a system, as measured through one of its elements or attributes over a specific period of time. However, the geomorphological concept of equilibrium has many confusing meanings [1], e.g., see below plus, quasi-equilibrium, time independence, and its semantics have been seemingly lost in a vast array of papers, e.g., the authors of [2–5] showed that catastrophe theory even suggests that any system might have numerous equilibrium states.

Equilibrium theory arises from Newton's laws of motion ($F = MLT^{-2}$) and refers to where an object's velocity is constant (no acceleration), or if an object is stationary (at rest) and any force acting on it has its vector sum as zero, i.e., force and reaction are balanced and the system's properties are unchanged over time. This is static equilibrium. Dean ([6], p.399), referring to beach dynamic equilibrium, defined it as "the tendency for beach geometry to fluctuate about an equilibrium, which also changed through time, but much more slowly." He added that our ability to predict quantitatively these changes is likely to remain poor for the coming decades.

A system can also exhibit other states:

- Steady: where the average system's condition trajectory is unchanged through time.
- Stable: where a system has a tendency to return to the same equilibrium state once it experiences disturbance.
- Unstable: where a system returns to a new equilibrium, post any disturbance.
- Static: where reactive forces and moments must balance the externally applied forces and moments.
- Metastable: where additional energy must be introduced before an object can reach true stability.

Further readings about these states may be found in [7,8].

Additionally, Renwick [9] has argued that if one has equilibrium (absence of a discernible trend over the period under observation), then its inverse is a disequilibrium state, i.e., a landform that tends towards equilibrium, but has not had sufficient time to reach it, i.e., a decaying state. The end result is confusion. It is not only with respect to beaches that the term is frequently used; care should be taken in usage of the term, as it has been strewn around the literature in many guises, e.g., equilibrium shoreline evolution models [10]; equilibrium types in planforms of bay beaches [11]. The latter's findings ranged from "dynamic" planforms where there is a constant sediment throughput to maintain beach stability, to a "static" position, as a result of a reduced/ceasing of an updrift sediment supply. Jackson and Cooper ([11], p.112), further commented that "Static equilibrium models represented a convenient yardstick with which to ascertain a particular shoreline's current stability status." However, they urged caution to the approach in identifying equilibrium and non-equilibrium shorelines, mentioning reliance on contemporary beach morphometrics as an input, and omission of other dynamic variables (secondary wave motions, tidal, and river currents).

Its geomorphological origin can be traced to Gilbert's (1877) classic work on sediment flux at the drainage basin scale in the Henry Mountains, USA [12], later quantified by Ahnert [13], relating to rock erosion and resistance, the concept of a systems approach in geomorphology and negative feedback. It is a process-based approach, and most discussion of the term has been in reference to hydrological processes and concepts. Gilbert [12] argued that all streams worked towards a graded condition (dynamic equilibrium) where the net effect of the stream is neither erosion nor deposition. Willgoose et al. [14] expanded this arguing that within river catchments episodic fluctuations can occur, but over time in dynamic equilibrium, a balance exists between uplift and erosion.

Equilibrium in any system has to be time dependent but the particular attributes must be defined over a particular period, but how is the scale defined? A classic paper on the fundamental importance of scale was that of Schumm and Lichty [15]. They showed that input/output relationships could change within any timespan, leading to the concept of fast and slow variables. Beattie and Oppenheim [16], dealing with thermodynamics, argued that processes can produce different equilibrium states that take place so slowly that no discernible displacement in the presumed equilibrium state can detected. Is this true in geomorphology and particularly in a beach environment? Usage of the equilibrium term as being scale-dependent is perhaps questionable, as it depends on the observer's view of the term. Howard ([17], p.71) was of the opinion that, "Equilibrium systems are generally not applicable to physical systems that exhibit oscillatory, threshold, hysteretic, multivalued or strongly unpredictable responses to constant or slowly changing inputs." He further stated that a change of time/scale could make a system giving complex responses be predictable or in equilibrium. Bracken and Wainwright [18] even posed the question whether equilibrium was a myth or a metaphor, arguing for the latter? Is the concept testable? Richards [19] even suggested that it was impossible to measure equilibrium.

For beaches, which are the concern of this paper, equilibrium infers a lack of change in a system (a component set gathered into a whole), as inputs and outputs remain in balance. If changes do occur moving it to a new position (dynamic equilibrium), then feedbacks (an output causing system changes to the inputs) will allow for correction (Figure 1). Feedbacks can be positive, i.e., exacerbating the size of the systems normal elements through time, or negative that can dampen or reverse the size of the systems elements/attributes. It is assumed that self-regulation occurs. Any beach can be thought of as an open system, i.e., energy/matter relating to the sediment budget can arrive and depart and if these are in balance then equilibrium is assumed to exist [20], e.g., beach sediment arrival equating to sediment removal. When this is broken, e.g., not only by human intervention (damming rivers, coastal structures, etc.), but also by the natural system itself, as occurred before the appearance of man on Earth, it is argued that the system will change in order to bring it back to equilibrium.

If one analyses the above closely, three key points emerge:

- The number of items involved in the sediment budget is so high that is statistically impossible for the balance to close at zero.
- Input and output volumes for each item change continuously and even if equilibrium is reached, it lasts one moment.
- Almost all the processes are not feedback regulated, which is the only chance for natural systems to be stable.

Does this infer that beaches can never be stable?

Among the forms that make up the Earth's surface, beaches are certainly those subject to the most rapid variations. A storm, even a modest one, is enough for the beach to assume a different morphology. The shape and position of the shoreline varies, as does the slope of the swash zone, the seabed, the position, size, and number of submerged bars. Moreover, from a granulometric viewpoint, the changes are significant and frequent, e.g., a small hole dug on the shoreline allows one to see more or less coarse sand levels that have been deposited in the previous days in different marine weather conditions.

Taking into consideration the whole beach (emerged and submerged part), these variations, both morphological and granulometric, are mainly due to movement of sand to and from the shore, and the overall volume of sediments that make up the beach does not vary as consistently as the changes in its morphology might suggest. For a short time scale, the beach is subject to feedback processes that make it assume the form that best dissipates wave energy, and as this changes moment by moment, even the beach adapts and changes. However, in these short periods some sediment can enter/exit, but does this infer that a beach is stable? Ahnert's ([21], p.322) comment that, "the huge concentration of energy in littoral processes means that all artificial disturbances of the natural dynamic equilibrium (sic) on the seacoast have very rapid and often unforeseen effects," is particularly pertinent here.

Beach stability is a concept frequently present in research papers, official documents and the press. Generally, claiming that beaches are not stable relates to concern for coastal sectors experiencing erosion, as accretion, in most cases, is not a problem. For the press and general public out of the three possible conditions for a beach, i.e., accretion, erosion, and stability, the latter is frequently considered as the natural one and the two-remaining are attributed to disturbances governed by anthropogenic actions, although beaches appeared, disappeared, grew and reduced before the appearance of *Homo sapiens*.

Hourly, daily, and seasonally changes in dry beach width are recognized in every coast, due to tide, air pressure, and waves, but these are considered as "normal" oscillations around longer-term "stable" conditions. Their magnitude order can range from few centimeters in micro-tidal sheltered coasts, e.g., many Mediterranean beaches, to hundreds of meters in macro-tidal exposed environments, e.g., the Glamorgan Heritage Coast, Wales, UK. Many papers/books refer to the sweep zone or envelope: a large sweep zone delineating an unstable beach, a very narrow one a stable (equilibrium) beach.

Bogaert et al. [22] distinguished between behavior and evolution. However, it is not the shoreline position which determinates beach condition but the beach sediment budget, calculated for a selected coastal segment and extending from the first dune (or anthropogenic structure) to the depth of closure [23]. The problem of high-frequency beach morphological changes around a hypothetical equilibrium has frequently been resolved by referring to the concept of dynamic equilibrium [6], but this status cannot be attributed to a beach as it refers to a continuous process which does not change the shape, temperature, or chemical constitution of an object. In chemistry, dynamic equilibrium is when, e.g., substances transition between the reactants and products at equal rates, meaning there is no net change, and can only occur in reversible reactions, i.e., when the rate of the forward reaction is equal to the rate of the reverse reaction.

The equilibrium concept could be applied to pocket beach rotation [24], where grains eroded on one side are deposited on the other one, but also in this case the balance between input and output is questionable. What is under discussion here is the existence of that hypothetical equilibrium around which the beach should fluctuate. The approach to this problem in this paper starts with processes that form the beach sediment budget looks at how they can modify sediment input and output from the beach, considers whether related processes are feedback regulated or not, and, finally, to try to find a functional definition of beach stability (if any).

A stable beach is one that oscillates "slightly" around a mean, as against a high oscillatory component that can spill into an erosion/deposition mode, i.e., an unstable beach has the potential to enter this phase. The time factor must be long enough to dampen out any irregularities due to transient storm activity. Many studies rely on the estimation done via interviews by Bird [25] who found that over 70% of shorelines are retreating because of climate change related processes, and this trend is increasing [26]. If these shorelines are in an eroding mode, then they are by definition unstable [26]. It should be noted that the figure quoted should refer to beaches that have been studied, as all beaches in the world were not investigated. Alternatively, Luijendijk et al. [27], through machine learning and image processing techniques on 1.9 million historical Landsat images, carried out global scale studies concluding that 24% of the beaches are eroding at a rate >0.5 m/year. Applying the same method, 48% of world beaches resulted as stable, but within a range of ±0.5 m/year.

2. Beach Sediment Budget

Each stretch of beach, even if delimited by coastal structures, such as harbors, jetties, groins, and detached breakwaters, is not a closed system where the amount of initial free energy is less easily available as the system moves towards a state with maximum entropy [20], as it might appear from previous descriptions, but it is crossed by a significant flow of entering and exiting material, some entering and some exiting, all contributing to a beach sediment budget [28]. It comprises multiple input and output items, each of them directly influenced by several factors, which, in their turn, respond to other forces, and none of these is stable over the time. In any beach the volume of input and output items can be many orders of magnitude different, and continuously changing, to make it statistically impossible for the statement to be equal.

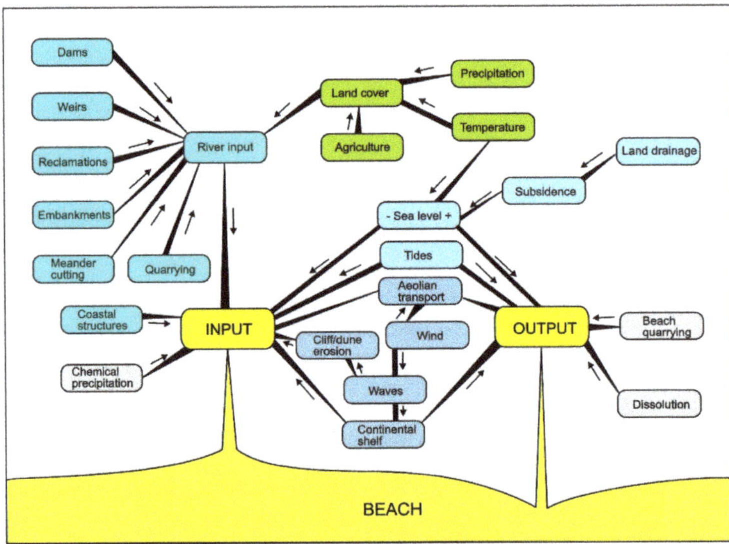

Figure 1. Schematic beach sediment budget Input and Output. Arrows show the direction of the process; connecting lines thickness increases and decreases depending on the effect (positive vs. negative) that each item has on the following one (modified from the work in [29]).

2.1. Riverine Input

Most world beaches owe their origin and survival to the sedimentary contribution from rivers (more than 90% according to Pethick [30] and 75% according to Best and Griggs [31] for the Santa Monica littoral cell). It is precisely due to the variations in river input that many of the shoreline oscillations that occurred in historical times are still in progress today. There are numerous natural and anthropogenic factors which, more or less directly, influence river input; some are easily identifiable, and quantification of their effects is rather simple, others are difficult to study and of uncertain effect.

2.1.1. Changes in Rainfall

Certainly, rainfall regime variations directly influence soil erosion rates [32] and modify transport capacities of rivers, so much so that an increase in rainfall should result in an increase in sedimentary input and a progradation of beaches. This, in general, can be considered true, but there are certainly cases in which the increase in rainfall determines an increase in vegetation cover which, in turn, reduces erodibility of the soils and surface runoff, so much so as to lead to a reduction of river input. A similar effect is most likely in semi-arid areas [33], where soil erosion rates are high despite low rainfall [34].

2.1.2. Variations in Vegetation Cover

Vegetation cover is also influenced by temperature (Figure 1); an increase in this, particularly in Mediterranean areas, can cause a reduction in vegetation with a consequent increase in erodibility. However, this temperature variation should lead to an upward shift of the tree vegetation limit in mountains [35] with a consequent reduction in erosion rates in the upper parts of catchment areas where, due to the greater relief, the production of sediments is greater [36].

Variations of this type certainly occurred in the Holocene and may have modified, even significantly, the sedimentary balance of some coasts. Unfortunately, studies of this nature are extremely scarce due to the objective difficulties of appreciating a phenomenon that is largely overshadowed by a much more relevant one that in turn is linked to variations of the vegetation cover induced by anthropic intervention. Forests cutting to convert land to pasture and agriculture have been recognized to be the main cause of coastal progradation in historical time [37–39].

Conversely, agricultural abandonment can lead to forest growth with a reduction of soil erosion [40–42], but where this favored collapse of terrace stone walls, as in many parts of the Mediterranean countries, this process increased [43].

2.1.3. Hydraulic Works

Among interventions carried out on rivers, it is not only deforestation that increases fluvial input; modifications of water courses and embankments construction have a similar effect. Cutting meanders, carried out to prevent alluvial plain flooding, imposes a shorter path, and therefore a greater slope, favoring sediment transport to the sea. From 1338 to 1771, the River Arno course from Pisa to the sea was shortened by approximately 5 km (from 12 to 7 km) by cutting three meanders, which almost halved its slope. On the same river, upstream and downstream from the town, embankments were built by the Romans together with reclamation works, in order to transform the marshy area into an agricultural area, as centurion footprints still show [44]. Together with the watershed deforestation this helped the river to build its delta and extend its course for 7.5 km (which reduces the slope). However, most anthropogenic interventions carried out on river channels has led to a reduction in sedimentary input: the construction of "diversions" to introduce the "slurries" into reclamation basins [41] and that of the bridles along river channels [45] have often caused a drastic reduction in the amount of sediment transported to the sea.

After reclamation, rivers can be redirected to the sea, enriching the sediment input, but the same can be done by diverting river course previously emptying into a lagoon in a natural way. This has been carried out with rivers on the western Veneto coast to prevent

Venice Lagoon siltation and expanding beaches near the new river mouth [46]. Additionally, weirs can be removed, releasing sediments they trapped [47], but more important, and well studied, is the effect of dams on riverine input to the coast [48].

From large dams, e.g., the Nile River at Aswan [49] and that of the Seven Gorges on the Yangtze River [50], to smaller ones, this impact has been well documented, e.g., rivers emptying into the Catania Gulf, Italy [51]. More than 100 $\times$ 10^9 tons of sediment has been sequestered in reservoirs constructed largely within the past 50 years [52]. Against this, several dam removals have been performed during the last few decades and channel evolution documented [53,54], but none had the tremendous impact on the coast as that of the Elwha River [55]. Here, a two years post-removal monitoring showed that 2.2 million m^3 of sand and gravel were deposited on the seafloor offshore at the river mouth [56]. However, each intervention does not have an instantaneous impact on the coast and its effect increases and reduces over time, never maintaining the same effect.

2.1.4. River Bed Quarrying

One of the most effective causes of reduction in sediment river input to the coast is riverbed extraction of aggregates [57]. This activity was conducted on an occasional basis and with modest means up to the last century and has become a real "industry" in recent decades, driven by the demand for construction materials for building and communication routes [58]. From river beds, each year materials have been removed with volumes far greater than the bed load of the rivers themselves, so much so that their level has been lowered by several meters, which has triggered serious problems of instability in banks and bridges. For example, ten times for the Po River [59], Italy; forty times according to for the Vembanad Lake catchments, India [60]. This operation was also favored by the fight against floods, as widening and deepening of river beds leads to an increase in the hydraulic variables. The fallout area of these interventions, the coastal strip, is too far, physically and politically, from populations affected by the floods and the fight against extractions has not yet been won everywhere. Damage caused by this activity is such that, even where mining has been banned for more than twenty years (e.g., Cecina River, Tuscany), the morphology of the rivers has not yet assumed the natural configuration and deep holes are still evident in their longitudinal and transverse profiles; actually, approximately 4 $\times$ 10^6 m^3 of aggregates was extracted on the 40 km terminal river course from the 1960s to early 1970s [61].

2.2. Coastal Processes

2.2.1. Beach and Dune Mining

Aggregates from beaches and dunes were traditionally used for construction in all coastal settlements, until its negative effect on coastal evolution was perceived; nevertheless, this activity is still, legally or illegally, carried out in many countries. Portobello beach, the 19th century fashionable watering place near Edinburgh, UK, had been the quartz rich sand quarry for glassware manufacturing from 1834 to mid-1930s, until all the white sand was lost, the promenade collapsed, and tourists disappeared [62]. At Poetto, the urban beach of Cagliari, Italy, approximately 2 million cubic meters of sediments were quarried on the beach, mostly for reconstruction of the town after WWII bombing, and this had been the main cause of the beach erosion [63]. In China, during the 1980s, 4000 million tons of beach sands was removed annually [57]. Dune mining along the 18 km of Monterey Bay from 1940 to 1984 was 128,000 m^3/year [64], i.e., approximately 7 m^3/m/year; when mines were closed the erosion rate decreased, but not significantly.

2.2.2. Relative Sea Level Change

Coastal plains have always been subject to subsidence, as sediments recently deposited are subject to compaction. This process is faster on river deltas, where deposition rates are higher, but can reach massive values where oil, gas, and water extraction is carried out. The high coastal erosion rate found at Ravenna, Italy, is mostly induced by oil and gas extraction both inland and offshore that produced a 40 mm/year subsidence in 1970–1977, the period of most intensive activity [65]. Satellite interferometry measured a value of 30 mm/year at the Gudao Oilfield, on the Yellow River Delta [66], and in the Piombino alluvial plain, natural subsidence rate shifted from 1 mm/year in the Holocene to the present 10 mm/year, as a consequence of water pumping for agriculture and a steel mill [67].

Sea level rise, as a cause of sediment loss, has its rationale in the Bruun Rule [68] which was firstly enthusiastically accepted, and later strongly rejected; it passed through several modifications, see, e.g., in [69–72], and is related to those sediments deposited near the depth of closure that subsequently exit the active beach following the "apparent deepening" caused by the rise in sea level. Although suggestions have been made that it is time to abandon the Bruun Rule [73], no operative alternatives have been proposed, but it is evident that if water depth increases, sediments moved to the limit of the active beach enter the "inactive" part of the profile. However, sea level rise can also induce sediment shifting to the coast, as occurred during the last 18,000 years, when barrier islands emerged and moved ashore [74,75], somehow contradicting the Bruun Rule. This input had been significant during the Holocene Sea Level Rise, but there have been no studies to investigate if it is still operating.

2.2.3. Aeolian Transport

Wind, with its direct effect on sediment erosion/transport/deposition, influences the "fine sediment" budget although wind drifted gravel has been documented by Zenkovich [76]. Aeolian processes are responsible both for input and output in the beach budget, although the former only has been generally considered, except for alongshore transport [77]. Offshore transport is evident from satellite images, but almost exclusively concerns silt-sized sediments coming from inland areas and not the beach itself. Inland transport has been documented in Neolithic times (burial of Skara Brea, Orkney [78]); historical times, e.g., abandonment of the Roman Via Julia, Wales after the year 1344 [79] and burial of the church at Skagen [80]; and in the present time, examples can be found in Spain [81], South Africa [82], and Australia [83].

2.2.4. Wave Energy

Waves influence the beach sediment budget and are the predominant process: waves and associated storm surges, move sediment longshore, offshore and onshore; the latter direction seldom produces sediment deficit unless sand is deposited in urbanized areas and not returned to the beach, e.g., if polluted by traffic [84]. Cross-shore [85] and long-shore [86] variability in sediment transport rate is well-known and documented. Net sediment long-shore transport ranges from few to millions cubic meters per year among the different coastal sectors. But the same magnitude order differences can be found locally at different times. Episodicity in long-shore sediment transport was studied by, e.g., Seymour et al. [87] who found that in one day sediment transport can be more than 600 times the mean daily net transport.

Shi-Leng and LiuTeh-Fu [88], studying a sector of the Mauritania coast, found that long-term variation in sediment transport follow a Gumbel distribution, but it is not necessary to wait for extreme events to have huge volumes of sediments moved along the coast: the net long-shore transport near Oregon Inlet (Outer Banks, NC, USA) is between one-half million and one million m^3/year [89]. Along the Mediterranean coast, a "protected sea environment" according to the Davies classification [90], a net sediment transport of more than 400,000 m^3 was calculated for a point near the harbor of Ashkelon, Israel [91].

With respect to long-shore transport, Raynor [92] introduced the term "cascade of uncertainties" involving the triple interaction of air, sea, and sediment into geomorphological literature. This is very pertinent when echoed in the concept of beach equilibrium. Pilkey and Cooper ([93], p.579) were of the opinion that "Sand volumes perhaps should be expressed as broad categories such as small, intermediate or large in recognition of the fact that meaningful determination of net annual transport of sand is probably impossible in this complex, dynamic and changing natural system." A case study for the Northwest coast of Portugal [94], showed that the annual littoral long-shore transport values exhibit a large variability with a maximum of 2.24 million m^3 $year^{-1}$, which exceeds the long-term mean magnitude by 105%, and a minimum, 108,000 m^3 $year^{-1}$, 10 times less than the mean value. Taborda et al. ([94], p.466) pointed out that, "The long-shore transport estimates reported in the literature for this coastal stretch were made through different techniques from cartographic comparison with mathematical modeling, whose results led to a wide range of values from 200,000 (Abecasis 1955) to 3.5 million m^3 $year^{-1}$ (Teixeira 1994), although most values converge toward a mean value of around 1 million m^3 $year^{-1}$."

Storms are known to have strongly modified the coast in specific periods of the Middle ages, e.g., the 1634 storm surge divided the German island of Strand in two parts, now known as Pellworm and Nordstrand islands [95]. In South Wales, UK, extreme storm events have been recorded in monastic records since the 3rd century culminating in the 14th and 15th centuries that saw much coastal erosion along with washovers. The storm events peaked in the 16th century, when they became especially ferocious causing much flooding and massive washover sand amounts formed the coastal dune fringe of South Wales [96,97]. Extreme events have been studied over a 6000-year-long period in Australia [98] surveying and dating beach ridges, showing the importance of the association of wave height and storm surge to build up those morphologies.

Climate change occurring during the last decades has modified storm frequency and intensity, e.g., along the Atlantic coastlines of Europe [99] and future scenarios forecast a further increase in storminess. This is an additional variable to the future beach sediment budget. Changes in wave directional distribution will modify long-shore transport direction and possibly invert its resultant direction, shifting convergence points, or completely modifying coastal cell geometry.

2.2.5. Biogenic Production and Chemical Precipitation, e.g., Inside Posidonia Prairies, Mangroves

Although most coastal sediments come from inland erosion, in some areas the beach comprises a consistent percentage of shell fragment and skeletons of marine animals. In Western Australia, these components are in the majority [100], but the same can be found in mid Latitude coasts, such as, Sardinia, where on the northern granite coast, the beach of Pelosa has more carbonate sand than quartz [101], or even in Scotland, where Coral Beach, Hebrides, has an abundance of calcareous seaweed [102]. Their production is influenced by water temperature, nutrients richness, and turbidity, all of which are extremely variable. Since historical times, this material was quarried to amend acid soils, e.g., in Scotland [103].

2.2.6. Hard Rock Coasts

Contrary to what one is led to believe, erosion of high coasts generally provides a very low sedimentary input (less than 5% according to Inman [104]), and those models of coastal evolution, which see the retreat of promontories and the filling in of the gulfs to the complete rectification of the coast, appear to be linked to a Davisian concept of coastal geomorphology by now largely outdated. The Cycle of Normal Erosion by Davies held sway for many years and hindered acceptance of the Gilbertian viewpoint that came to the fore again in the evolution of quantitative geography in the 1950s.

Nevertheless, rock coast erosion can be locally significant and constitute the main sediment source not only for pocket beaches, but also for significant segments of open coasts. Erosion is very dependent upon lithological setting and can be very high in favorable conditions, e.g., soft rocks strata. On some sectors of the Algarve coast, a 10 to 50 m high soft rock cliff is retreating for 1–2 m/year [105], but the highest cliff retreat rate was measured on pyroclastic deposits on Nishinoshima Island (Japan), with 80 m/year [106]. In East Anglia, UK, at Dunwich, coastal retreat has been 2 km since the Roman time, with 13 parishes disappeared after XII century, the last one in 1919 [107]; this sediment input feeds the coast down to Orford Ness, more than 30 km south. Obviously, sediment input from a rocky coast is extremely discontinuous [108] and beaches taking advantage of this follow an uneven evolution [109]. As protection of Fairlight Cove, Sussex, UK was being undertaken, French [103,110] noted that "cliff stabilization represents a net loss of sediments to the coastal budget", as an input of 9750 m^3/year was interrupted.

2.2.7. Coastal Structures

Sediment offshore dispersion is favored by coastal structures, both in harbor and with shore protection projects. These, not only modify long-shore sediment distribution, favoring or penalizing different sectors, but through wave reflection, especially with shore parallel structures: [111,112] can induce topographically controlled rip currents through the structures' gaps [113]. Structures orthogonal or oblique to the coast [114,115] favor the shift of sediments to deeper areas, where they can be lost or from where they hardly return to the near-shore further downdrift [116]. This is well known and considered by researchers, but here we want to highlight the variability of their effect.

In front of shore parallel structures, beach profiles gradually lower [117,118], so wave energy dissipation reduces, and the structure becomes hit with increasing intensity, under a positive feedback-regulated process. However, when depth is so large that no shoaling processes are present and waves are fully reflected, no impact on sediment transport is to be expected; but this terminal condition is hard to be achieved. If this happens, a contributing cause is a regional negative sediment budget, e.g., at Marina di Pisa, Italy, where breakwaters located at an initial depth of 2.5 m, now have a depth of up to 7.0 m at the seaward side [112].

With respect to groins, trapping capacity reduces with time, until sediments bypass the obstacle; at this point, offshore dispersion is increased, although some material can overpass the structure (if it does not extend over the depth of closure). Therefore, updrift accumulation, offshore dispersion and input to the downdrift sectors change over time [119,120].

3. Conclusions

Most of the above examples dealt with in this paper are not linked by feedback processes, e.g., an eroding beach cannot send a message upstream to ask the river for more sediment. Without feedback any equilibrium is impossible not only in nature, but also in, for example, economics or human sciences. The beach is subject to feedback-regulated processes [121], that act within the beach and do not operate linking it with the external environment (atmosphere, catchment, etc.). Beach slope changes under the attack of different waves (to better dissipate wave energy), bars formed at the breaker line induce more waves to break there (a positive feedback) but waves can demolish the bar if it is too elevated (negative feedback) giving it a well-defined height.

Even without considering variations in each of these input and output items, each of which involves from few to millions of cubic meters of sediments a year, it is statistically impossible that their algebraic sum is zero, or even not far from zero. When a beach is classified to be in equilibrium, it almost certainly means that data could not reveal changes that were of the same order of magnitude; and one may question the data accuracy.

For the Tuscany coast a stable classification refers to sectors where the mean shoreline displacement of two surveyed shorelines was between +/−5 m. As far as the real sedimentary budget is considered, i.e., the sand volume change, the surveys accuracy is generally unsuitable to monitor actual values. Echo sounding accuracy is between 5 and 10 cm, and in the worst conditions depth changes of +/−20 cm are not significant [122]. If the depth of closure is 1 km offshore, 200 m^3 per meter of coast is the expected error, which equals a medium artificial nourishment project.

Assessing the possibility for a beach to be in equilibrium is contentious and socially dangerous. To give stakeholders this idea makes them ask for this stability and turns away the possibility of allowing the beach to achieve its due resilience. In many papers, "stability" has been used, but clearly exhibiting such a beach evolution trend refers to cases where observed changes are within any measurement error.

Many coastal geomorphic features are relatively stable and form changes take place slowly; others are less stable and have comparatively rapid change in response to dynamic environmental factors [123]. Quickly changing trends are a major problem for the present, and for the temporal dimension, which we usually use to build the knowledge, explanations, and predictions [44,124].

Unconsolidated beach materials respond rapidly to changes in the dynamic environment. Comments made in many books and documents of the last century, were to the effect that if a beach for an adequate period has been subject to environmental forces, then the profile will respond to both long-term and short-term changes, which tends to restore an equilibrium profile. The view was that equilibrium—the amount of sediment deposited by waves and currents—will be balanced by the amount removed, with many researchers even producing equations for the slope of natural profiles [125]. Cooper and Pilkey ([126], p. 605) concluded an overview of long-shore transport modeling, assessing, "that our present understanding permits only a qualitative estimate of direction of long-shore drift and identification of some of the controls"; but are the controls reliable?

So what is beach stability and how is it (if at all) entwined with dynamic equilibrium, or has the term dynamic been corrupted and used as a useful blanket adjective? As has been shown, it is virtually impossible for subsequent beach profiling to cover the exact same profile, so the stability (equilibrium) concept must be questioned. A beach is certainly a dynamic entity as it constantly changes shape as sediment is moved around. Is dynamic equilibrium simply an idea for a beach striving to assume an ideal shape? Is stability neither erosion nor deposition? Is an unstable beach one undergoing strong erosion only—if so what about strong deposition? Subjectivity of input selection, survey accuracy and also the potential for geological, dynamic, and sedimentological constraints reflect the uncertainty in the quest for the elusive conundrum of equilibrium. Is it akin to Lewis Carroll's [127] snark, is it a Boojum (a particularly nasty type of snark, which if glimpsed by a person, he/she disappears, so perhaps the term might disappear from the literature), or an allegory of an attempt to discover an absolute measure that is doomed to failure? The concept is so theoretical as to be functionally irrelevant.

To end, we believe that the word "stability" should not be used for beaches, unless it is accompanied by a value of the accuracy with which this condition is assessed and the time period to which it refers.

Funding: This research received no external funding.

Conflicts of Interest: The authors declare no conflict of interest.

References

1. Kennedy, B.A. Dynamic equilibrium. In *The Dictionary of Physical Geography*; Goudie, A., Ed.; Oxford: Oxford, UK, 1985; pp. 142–143.
2. Kennedy, B.A. Hutton to Hutton—views of sequence, progression and equilibrium in geomorphology. *Geomorphology* **1992**, *5*, 231–250. [CrossRef]
3. Phillips, J.D. Nonlinear dynamical systems in geomorphology; resolution or revolution. *Geomorphology* **1992**, *3*, 219–229. [CrossRef]
4. Thorn, C.E.; Welford, M.R. The equilibrium concept in geomorphology. *Ann. Assoc. Amer. Geog.* **1994**, *84*, 666–696. [CrossRef]
5. Graf, W.I. Mining and channel responses. *Assoc. Amer. Geog* **1979**, *69*, 262–275. [CrossRef]
6. Dean, R.G. Dynamic equilibrium. In *Encyclopaedia of Coastal Science*; Schwartz, M., Ed.; Springer: Berlin/Heidelberg, Germany, 2005; pp. 399–400.
7. Huggett, R.J. A history of the systems approach in geomorphology. *Géomorphol.: Relief Process. Environ.* **2007**, *2*, 145–158. [CrossRef]
8. Gregory, K.J.; Lewin, J. *The Basics of Geomorphology, the Key Concepts*; Sage: Thousand Oaks, CL, USA, 2014; p. 248.
9. Renwick, W.H. Equilibrium, Disequilibrium, and Nonequilibrium Landforms in the Landscape. *Geomorphology* **1992**, *5*, 265–276. [CrossRef]
10. Jaramillo, C.; Sánchez-García, E.; Jaru, M.S.; González, M.; Palomar-Vasquez, J.M. Subpixel satellite-derived shorelines as valuable data for equilibrium shoreline evolution models. *J. Coast. Res.* **2020**, *36*(6), 1215–1228. [CrossRef]
11. Jackson, D.W.T.J.; Cooper, J.A.G. Application of the equilibrium planform concept to natural beaches in Northern Ireland. *Coast. Eng.* **2010**, *57*, 112–123. [CrossRef]

12. Gilbert, G.K. Report on the geology of the Henry Mountains: US Geogr. and Geol. Survey of the Rocky Mountain Region (Powell Survey): US Govt. Printing Office, 160 p. 1909, The convexity of hilltops: Lour. *Geology* **1877**, *17*, 344–350. [CrossRef]
13. Ahnert, F. An approach to dynamic equilibrium in theoretical simulations of slope development. *Earth Surf. Process. Landf.* **1987**, *12*, 3–15. [CrossRef]
14. Willgoose, G.; Bras, R.L.; Rodriguez-Iturbe, I. The relationship between catchment and hillslope properties: Implications of a catchment evolution. *Geomorphology* **1992**, *5*, 21–37. [CrossRef]
15. Schumm, S.A.; Lichty, R.W. Time, space and casualty in geomorphology. *Amer. Jn. Sci.* **1965**, *263*, 110–119. [CrossRef]
16. Beattie, J.A.; Oppenheim, I. *Principles of Thermodynamics*; Elsevier Scientific Publishing Company: Amsterdam, The Netherlands, 1979; p. 329.
17. Howard, A.D. Equilibrium models in geomorphology. In *Modelling Geomorphological Systems*; Anderson, M.G., Ed.; John Wiley and Sons: Hoboken, NJ, USA, 1988; pp. 49–72.
18. Bracken, L.J.; Wainwright, J. Geomorphological equilibrium; myth or metaphor? *Trans. Brit. Geog.* **2006**, *31*, 167–178. [CrossRef]
19. Richards, R. *Rivers, form and Processs in Alluvial Channels*; Methuen: London, UK, 1982; p. 358.
20. Chorley, R.J.; Kennedy, B.A. *Physical Geography: A Systems Approach*; Prentice Hall: London, UK, 1971; p. 370.
21. Ahnert, F. *Introduction to Geomorphology*; Arnold: London, UK, 1996; p. 360.
22. Bogaert, P.; Montreuil, A.L.; Chen, M. Predicting volume change for beach intertidal systems: A space-time stochastic approach. *J. Mar. Sci. Eng.* **2020**, *8*, 901. [CrossRef]
23. Hallermaier, R.J. Use for a calculated limit depth to beach erosion. *Coast. Eng.* **1978**, *1978*, 1493–1512.
24. Short, A.D.; Masselink, G. Embayed and structurally controlled beaches. In *Handbook of Beach and Shoreface Morpho-Dynamics*; Wiley: Chichester, UK, 1999; pp. 230–250.
25. Bird, E.C.F. *Coastline Changes; A Global Review*; Wiley: Chichester, UK, 1985; p. 219.
26. Bird, E.C.F. *Encyclopedia of the World's Coastal Landforms*; Springer: Berlin/Heidelberg, Germany, 2010; p. 1516.
27. Luijendijk, A.; Hagenaars, G.; Ranasinghe, R.; Baart, F.; Donchyts, G.; Aarninkhof, S. The State of the World's Beaches. *Sci. Rep.* **2018**, *8*, 6641. [CrossRef]
28. Rosati, J.D. Concepts in Sediment Budgets. *J. Coast. Res.* **2005**, *21*, 307–322.
29. Pranzini, E. Cause naturali ed antropiche nelle variazioni del bilancio sedimentario del litorali. *Riv. Geogr. It* **1995**, *1*, 47–62.
30. Pethick, J. *An Introduction to Coastal Geomorphology*; Arnold: London, UK, 1984; p. 260.
31. Best, T.C.; Griggs, G.B. A sediment budget for the Santa Cruz littoral cell. In: From Shoreline to the Abyss. *Soc. Econ. Palaeontol. Mineral.* **1991**, *46*, 35–50.
32. Nastos, P.T.; Evelpidou, N.; Vassilopoulos, A. Does climatic change in precipitation drive erosion in Naxos Island, Greece? *Nat. Hazards Earth Syst. Sci.* **2010**, *10*, 379–382.
33. Fensham, R.J.; Fairfax, R.J.; Arecher, S.R. Rainfall, land use and woody vegetation cover change in semi-arid Australian savanna. *J. Ecol.* **2005**, *93*, 596–606. [CrossRef]
34. Langbein, W.B.; Schumm, S.A. Yield of sediment in relation to mean annual precipitation. *Eos Trans. Am. Geophys. Union* **1958**, *39*, 1076–1084. [CrossRef]
35. Holtmeier, F.-K.; Broll, G. Sensitivity and response of northern hemisphere altitudinal and polar treelines to environmental change at landscape and local scales. *Glob. Ecol. Biogeogr.* **2005**, *14*, 395–410. [CrossRef]
36. Milevski, I. Morphometric elements of terrain morphology in the Republic of Macedonia and their influence on soil erosion. In Proceedings of the International Conference Erosion and Torrent Control as A Factor in Sustainable River Basin Management, Belgrade, Serbia, 25–28 September 2007; pp. 25–28.
37. Hughes, J.D.; Thirgood, J.V. Deforestation, erosion, and forest management in ancient Greece and Rome. *J. For.* **1982**, *29*, 60–75. [CrossRef]
38. Innocenti, L.; Pranzini, E. Geomorphological evolution and sedimentology of the Ombrone River delta (Italy). *J. Coast. Res.* **1993**, *9*, 481–493.
39. Han, Z.; Dai, Z. Reclamation and river training in the Quintang estuary. In *Engineered Coasts*; Chen, J., Eisma, D., Hotta, K., Walker, H.J., Eds.; Kluwer Academic Press: Dordrecht, The Netherlands, 2002; pp. 121–138.
40. Cipriani, L.E.; Pranzini, E.; Rosas, V.; Wetzel, L. Landuse changes and erosion of pocket beaches in Elba Island (Tuscany, Italy). *J. Coast. Res.* **2011**, *SI64*, 1774–1778.
41. Frangipane, A.; Paris, E. Long-term variability of sediment transport in the Ombrone River basin (Italy). *Iahs Publ. -Ser. Proc. Rep. -Intern Assoc Hydrol. Sci.* **1994**, *224*, 317–324.
42. Nyssen, J.; Van den Branden, J.; Spalević, V.; Frankl, A.; Van de Velde, L.; Čurović, M.; Billi, P. Twentieth century land resilience in Montenegro and consequent hydrological response. *Land Degrad. Dev.* **2014**, *25*, 336–349. [CrossRef]
43. Arnaez, J.; Lasanta, T.; Errea, M.P.; Ortigosa, L. Land abandonment, landscape evolution, and soil erosion in a Spanish Mediterranean mountain region: The case of Camero Viejoj. *Land Degrad. Develop.* **2011**, *22*, 537–550. [CrossRef]
44. Pasquinucci, M.; Mecucci, S.; Morelli, P. Territorio e popolamento tra i fiumi Arno, Cascina ed Era: Ricerche archeologiche, topografiche e archivistiche. In Proceeding of the Atti 1° Congresso Archeologia Medioevale, Pisa, Italy, 29–31 May 1997.
45. Rinaldi, M. Recent channel adjustments in alluvial rivers of Tuscany, Central Italy. *Earth Surf. Process Landf.* **2003**, *28*, 587–608. [CrossRef]
46. Zunica, M. *Le Spagge del Veneto. Consiglio Nazionale Delle Ricerche*; Tipografia Antoniana: Padova, Italy, 1971; p. 146.

47. Im, D.; Kang, H.; Kim, K.-H.; Choi, S.-U. Changes of river morphology and physical fish habitat following weir removal. *Ecol. Eng.* **2011**, *37*, 883–892. [CrossRef]
48. Walling, D.E. *The Impact of Global Changen Erosion and Sediment Transport by Rivers: Current Progress and Future Challenges*; UNESCO World Water Assessment Programme: Paris, France, 2009; p. 30.
49. Torab, M.; Azab, M. Modern shoreline changes along the Nile delta coast as an impact of construction of the Aswan high dam. *Geogr. Tech.* **2007**, *2*, 69–76.
50. Liu, P.; Li, Q.; Li, Z.; Hoey, T.; Liu, Y.; Wang, C. Land Subsidence over Oilfields in the Yellow River Delta. *Remote Sens.* **2015**, *7*, 1540–1564. [CrossRef]
51. Amore, C.; Giuffrida, E. L'influenza dell'interrimento dei bacini artificiali del F. Simeto sul litorale del Golfo di Catania. *Boll. Soc. Geol. Italy* **1984**, *103*, 731–753.
52. Syvitski, J.P.M.; Vörösmarty, C.J.; Kettner, A.J.; Green, P. Impact of Humans on the Flux of Terrestrial Sediment to the Global Coastal Ocean. *Science* **2005**, *308*, 376–380. [CrossRef]
53. O'Connor, J.E.; Duda, J.J.; Grant, G.E. 1000 dams down and counting. *Science* **2015**, *348*, 496–497. [CrossRef]
54. Harrison, L.R.; East, A.E.; Smith, D.P.; Logan, J.B.; Bond, R.M.; Nicol, C.L.; Williams, T.H.; Boughton, D.A. River response to large-dam removal in a Mediterranean hydroclimatic setting: Carmel River, California, USA. *Earth Surf. Process. Landf.* **2019**, *43*, 3009–3021. [CrossRef]
55. East, A.E.; Pess, G.R.; Bountry, J.A.; Magirl, C.S.; Ritchie, A.C. Large-scale dam removal on the Elwha River, Washington, USA: River channel and floodplain geomorphic change. *Geomorphology* **2015**, *246*, 687–708.
56. Warrick, J.; Bountry, J.; East, A.; Magirl, C.S.; Randle, T.J.; Gelfenbaum, G.; Ritchie, A.C.; Pess, G.R.; Leung, V.; Duda, J.J. Large-scale dam removal on the Elwha River, Washington, USA: Source-to-sink sediment budget and synthesis. *Geomorphology* **2015**, *246*, 729–750. [CrossRef]
57. Dongxing, X.; Wenhai, Y.; Guiqiu, W.; Jinrui, C.; Fulin, L. Coastal erosion in China. *Acta Geogr. Sin.* **1993**, *60*, 468–476.
58. Coltorti, M. Human impact in the Holocene fluvial and coastal evolution of the Marche region, Central Italy. *Catena* **1977**, *30*, 311–335. [CrossRef]
59. Bondesan, M.; Dal Cin, R. *Rapporti fra Erosione Lungo i Litorali Emiliano-Romagnoli e del Delta del Po e Attività Estrattiva Negli Alvei Fluviali. Cave e Assetto del Territorio*; Italia Nostra—Regione Emilia-Romagna: Rome, Italy, 1975; pp. 127–137.
60. Padmalal, D.; Maya, K.; Sreebha, S.; Sreeja, R. Environmental effects of river sand mining: A case from the river catchments of Vembanad lake, Southwest coast of India. *Environ. Geol.* **2008**, *54*, 879–889. [CrossRef]
61. Bartolini, C.; Berriolo, G.; Pranzini, E. Il riassetto del litorale di Cecina. *Porti Mare Territ.* **1982**, *4*, 79–87.
62. Duck, R. *On the Edge: Coastlines of Britain*; Edinburgh University Press: Edinburgh, UK, 2015; p. 222.
63. Pranzini, E. Protection projects at two recreational beaches: Poetto and Cala Gonone beaches, Sardinia, Italy. In *Beach Management*; Williams, A., Micalleff, A., Eds.; Earthscan: London, UK, 2009; pp. 287–306.
64. Thornton, E.B.; Sallengerb, A.; Conforto Sestoc, J.; Egleyd, L.; McGeee, T.; Parsons, R. Sand mining impacts on long-term dune erosion in southern Monterey Bay. *Mar. Geol.* **2006**, *229*, 45–58. [CrossRef]
65. Vicinanza, D.; Ciavola, P.; Biagi, S. Progetto sperimentale di iniezione d'acqua in unità geologiche profonde per il controllo della subsidenza costiera: Il caso di studio di Lido Adriano (Ravenna). *Studi Costieri* **2008**, *15*, 121–138.
66. Luo, X.X.; Yang, S.L.; Wang, R.S.; Zhang, C.Y.; Li, P. New evidence of Yangtze delta recession after closing of the Three Gorges Dam. *Sci. Rep.* **2017**, *7*, 41735. [CrossRef]
67. Bartolini, C.; Palla, B.; Pranzini, E. Studi di geomorfologia costiera: X—Il ruolo della subsidenza nell'erosione litoranea della pianura del Fiume Cornia. *Boll. Soc. Geol. It.* **1988**, *108*, 635–647.
68. Bruun, P. Sea level rise as a cause of shore erosion. *J. Waterw. Harb. Div.* **1962**, *88*, 117–130.
69. Schwartz, M.L. The Bruun Rule—twenty years later. *J. Coast. Res.* **1987**, *3*, ii–iv.
70. Dubois, R.N. A re-evaluation of Bruun's rule and supporting evidence. *J. Coast. Res.* **1992**, *8*, 618–628.
71. Davidson-Arnott, R.G.D. A conceptual model of the effects of sea level rise on sandy coasts. *J. Coast. Res.* **2005**, *21*, 1166–1172. [CrossRef]
72. Ranasinghe, R.; Callaghan, D.; Stive, M.J.F. Estimating coastal recession due to sea level rise: Beyond the Bruun Rule. *Clim. Chang.* **2012**, *110*, 561–574. [CrossRef]
73. Cooper, J.A.G.; Pilkey, O.H. Longshore drift: Trapped in an expected universe. *J. Sediment. Res.* **2004**, *74*, 599–606. [CrossRef]
74. van Straaten, L.M.J.U. Coastal barrier deposits in south and north Holland, in particular in the area around Scheveningen and Ijmuiden. *Geol. Sticht. Med.* **1965**, *17*, 41–87.
75. Shepard, F.P. 35,000 years of sea level. In *Essay in Marine Geology*; Clement, T., Ed.; University Southern California Press: Los Angeles, CL, USA, 1963; pp. 1–10.
76. Zenkovich, V.P. *Processes of Coastal Development*; Oliver & Boyd: Edinburgh, UK, 1967; p. 738.
77. Psuty, N.P. Foredune mobility and stability. Fire Island, New York. In *Coastal Dunes. Form and Processes*; Nordstrom, K.F., Psuty, N., Bartel, B., Eds.; Wiley: Chichester, UK, 1990; pp. 159–176.
78. Linklater, E. *Orkney and Shetland*; Robert Hale: London, UK, 1971; 272p.
79. Steers, J.A. *The Sea Coast*, 3rd ed.; Collins: London, UK, 1962; 292p.
80. Clemmensen, L.B.; Glad, A.C.; Hansen, K.W.T.; Murray, A.S. Episodes of aeolian sand movement on a large spit system (Skagen Odde, Denmark) and North Atlantic storminess during the Little Ice Age. *Bull. Geol. Soc. Den.* **2015**, *63*, 17–28.

81. Vallejo, I.; Ojeda, J. El sistema de dunas activas del Parque Nacional de Doñana. In *La dunas en España. Enquadernaciones Martinez;* Sanjaume, E., Garcia, E.J., Eds.; CandelaInk: Cadiz, Spain, 2011; pp. 427–444.
82. Illenberger, W.K.; Rust, I.C. A sand budget for the Alexandria coastal dune field, South Africa. *Sedimentology* **1988**, *35*, 513–521. [CrossRef]
83. Hesp, P.A.; Short, A.D. Barrier morphodynamics. In *Handbook of Beach and Shore Morphodynamics;* Short, A.D., Ed.; John Wiley & Sons: Chichester, UK, 1999; pp. 307–333.
84. Choi, J.Y.; Jeong, H.; Choi, K.Y.; Hong, G.H.; Yang, D.B.; Kim, K.; Ra, K. Source identification and implications of heavy metals in urban roads for the coastal pollution in a beach town, Busan, Korea. *Mar. Pollut. Bull.* **2020**, *161*, 111724. [CrossRef] [PubMed]
85. Larson, M.; Kraus, N. Prediction of cross-shore sediment transport and temporal scales at different spatial. *Mar. Geol.* **1995**, *126*, 111–127. [CrossRef]
86. Chowdhury, P.; Behera, M.R. Effect of long-term wave climate variability on longshore sediment transport along regional coastlines. *Prog. Oceanogr.* **2017**, *156*, 145–153. [CrossRef]
87. Seymour, R.J.; Castel, D. Episodicity in longshore sediment transport. *J. Waterw. Portcoastalocean Eng. J. Waterw. Port Coast. Ocean Eng.* **1985**, *111*, 542–551. [CrossRef]
88. Shi-Leng, X.; Teh-Fu, L. Long-term variation of longshore sediment transport. *Coast. Eng.* **1987**, *11*, 131–140. [CrossRef]
89. Inman, D.L. Littoral cells. In *Encyclopaedia of Coastal Science;* Schwartz, M.L., Ed.; Springer: Dordrecht, The Netherlands, 2005; pp. 594–599.
90. Davies, J.L. *Geographical Variations in Coastal Development;* Oliver and Boyd: Edinburgh, UK, 1977; p. 212.
91. Perlin, A.; Kit, E. Longshore sediment transport on Mediterranean coast of Israel. *J. Waterw. Portcoastalocean Eng. J. Waterw. Port Coast. Ocean Eng.* **1999**, *125*, 80–87. [CrossRef]
92. Raynor, S. Prediction and other approaches to climate change policy. In *Prediction: Science;* Sarewitz, D., Pielke, R.A., Jr., Byerly, R., Jr., Eds.; Decision Making and the Future of Nature: Washington, DC, USA, 2000; pp. 269–296.
93. Pilkey, O.H.; Cooper, J.A.G. Longshore transport volumes: A critical review. *J. Coast. Res.* **2002**, *36*, 572–580. [CrossRef]
94. Nobre Silva, A.; Taborda, R.; Bertin, X.; Dodet, G. 2012 Seasonal to Decadal Variability of Longshore Sand Transport at the Northwest Coast of Portugal. *J. Waterw. Portcoastalocean Eng. J. Waterw. Port Coast. Ocean Eng.* **2009**, *138*, 464–472. [CrossRef]
95. Meier, D. The Historical Geography of the German North-Sea Coast: A Changing Landscape. *Die Küste* **2008**, *74*, 18–30.
96. Trevelyan, M. *Llantwit Major: Its History and Antiquities;* John E. Southhall: Newport, UK, 1910; p. 124.
97. Davies, P.; Williams, A.T. The enigma of Colhuw Port. *Geogr. Rev.* **1991**, *81*, 257–262. [CrossRef]
98. Nott, J.; Smithers, S.; Walsh, K.; Rhodes, E. Sand beach ridges record 6000 year history of extreme tropical cyclone activity in north-eastern Australia. *Quat. Sci. Rev.* **2009**, *28*, 1511–1520. [CrossRef]
99. Lozano, I.; Devoy, R.J.N.; May, W.; Andersen, U. Storminess and vulnerability along the Atlantic coastlines of Europe: Analysis of storm records and of a greenhouse gases induced climate scenario. *Mar. Geol.* **2004**, *210*, 205–225.
100. Short, A.D.; Woodroffe, C.D. *The Coast of Australia;* Cambridge University Press: New York, NY, USA, 2009; p. 288.
101. Devoti, S.; Silenzi, S.; Amici, I.; Aminti, P.; Amodio, M.; Bovina, G.; Callori Vignale, C.; Cappietti, L.; Chiocchini, O.; Di Gregorio, F.; et al. *Il Sistema Spiaggia-Duna Della Pelosa (Stintino);* Quaderno: Ispra, Italy, 2010; p. 288.
102. Whittow, J.B. *Geology and Scenery in Scotland;* Penguin Books: Harmondsworth, UK, 1977; p. 362.
103. Entwistle, J.A.; Abrahams, P.W.; Dodgshon, R.A. Multi-element analysis of soils from Scottish historical sites. Interpreting Land-use history through the physical and geochemical analysis of soil. *J. Archaeol. Sci.* **1998**, *25*, 53–68. [CrossRef]
104. Inman, D.L. Shore Processes. In *Encyclopaedia of Science and Technology;* McGraw Hill: New York, NY, USA, 1960; pp. 299–306.
105. Dias, J.M.A.; Neal, W.J. Sea Cliff Retreat in Southern Portugal: Profiles, Processes, and Problems. *J. Coast. Res.* **1992**, *8*, 641–654.
106. Mogi, A.; Tsuchide, M.; Fukushima, M. Coastal erosion of a new volcanic island Nishinoshima. *Geogr. Rev. Jpn.* **1980**, *53*, 449–462. [CrossRef]
107. Morris, B. In defence of oblivion: The case of Dunwich, Suffolk. *Int. J. Herit. Stud.* **2014**, *20*, 196–216. [CrossRef]
108. Hapke, C.; Plant, N. Predicting coastal cliff erosion using a Bayesian probabilistic model. *Mar. Geol.* **2010**, *278*, 140–149. [CrossRef]
109. Priest, G.R. Coastal Shoreline Change Study Northern and Central Lincoln County, Oregon. *J. Coast. Res.* **1999**, *28*, 140–157.
110. French, P.W. Coastal defences. In *Processes, Problems and Solutions;* Routledge: London, UK, 2001; p. 366.
111. Silvester, R.; Hsu, J.R.C. *Coastal Stabilization. Innovative Concepts. Englewood Cliffs;* PRT: Prentice Hall, NJ, USA, 1993; p. 539.
112. Anfuso, G.; Pranzini, E.; Vitale, G. An integrated approach to coastal erosion problems in northern Tuscany (Italy): Littoral morphological evolution and cells distribution. *Geomorphology* **2011**, *129*, 204–214. [CrossRef]
113. Aagaard, T.; Masselink, G. The Surf Zone. In *Handbook of Beach and Shore Morphodynamics;* Short, A.D., Ed.; John Wiley & Sons: Chichester, UK, 1999; pp. 72–118.
114. Evert, C.H. Beach behaviour in vicinity of groins -two New Jersey field experiments. *Proc. Coast. Struct.* **1979**, *2*, 853–857.
115. Muir Wood, A.M.; Fleming, C.A. *Coastal Hydraulics;* The Macmillan Press Ltd.: London, UK, 1969; p. 280.
116. Allen, J.R. Nearshore Sediment Transport. *Geogr. Rev.* **1988**, *78*, 148–157. [CrossRef]
117. Bernatchez, P.; Fraser, C. Evolution of coastal defence structures and consequences for beach width trend, Québec, Canada. *J. Coast. Res.* **2012**, *28*, 1550–1566. [CrossRef]
118. Neshaei, M.L.; Holmes, P.; Salimi, M.G. A semi-empirical model for beach profile evolution in the vicinity of reflective structures. *Ocean Eng.* **2009**, *36*, 1303–1315. [CrossRef]

119. Lima, M.; Coelho, C.; Veloso-Gomes, F.; Roebeling, P. An integrated physical and cost-benefit approach to assess groins as a coastal erosion mitigation strategy. *Coast. Eng.* **2020**, *156*, 103614. [CrossRef]
120. Tereszkiewicz, P.; McKinney, N.; Meyer-Arendt, K.J. Groins along the northern Yucatan coast. *J. Coast. Res.* **2018**, *34*, 911–919. [CrossRef]
121. King, C.A.M. Feedback relationships in geomorphology. *Geogr. Ann.* **1970**, *52A*, 147–159. [CrossRef]
122. Pranzini, E.; Rossi, L. The Role Of Coastal Evolution Monitoring. In *Coastal Erosion Monitoring. A Network of Regional Observatories*; Cipriani, L.E., Ed.; Nuova Grafica Fiorentina: Florence, Italy, 2013; pp. 11–55.
123. Mossa, J.; Meisburger, E.P.; Morang, A. *Geomorphic Variability in the Coastal Zone, Coastal Geology and Geotechnical Program*; Technical Report CERC-92-4; Army Corps of Engineers: Washington, DC, USA, 1992; pp. 20314–21000.
124. Furlani, S.; Ninfo, A. Is the present the key to the future? *Earth-Sci. Rev.* **2015**, *142*, 38–46. [CrossRef]
125. Dean, R.G.; Maurmeyer, E.M. Models for Beach Profile Response. In *CRC Handbook of Coastal Processes and Erosion*; Komar, P.D., Ed.; CRC Press Inc.: Boca Raton, FL, USA, 1983; pp. 151–167.
126. Cooper, J.A.G.; Pilkey, O.H. Sea-level rise and shoreline retreat: Time to abandon the Bruun Rule. *Glob. Planet. Chang.* **2004**, *43*, 157–171. [CrossRef]
127. Carroll, L. *The Hunting of the Snark*; Macmillan and Co.: London, UK, 1876; p. 84.

Article

Influence of Wave Climate on Intra and Inter-Annual Nearshore Bar Dynamics for a Sandy Beach

Nataliya Andreeva [1,*], Yana Saprykina [2], Nikolay Valchev [1], Petya Eftimova [1] and Sergey Kuznetsov [2]

[1] Institute of Oceanology, Bulgarian Academy of Sciences, 40 Parvi May Blvd., P.O. Box 152, 9000 Varna, Bulgaria; valchev@io-bas.bg (N.V.); eftimova@io-bas.bg (P.E.)

[2] Shirshov Institute of Oceanology, Russian Academy of Sciences, Nahimovskiy Prospekt, 36, Moscow 117997, Russia; saprykina@ocean.ru (Y.S.); kuznetsov@ocean.ru (S.K.)

* Correspondence: n.andreeva@io-bas.bg

Abstract: The study investigates cross-shore outer sand bar dynamics in an open-coast non-tidal beach at the Bulgarian Black Sea due to wave climate. On seasonal to short-term (1–2 years) time scale, monthly field measurements of the outer bar profiles were related to respective modeled nearshore wave data. Hereby, seaward-shoreward bar migration was examined depending on the wave forcing, wave non-linearity, wave transformation scenarios, storms and direction of wave incidence. Analysis revealed that intra-annually highly non-linear waves were responsible for outer bar displacement, while the direction of migration depended on wave period, duration of conditions with wave steepness >0.04, angle of approach and total duration of storms. Short-term bar evolution was mainly governed by wave height and storms' parameters as the angle of approach and duration. The correlation between the outer bar location and wave height annual variations initiated the first for the explored Black Sea region examination of possible connection between wave height's temporal fluctuations and the variability of climatic indices the North Atlantic Oscillation (NAO), the Atlantic Multi-decadal Oscillation (AMO), the East Atlantic Oscillation (EA), the Arctic Oscillation (AO), the East Atlantic-Western Russia (EA/WR) and the Scandinavian (SCAND) patterns. According to the results the inter-annual outer bar location may vary depending on periods of maximum annual wave fluctuations, which in turn predominantly depend on indices the EA (4–5, 10–11, 20–30 years), the EA/WR (2–4, 9–13 years) and the NAO (15 years).

Keywords: wave climate; nearshore bars; field data; cross-shore bar migration; wave conditions; wave non-linearity; storms; teleconnection patterns; Black Sea

Citation: Andreeva, N.; Saprykina, Y.; Valchev, N.; Eftimova, P.; Kuznetsov, S. Influence of Wave Climate on Intra and Inter-Annual Nearshore Bar Dynamics for a Sandy Beach. *Geosciences* **2021**, *11*, 206. https://doi.org/10.3390/geosciences11050206

Academic Editors: Jesus Martinez-Frias and Gianluigi Di Paola

Received: 16 February 2021
Accepted: 5 May 2021
Published: 8 May 2021

Publisher's Note: MDPI stays neutral with regard to jurisdictional claims in published maps and institutional affiliations.

1. Introduction

The nearshore environments of mild-sloped sandy beaches are often characterized by the presence of a variety of morphological features, as most common are sand bars. Typically, they are shore-parallel, crescentic or irregular shoals [1]. Sand bars are often found at depths less than 10 m and within or just seaward of the surf zone, whose width is influenced by variations in the incident wave climate. Over the years, many mechanisms have been proposed to explain bar formation [2], which eventually were placed into three main groups [3]: breakpoint-related mechanisms [4,5], infra-gravity wave-related mechanisms [6,7] and self-organizational mechanisms [8,9]. The presence of sand bars in the coastal zone is of significant importance for sandy beach morphodynamics since they act as a storage of sediments and as a natural protection by dissipating high wave energy during storms by breaking, thus limiting coastal erosion and flooding hazards [10,11]. These morphological features may exert a significant impact on the nearshore hydrodynamics due to their cross-shore and long-shore geometry and location [12,13]. Moreover, studying the sand bar dynamics allows identification of important physical processes that control coastal evolution, and thus enrich the understanding and knowledge about sediment transport in the coastal zone [14].

Nearshore bathymetry varies on a spectrum of time scales ranging from inter-annual (longer than a year) to intra-annual (equal to or shorter than 1 year). Intra-annual time scales include the seasonal cycle and shorter fluctuations associated with passing storms and weak to calm wave conditions [1]. The strong seasonal variability of wind and wave climate drives sandy beaches to exhibit strong seasonal morphodynamic cycles [15]. They are associated with sediments being eroded from the foreshore during high waves, forming storm or winter beach profile [16] followed by recovery during lower energy conditions (swell or summer profile). Typically, these seasonal cycles in beach morphodynamics are reflected in the submerged beach profile with "sandbars building and migrating offshore during storm conditions, and deflating and migrating onshore during low-energy swell conditions" [15]. Cross-shore sand bar migration, resulting in beach profile changes is associated with "imbalance between the cross-shore sediment transport driven by wave non-linearities (orbital velocity skewness and asymmetry), undertow and the gravitational downslope effects" [17]. The offshore sandbar migration is attributed to breaking of large storm waves over the bar with a dominant offshore bed return flow-induced sediment transport. On the other hand, its slow onshore migration takes place in between storms due to a dominant onshore sand transport driven by wave non-linearities during weakly to non-breaking wave regimes across the bar [17,18]. Meanwhile, frequent changes in wave intensity may exert minor bar displacements reflecting its relative stability in the nearshore zone [19]. In the last decades the seasonal sandbar migration has been studied using in situ measurements and numerical modeling in [15,20–25]. Despite that, the nearshore morphological evolution on seasonal time scales is still poorly understood since field observations have been limited due to difficulties and expense of working in nearshore environments [15,23]. Over the years, nearshore morphodynamics on different coastal stretches along the Bulgarian Black Sea coast has been investigated via field measurements and numerical simulations. On inter-annual time scale in [26,27], on seasonal and monthly basis in [28,29], and due to individual storms in [30–35].

The inter-annual variability of sand bar migration has not been sufficiently studied due to lack of long-term continuous field observations in various coastal environments. At present, sand bar migration is much better investigated and understood at the medium-term time spans, i.e., less than 10 years. It is reported that there are migration cycles (return periods) of about 3–4 years [11,36] and 2–5 years [37]. It is assumed that the decrease of the return period is associated with more energetic wave conditions [36]. According to an analysis of bathymetric data collected for 30 years on the Terschelling coast the long-term evolution of sand bars has characteristic stages: generation, seaward migration and decay, and this cycle has periods up to 10–20 years [38]. Investigations of long-term bar behavior based on nearshore profiles' measurements performed for nearly 30 years along the Holland coast have revealed strong alongshore bar variability [39]. Regions of homogeneous Large-Scale Coastal Behavior (LSCB-regions) controlled by different hydrodynamics, or sediment and morphological structure were selected. The length sizes of the LSCB-regions are from 5 km up to 42 km. The change between LSCB-regions occurs sharply at distances shorter than 2 km. Such alongshore variability also does not allow determination of clear patterns of long-term variability trends on barred shores. In the model proposed in [40] a short-term (months) dynamical processes in the coastal zone are used to predict medium-term bar displacements. The model reproduces well enough the seaward bar migration, but do not explain bar generation stage, because of the influence of wave climate on the medium-term bar migration and because the nearshore sediment transport is not completely understood.

The first study that has associated the sand bar migration with changes in the climatic indices (the North Atlantic Oscillation in particular) is [41]. They confirmed the existence of such connection based on 15-year time series of Argus video images on the evolution of the submerged profile in the Perranporth beach. In [42], on the example of the Black Sea, it was shown that coastal wave climate is rather heterogeneous and different parts of the coast can correlate in a different manner with variations in the climatic indices, i.e., having different periods of long-term fluctuations. This can also cause heterogeneous variability in the bar's migration along the coast.

Thus, the primary objective of the study is to investigate inter- and intra-annual cross-shore outer bar migration and their dependence on the regional wave climate on the example of an open-coast non-tidal beach at the western Black Sea coast. Description of the study site is given in Section 2. Section 3 introduces data sets of beach profile field measurements (Section 3.1), numerically modeled wave forcing (Section 3.2) and their statistical processing. Results are discussed in Section 4. Section 4.1 presents examination of the intra-annual bar evolution on the grounds of variability of wave forcing and non-linearity, storms and direction of wave incidence. Section 4.2 explores the inter-annual bar evolution and reveals main periods of its middle and long-term variability due to changes in the main climatic indices affecting the wave climate of the coastal region under consideration. The conclusions follow in Section 5.

2. Study Site

The study site is located in the southern part of Kamchia-Shkorpilovtsi beach, which is the largest and longest (≈14 km) sandy beach along the Bulgarian Black Sea coast with well-developed dunes and the presence of two rivers—Kamchia and Fandakliyska (Figure 1).

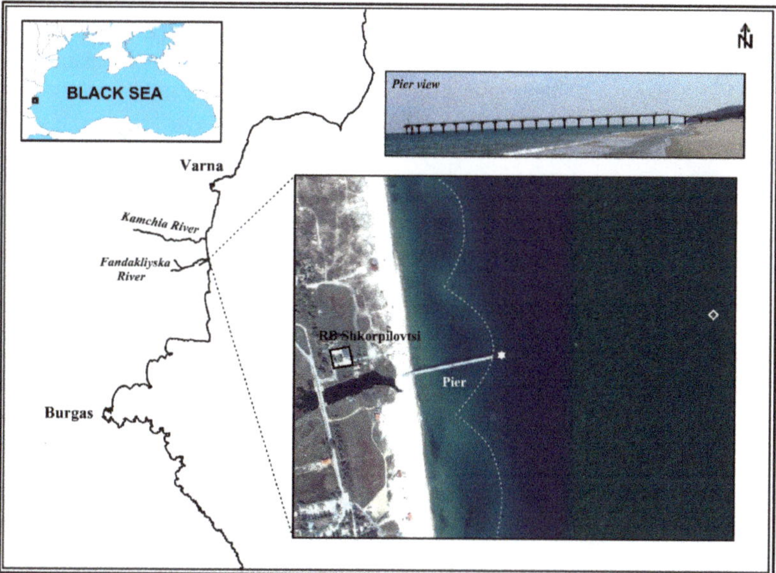

Figure 1. Description and location of the study site. The satellite view shows the location of the field site on the Bulgarian Black Sea coast and indicates the positions of the research station 'Shkorpilovtsi', the pier, and the state of the crescentic bars (dashed grey line) along the coastline on 24th September 2009; white diamond—extraction point of the climatic wave data, depth 17 m; white star—same for wave time series, depth 4.4 m; data sources: the Black Sea basin map was taken from free maps platform [43] and the satellite image was acquired by Maxar Technologies, USA [44].

The coastal area is open to the winds and waves of the eastern half. Strong seasonal variability is the most remarkable feature of the wind and wave climate. Winds from the northeast quarter are the most frequent in the western Black Sea. Having the largest fetch, they trigger the most severe storms. The northeastern winds prevail over the study area. In general, the southeastern winds are less significant in terms of storm intensity. Following the wind pattern waves approach the shore mostly from the northeast, east and southeast as the impact of waves coming from northeast and east is much more pronounced. Thus, a large amount of the waves approach the shore in the normal generating intensive cross-shore water circulations as compared to relatively weak long-shore wave-driven currents [45]. Based on wave hindcast data the maximum significant wave height can reach up to 5 m at depths of about 20 m [46]. The beach is formed due to the accumulation of erosive and fluvial sediments and has a positive sediment budget since it is constantly fed by sediments from both rivers, as well as from a well-developed ravine and gully system [47]. Morphological conditions of the adjacent coastal zone are rectilinear shoreline with almost parallel isobaths and nearshore crescentic bars (Figure 1).

The study site area near the mouth of Fandakliyska River is equipped with two field research facilities: a research station 'Shkorpilovtsi' and a 230 m-long pier constructed perpendicularly to the shore, thus covering the most dynamic part of the coastal zone ($\approx$200 m). It rises 7 m above the mean sea level and water depths at its sea-end vary between 4–5 m (Figure 1). Therein, at depths 0–10 m the mean slope of the seabed is 0.02–0.025, while for depths 10–20 m it is 0.006 [48]. Reflection coefficient estimated for different parts of the shore is of the order of 10^{-2}–10^{-3}, which means that the reflection of the waves at this dissipative coast has practically no effect on their heights [49]. The shoreface around the pier usually features one inner sand bar at distances 40–60 m and depths 0.5–2 m [50] and an outer crescentic sand bar at depths 3–5 m and distances 110–200 m. On the upper submerged profile (depths <2.5 m), over 95% of bed sediments are composed of coarse to medium-sized sands (0.3–0.76 mm), which mostly consist of quartz (96%), while the remaining 4% are represented by fine shell particles ($CaCO_3$). As depth grows, their content decreases, and at 8–10 m depth over 90% of the sediments consist of particles less than 0.25 mm in size [45,47].

3. Materials and Methods

3.1. Field Data: Cross-Shore Profiles

The morphodynamics of the outer bar was analyzed using discrete measurements of beach profile depths collected along the pier. In the last decades, such measurements were made during a number of international field experiments 'Shkorpilovtsi 2007, 2016, 2018' and within the MICORE project in 2008–2010. Each survey extended from a base point marked as zero at the shore-end of the pier out to typical depths of 4–5 m. The base point was a reference marker in the determination of the offshore distance and it does not coincide with elevation zero of the beach profiles. The data were collected with either a metal pole or a rope lot, as both devices have markings at every 10 cm. The accuracy of both measuring methods is approximately commensurable, as reported accuracy for the metal pole is about 10 cm [51]. The measurements were made at points of known coordinates distanced at every 2 m by lowering the metal pole (rope lot) from the pier deck. In 2008–2010 measurements were acquired only with the metal pole. In 2007, 2016 and 2018 cross-shore profiles were surveyed on a daily basis for about a month, usually in October [50]. In 2008–2010 measurements were done predominantly once a month [32].

The dataset collected in 2008–2010 was used to study the intra-annual morphodynamics of the outer bar. During these monthly measurements, the summer surveys were less frequent, while in winter the regularity depended primarily on meteorological forecasts, giving preference to calm seas or very weak swell. In total, 33 cross-shore profiles were measured: one profile in November 2008, 25 profiles in January-December 2009, six profiles in January–April 2010, and one profile in August 2010. For this period, the average shoreline position was approx. 29 m away from the pier base point. The profiles were subjected to additional processing. The mean slope of each profile was determined as $\tan\beta$ [52] taking into account only its submerged part, i.e., from elevation zero down to the last measured depth. Thus, for 2008 $\tan\beta = 0.034$, while the average $\tan\beta$ for 2009 and 2010 is 0.039 and 0.045, respectively. Preliminary analysis of the transects revealed that in 2008–2010 the outer bar was localized at distances 130–220 m off the shore at depths 3.5–4.5 m. Having this in mind, profile depths from elevation zero down to a 100 m-offshore distance were removed in order to ease the determination of bar characteristics. The remaining data were subjected to smoothing by Moving Average method with factor 5 for correction of potential errors due to applied measuring technique. The smoothed data series were used to define the deepest shoreward point (fore-bar depth) and the shallowest shoreward value (bar crest) in order to isolate its entire form. It should be preliminary commented that offshore distance available for performing cross-shore measurements was limited by the pier's length. Thereby, in some cases, the last measured profile points with the smallest depths were considered as bar crests, e.g., data for March–July 2009. Outer bar characteristics used for analysis of its intra-annual dynamics were bar profile, bar crest offshore distance, depth above the bar crest (bar location depth) and bar height, determined by subtraction of fore-bar depth from the depth above the crest.

3.2. Modeled Data: Wave Parameters

Wave fields at the study site were simulated by the SWAN (Simulating Waves Nearshore) Cycle III spectral numerical wave model, which is based on wave action balance equation [53]. The regional wave climate was reconstructed by means of hindcasting for a period of 62 years. More specifically, the available hindcast data for 1948–2006, obtained using wind fields from the regional climate model REMO (Regional Atmosphere Model) [54] were complemented by simulations for 2007–2010 making use of the Global Forecast System's (GFS) analysis winds [55] to encompass the study period. Both sources of wind forcing are based on atmospheric reanalysis of the National Centers for Environmental Prediction (NCEP) having horizontal resolution of 0.5°. The model was run in a non-stationary mode as numerical simulations had 32 frequencies and 36 directional bands. Two nested SWAN regular grids were set to compute the wave fields in the Western Black Sea shelf and nearshore domain off Kamchia-Shkorpilovtsi beach with horizontal resolution of 1/30° and 400 m, respectively. Boundary condition for the SWAN runs were provided by the WAM (Wave Model) output. More details on both models set-up and validation can be found in [46,56]. Thus, long-term numerical simulations secured availability of multi-annual time series of various wave parameters with temporal resolution of one hour.

To analyze the role of regional wave climate on the inter-annual outer bar dynamics the annual 99th Quantiles (Q99) of significant wave height Hs and peak period Tp were calculated from 1948–2010 hindcast data for deep water location (depth 52 m) in front of the study site. For intra-annual analysis, two data sets were needed. First set served to establish how 2009–2010 wave conditions related to the regional wave climate. Therefore, multi-annual monthly Q99 of Hs and Tp were evaluated based on the full time span of the hindcast, while monthly Q99 of the same parameters were determined only for 2009 and 2010. In both cases the source data for quantiles estimation were extracted for location offshore the pier in a zone of weak transformation of waves at 17 m depth (Figure 1). Hourly time series of Hs, Tp and mean direction of wave propagation Dm in November 2008–August 2010 were made available to support detailed analysis. The second set of wave data aided assessment of 2009–2010 bar's dynamics in close vicinity to its location. To this end, initial sea states (wave parameters' time series and monthly Q99 values) were transformed from 17 to 4.4 m depth—a position located at the end of pier (Figure 1). The transformation was performed using the XBeach (eXtreme Beach) model [57] on a cross-shore profile located along the pier on a grid with variable cell sizes and resolution increasing from 7.3 m at the offshore boundary up to 1 m at shore. The input boundary conditions were provided from the SWAM modeling and the XBeach model was run in stationary mode without calculation of sediment transport and morphology update. Model output was post-processed to extract significant wave heights Hs, spectral peak wavelengths Lp and wave steepness Hs/Lp. According to recent studies [58] the efficiency of the SWAN model is very high in deep and transitional waters, and it should not be used for shallow water applications. The XBeach, on the other hand, is used for computation of nearshore hydrodynamics, as in stationary mode it resolves physical processes as wave propagation, directional spreading, shoaling, refraction, wave breaking, etc. [57]. The preference to work with the 99th quantile of wave parameters was based on the concept that the position of the nearshore bars is closely related to the position of the breaker line(s), which in turn depends on the severity of wave conditions [4]. Thus, when considering the outer bar, it is only expected to be in the breaker zone during severe storms [10], and as pointed out in [59] large offshore waves are required to induce the outer bar into activity and cause significant morphological change.

4. Discussion of Results

4.1. Intra-Annual Sand Bar Evolution

4.1.1. Morphodynamics of the Seabed Relief

As stated in the field site description the nearshore area around the pier features a double bar profile. Bars' morphodynamics over time could be easily traced in a cross-shore plane if presented in chronological order of all field measurements (Figure 2).

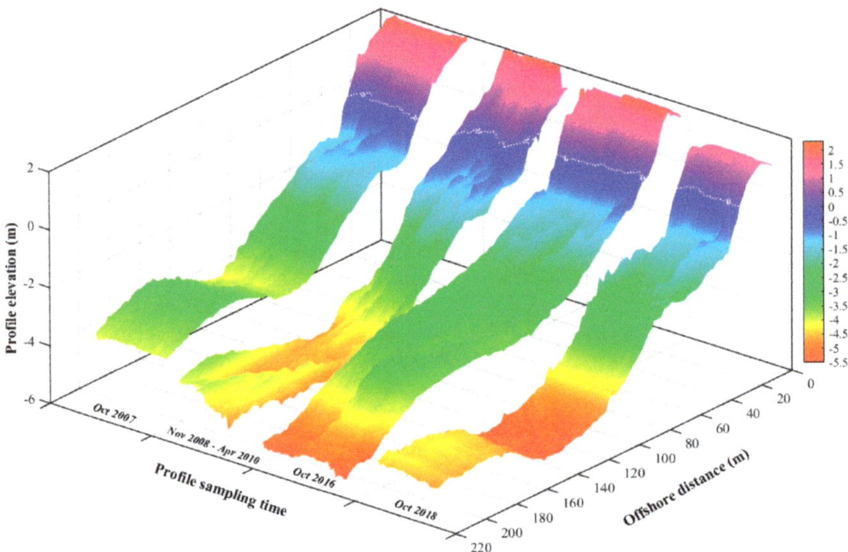

Figure 2. Coastal morphodynamics constructed from cross-shore profile measurements for all field experiments in October 2007, November 2008–April 2010, October 2016 and October 2018 at Kamchia-Shkorpilovtsi beach; white dashed line—contour of the mean sea level.

Considering the surveys in 2007, 2016 and 2018 the inner bar evolution could be observed at offshore distances 40–60 m, as during each of these field campaigns at least one storm event has occurred. In 2008–2010, this zone has widened up to 30–70 m. According to recent studies at the site, the inner bar has a storm origin and commonly exists in the surf zone on a time scale of a few days [60]. Depending on the wave regime, it develops during the storm's initial stage and eventually adjoins the shore while storm attenuates [61]. Analysis of wave and profile data from 2007 and 2016 confirmed the bar formation due to wave breaking by plunging and the combined effect of waves and undertow [50,62], while responsible for its onshore migration are processes of wave breaking switching from plunging to spilling and periodic energy exchange between non-linear wave harmonics [60,62].

The dynamics of the outer bar is quite remarkable, as well (Figure 2). It shows a variability concerning its form, its (a) symmetry, and its offshore and depth location, which is not only inter-annual but annual to seasonal (2008–2010), and even could be observed within a single storm event, as in October 2016. According to the measurements, over the years the outer bar is always present, and its evolution and displacement commonly occur within distances of 110–220 m at depths 3–5 m. The closest to the shore the sand bar was in 2007 and 2016 at distances 110–190 m and 120–210 m, respectively, and it was localized at depths 3–4 m, while in 2018 it was farthermost off the shore at 150–220 m (depths 4–5 m). On the grounds of measurements in 2016, the beach profile deformations were studied during different stages of the reported storm (Hs ≈ 4.5 m), considered as seasonal due to estimated return period of 1 year [61]. The results stated that the outer bar built during the stages of the maximum storm intensity, which led to the formation of a typical storm profile. Storm's attenuation stage, i.e., wave parameters' gradual decrease contributed to the natural process of the beach foreshore recovery, while a slight shoreward shift of the outer bar and increase of the depth above its crest were registered. Based on the same data, [50,62] reported that the growth and displacement of the outer bar depend on the breaking type and location of the breaking zone within the nearshore area. Large storm waves breaking by plunging and wave asymmetry variations with respect to the vertical axis [63,64] trigger initial and further outer bar development, while shifts of the breaking area by plunging entail respective bar crest relocations.

The availability of regular monthly seabed measurements in 2008–2010 gave opportunity for more detailed analysis of the outer bar dynamics on annual and seasonal basis (Figure 3).

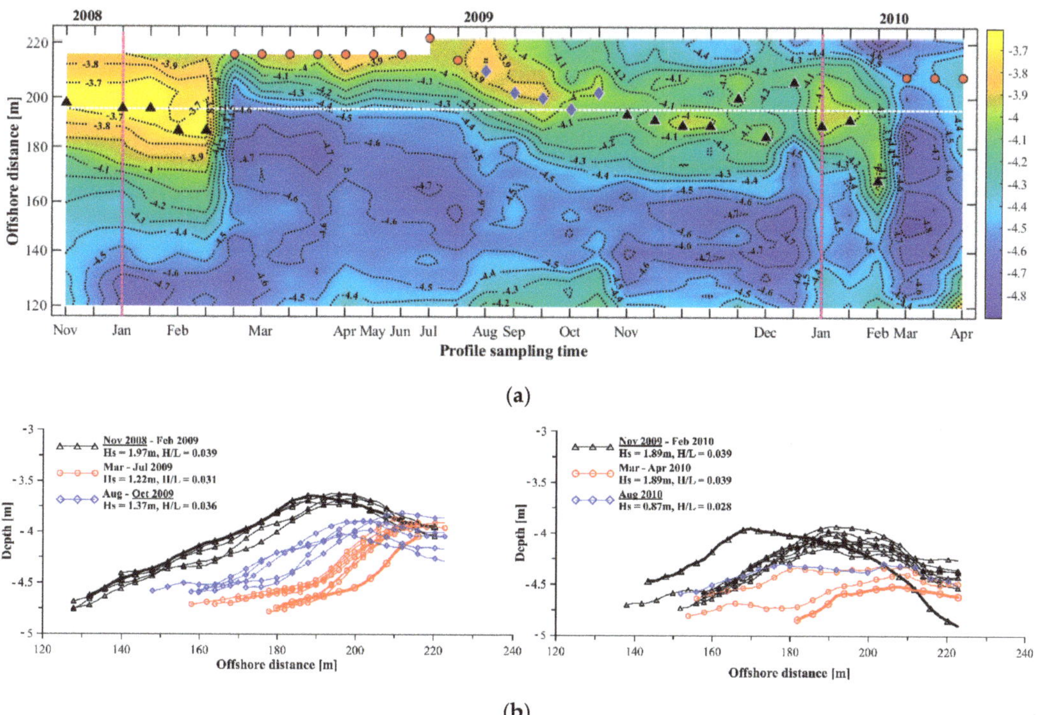

Figure 3. Intra-annual morphodynamics of the outer sand bar: (**a**) chronological seabed relief changes in November 2008–April 2010; colors and symbols of bar crest positions correspond to profile groups as described below in the caption text; vertical magenta lines—beginning of each year, horizontal white line—reference winter bar crest location defined from 2009 data; (**b**) outer bar profile envelopes of temporal variations (sweep zones) for November 2008–October 2009 (left panel) and November 2009–August 2010 (right panel), Hs and H/L—seasonally averaged monthly maximum values of the parameters at depth 4.4 m; black lines with triangles—winter pro-files, red lines with circles—summer profiles, and blue line with diamond—transitional profiles; heavy black line with triangles—last measured profile prior to the abrupt seaward bar shift, heavy red line with circles—first profile after the bar's displacement to its summer location.

As previously mentioned, during this period the sand bar migrated within depths ≈ 3.5–4.5 m and offshore distances 130–220 m. Changes of bed relief and crest positions show that the outer bar's intra-annual variations followed a certain repetitive seasonal pattern (Figure 3a). It was characterized by the bar's presence closer to the shore in winter at distances 180–200 m followed by an abrupt seaward shift to larger depths at distances ≈ 210–220 m for spring-summer season, and again a gradual return to its winter location.

It was evident that such bar crest dynamics was accompanied by changes in the bar's form (Figure 3b). After inspection of the measured transects in November 2008–August 2010 and taking into account the seasonal pattern, three groups of profiles were determined according to their similarity in shape:

- Winter profiles, corresponding to sand bar shoreward location in November 2008 (2009)–February 2009 (2010);
- Summer profiles, corresponding to its farthermost positions offshore as in March–July 2009 and March–April 2010;
- Transitional profiles, representing the gradual bar displacement (August–October 2009 and August 2010) back to its winter position in the nearshore zone.

The proposed "winter-summer" distinction is rather based on the seasonality of the outer bar profile evolution discussed herein, than on the classical concept of seasonal beach profiles—winter and summer ones [16]. Moreover, the possibility to distinguish the group of transitional profiles is connected with the wave regime seasonality in general, which is not an abrupt transition from summer low-energy regimes to winter high-energy ones, but rather a process of gradual wave energy increase in a form of a series of subsequent storms [65].

In the first seasonal cycle of bar migration in November 2008–October 2009 (Figure 3b, left panel) each of the introduced profile groups is very easily distinguished, keeping relatively uniform slightly asymmetric shapes in comparison to the next season in November 2009–August 2010. Regardless of its slight cross-shore migration the bar remains relatively high, as its height varies between 0.99–0.66 m without significant change of its location depth from ≈3.6 m to ≈3.9 m (Table 1).

Table 1. Outer bar average characteristics for different profile groups for November 2008–August 2010.

Profile Type	Winter Profiles		Summer Profiles		Transitional Profiles	
Seasons	Nov 2008–Feb 2009	Nov 2009–Feb 2010	Mar–Jul 2009	Mar–Apr 2010	Aug–Oct 2009	Aug 2010
Bar crest average offshore distance (m)	193	191	217	208	202	204 [1]
Bar crest average depth (m)	3.65	4.02	3.91	4.4	3.92	4.29 [1]
Average bar height (m)	0.99	0.6	0.78	0.35	0.66	0.29 [1]

[1] Real profile characteristics for August 2010.

During the second seasonal cycle in November 2009–August 2010 (Figure 3b, right panel; Table 1), regardless of partial availability of data, it is evident that the pattern of intra-annual bar evolution is being repeated, but from shoreward shifted location of origin. Even though, only the winter profiles are relatively uniform in shape, generally all profiles are more symmetric. At the beginning of the second cycle, the initial average bar height was 0.60 m and as the repetition of migration pattern proceeded the offshore bar relocation was accompanied by significant bar crest erosion, thus, increasing the average depth above the crest to 4.4 m. A common feature for both seasonal cycles is that in winter the sand bar has larger heights than in spring-summer season.

4.1.2. Bar Evolution due to Wave Parameters and Non-Linearity

The initial steps to analysis of the annual and seasonal bar evolution were to examine the dependence between bar crest locations in 2009–2010 and wave parameters such as significant wave height Hs, peak period Tp and wave steepness Hs/Lp (Figure 4).

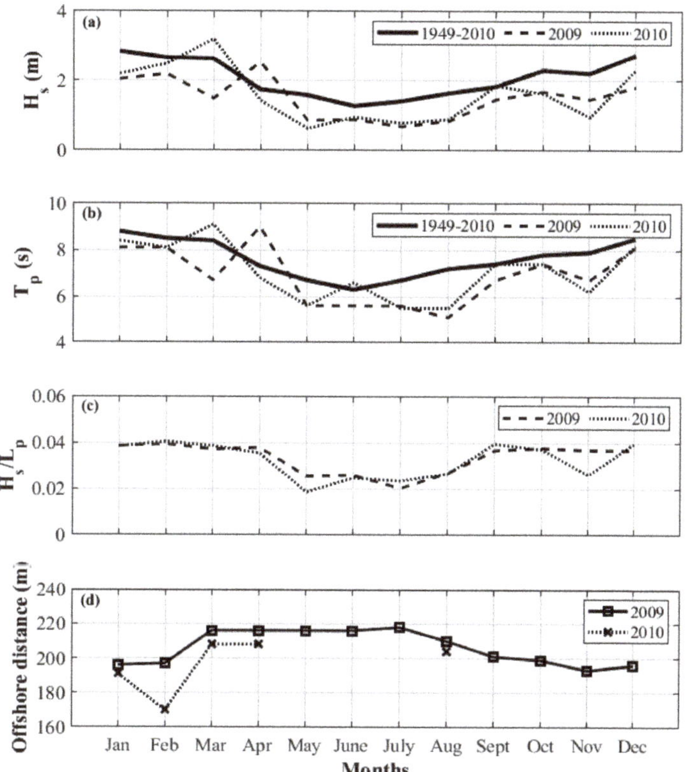

Figure 4. Intra-annual variations of: (**a**,**b**) monthly Q99 of significant wave heights Hs and peak periods Tp for 2009–2010 vs. annual monthly Q99 of the same parameters for 1948–2010 at depth 17 m; (**c**) wave steepness of transformed monthly Q99 waves for 2009–2010 at depth 4.4 m; (**d**) average monthly positions of the outer bar crests for 2009–2010.

The analysis comprised an estimate of 2009–2010 wave conditions in comparison to the regional wave climate and whether either of the years was more energetic than the others. To this end, monthly Q99 of Hs and Tp for 2009–2010 were compared to the same annual monthly Q99 for 1948–2010 (Figure 4a,b). Results show that in general, 2009–2010 wave parameters do not exceed the climatic ones. Exceptions concerning wave height are higher estimates for April 2009 and March 2010, and a commensurate value for September 2010. Wave periods follow the same tendency plus higher estimates for June 2010. As for the comparison between the years 2009 and 2010, it appears that on average wave heights are commensurable (1.5 m in 2009 vs. 1.61 m in 2010), but a bit longer wave periods (7.1 s) were present in 2010 as oppose to those in 2009 (6.9 s).

The comparison also shows that bar crest position depends on variations of all wave parameters throughout the year. It keeps the same displacement pattern during both years and a clear dependence is seen not only on wave height, but also on wave period and steepness. Winter storms with larger wave heights (periods) and Hs/Lp ≈ 0.04 erode the profile moving the sand bar offshore (Figures 3 and 4d). Summer storms with lower heights (periods) and Hs/Lp < 0.03 are not capable to induce shoreward bar migration, which is done in the autumn when steepness again increases up to 0.04. Thus, as per Figure 4d in November–February 2009 (same for 2010) the bar crests are found at distances 195–200 m, while at the end of the season the bar is moved offshore (≈220 m) and remains stable during March–July 2009 and March–April 2010. Shoreward bar's shift toward its winter location was initiated by storms in August–October 2009 and August 2010.

It is of interest to further examine which values of wave steepness (Hs/Lp) correspond to the positions of the bar crest in the coastal zone. For that purpose, crest offshore distances were set against steepness values of the transformed monthly Q99 waves for 2009–2010 (Figure 5).

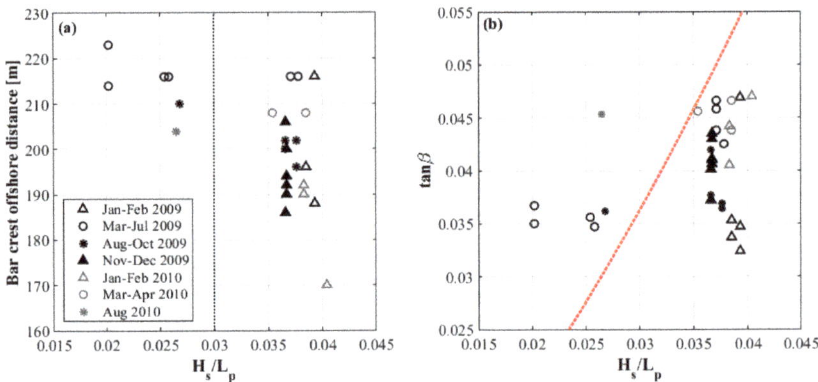

Figure 5. Dependence of steepness Hs/Lp of transformed monthly Q99 waves (depth 4.4 m) for 2009–2010 on (**a**) the offshore distances of the outer bar crest, black dashed line—wave steepness threshold, and (**b**) the mean slope tanβ, red dashed line—criterion for identification of non-linear wave transformation scenarios [$\tan\beta = 7(Hs/Lp)^{3/2}$]; black symbols—2009 values, grey symbols—2010; circles—seaward bar crest positions, triangles—shoreward bar crest locations, stars—bar crest displacements.

According to the presented results steepness 0.03 could be introduced as a threshold to support differentiation between bar's stable position and its relocations within the nearshore area. Thus, for Hs/Lp < 0.03, as in summer months May–August 2009 and August 2010, the sand bar was stable and located offshore, while for Hs/Lp $\geq$ 0.035–0.04 the bar crest was found either offshore ($\approx$220 m) or closer to the coastline, or migrating within distances 170–210 m (Figures 3 and 4). Additionally, wave steepness for January–February 2009 (same for 2010) was higher than for November–December 2009 due to lower wave heights at the end of 2009.

As next, it was necessary to examine for which steepness values the bar remains stable or is subjected to displacement. Another parameter affecting wave steepness and non-linear evolution of waves in the coastal zone is the mean slope (tanβ). Depending on the Irribaren number, non-linear wave transformation may proceed by four characteristic scenarios [49]. The scenarios are distinguishable for the periodicity of wave energy exchange between the first and the second non-linear harmonics, as the slope affects the number of periodic cycles and the growth of the second harmonics, which in turn influence the sediment transport along the profile [60].

Using the criterion for realization of the scenarios [49] a dependence was sought between the wave steepness and the mean slope (Figure 5b). As per criterion, for mean slope $\tan\beta > 7(Hs/Lp)^{3/2}$ waves in the nearshore zone transform by either of two scenarios, i.e., amplitudes of the second harmonics grow only very close to the shore (S1) or remain small along the entire coastal zone (S3). Results in (Figure 5b) state that for waves in May–August 2009 and in August 2010 scenarios S1 and S3 are applicable. According to the hindcast data these waves have Hs/Lp < 0.03, the average monthly Hs varies within 0.67–0.86 m and the average monthly Tp is between 5.1–5.6 s. Based on 2007 field data, the influence of scenario S3 on the morphology of the submerged profile was explored in [60] on a case of a wave regime with significant height 0.6 m and peak period $\approx$ 5 s measured at the pier's end. These results confirm the validity of a wave parameters' range introduced above. Moreover, for this scenario only the inner bar profile was subjected to deformations and significant shoreward relocation, while changes on the outer bar were minor. Thus, absence of large second harmonics and spatially small periods of energy exchange dictate weak non-linearity of waves over the outer bar, which contributes to its relative stability in the spring-summer months.

On the other hand, for $\tan\beta < 7(Hs/Lp)^{3/2}$ wave transformation proceeds according to one of the other two scenarios, i.e., amplitudes of the second harmonics reach their maxima within the coastal zone (S2) or second harmonics already have large amplitudes upon entering the coastal zone (S4). Such highly non-linear scenarios are valid for data pairs in January–April 2009 and September 2009–April 2010 (Figure 5b). According to hindcast data, steepness of these waves varies between 0.035–0.04, the average monthly Hs: 1.48–2.12 m and the average monthly Tp: 6.7–9.1 s. Once again on the basis of 2007 field data, the influence of scenario S2 on the evolution of the submerged beach profile was studied in [60,62] for a wave regime with significant height 1.1 m and period 7 s measured at the pier's end. They established that at the length of each full period of energy exchange between the first and the second non-linear harmonics erosion occurs on the seaward bar's slope with sediments being transferred to its shoreward front either for the inner or the outer bar. Furthermore, if the contribution of the undertow and the transition from S2 to S3 is considered, such periodic energy exchange can lead to bar's significant shift toward the shore and changes in its symmetry [60].

To draw a more precise conclusion about the waves capable to act on the bar's stability or displacement, threshold values for Hs and Tp were determined by setting the monthly Q99 Hs and Tp (depth 4.4 m) against the bar crest offshore distances. This resulted in definition of Hs $\approx$ 1 m and Tp $\approx$ 6 s. Having this in mind, it was assumed that waves responsible for cross-shore bar migration transform according to scenarios S2 or S4 and have steepness >0.03, Hs > 1 m and peak periods over 6 s.

Therefore, it might be suggested that the summer profiles and the relevant bar's stability are aided and contributed by the influence of weakly non-linear waves (scenarios S1 or S3), while winter and transitional profiles, as well as bar's migration are governed by strong non-linearity of waves represented by scenarios S2 or S4.

Next in line was to examine whether any combination of wave parameters render influence on the bar's movement. To this end, we considered the distribution of bar crest positions against the average monthly data of Hs, Tp and Hs/Lp (depth 4.4 m) for January 2009–August 2010. In addition, a reference bar crest distance of 196 m was determined, based on the most frequent profile measurements in winter 2009 (Figure 3a). The comparison showed that in January–April 2009 and September 2009–April 2010 for similar wave steepness values $\approx$ 0.04 the bar crest was localized either close or away from the shore with regard to the reference line (Figure 3a). As for the other wave parameters, it became evident that to a narrow interval of steepness values (0.035–0.04) corresponded a wide range of wave heights: 1.48–2.12 m, respectively periods: 6.7–9.1 s. Despite the time scale discrepancy in the compared data, it was concluded that regardless of the similar steepness values waves that cause the bar crest to shift seaward are higher and with longer periods than those moving it in the opposite direction, which is especially valid for January–April 2009 (2010)—Figures 3a and 4. However, these general findings do not justify in completion the observed intra-annual bar evolution. This implied the necessity to investigate the influence of individual storms and their characteristics on the cross-shore bar crest relocations during January–April 2009 and September 2009–April 2010.

4.1.3. Influence of Storms and Wave Incidence Angle on Bar Dynamics

The analysis was based on hourly time series (SWAN model output transformed to 4.4 m depth) of parameters Hs, Tp and Hs/Lp for November 2008–August 2010, which were chronologically collated with bar crest positions. It was obvious that discrepancies in the temporal resolution of modeled and measured data introduce difficulties in definition of direct causal connection between particular wave events and monthly measurements of the bar profiles with pre- or post-storm identity. Therefore, we applied the thresholds for scenarios with strong non-linearity (Hs > 1 m, Tp > 6 s and Hs/Lp > 0.03) to the time series in order to identify individual storms or sequences of storms that might be claimed responsible for the cross-shore bar migration.

Application of the thresholds showed that in summer months of 2009–2010 waves corresponded to scenarios S1 and S3 and wave conditions caused minor changes to bar height at its farthermost position at ≈ 220 m (Figure 3). However, influenced by the first autumn storms in September 2009 having characteristics valid for scenarios S2&S4 (Hs ≈ 1.5 m, Tp ≈ 6.5 s, Hs/Lp ≈ 0.04), the sand bar began to move towards the shore reaching its reference location of 196 m in November 2009. The following storm events (Hs ≈ 1.5–2 m, Tp ≈ 7–9 s, Hs/Lp > 0.04) caused minor back-and-forth bar crest shifts of 10–15 m. This state retained up to mid-storm season (January–February), when a storm occurred causing the bar crest initial shoreward displacement. For 2009, such a storm was in January, while for 2010 in February (Figure 3b). The subsequent heavy storm: in February for 2009 (Hs > 2 m, Tp ≈ 8 s, Hs/Lp ≈ 0.04) and in March for 2010 (Hs ≈ 2.2 m, Tp ≈ 9 s, Hs/Lp ≈ 0.045) caused the maximum bar crest offshore shift to its spring-summer location increasing significantly the depth above the crest (Figure 3a,b; Table 1).

Such approach contributed to the definition of individual storms that were deemed responsible for the shoreward/seaward shifts of the bar (Table 2). The hourly time series for each storm or sequence of storms were additionally processed, keeping for analysis only the data fulfilling the thresholds for non-linear scenarios, i.e., Hs > 1 m, Tp > 6 s and Hs/Lp > 0.03. Afterwards, wave parameters of each storm were averaged to a single value in order to make a comparison between wave steepness values, significant wave heights and peak periods for each storm group (Table 2).

Table 2. Average wave parameters fulfilling the thresholds for scenarios S2&S4 for storms responsible for outer bar displacements.

Storm Groups (SG)	Hs (m)	Tp (s)	Hs/Lp	Duration of Wave Conditions with Hs/Lp > 0.04 (h)
SG1: Storms moving the outer bar shoreward (January 2009, September–October–November 2009, January 2010)	1.59	6.7	0.036	66
SG2: Storms moving the outer bar seaward (February 2009, December 2009, February–March 2010)	1.7	7.2	0.037	121

It appeared that in 2009–2010 storm conditions for each group were rather similar, although waves in SG2 are slightly higher, steeper and with longer periods. The importance of wave height and period for shore profile changes at the site were studied by means of numerical modeling in [66,67]. They revealed that the typical scenario of wave action on the beach profile regardless of the season is characterized by erosion near the shoreline and accumulation of sediments seaward. The general trend is that the higher the storm waves and longer the periods the larger the shoreline erosion and the further the seaward accumulation of sediments. For same storm waves with shorter periods, the sediments move closer to the shore. However, a significant difference was noticed concerning the time duration of wave conditions for which Hs/Lp exceeded 0.04 (Table 2). For storms causing bar's seaward shift (SG2) these conditions lasted as twice longer (121 h) as those moving the bar closer to the coast (66 h). This suggests that not only steepness and wave period should be taken under consideration in analysis of bar behavior but the duration of these conditions as well.

To examine the role of wave incidence angle on bar migration hourly time series of mean direction of wave propagation Dm (depth 17 m) for November 2008–August 2010 were used. As previously, the time series were subjected to the threefold criterion for scenarios S2&S4 to analyze only the data fulfilling these conditions on three time scales:

- Storm season, November 2008–April 2009 and September 2009–April 2010;

- Storm groups, as per Table 2;
- Monthly storms, here, three months with the most severe wave conditions were selected from each storm group SG1&SG2.

 Results are presented in Figure 6 using rose charts for each time scale.

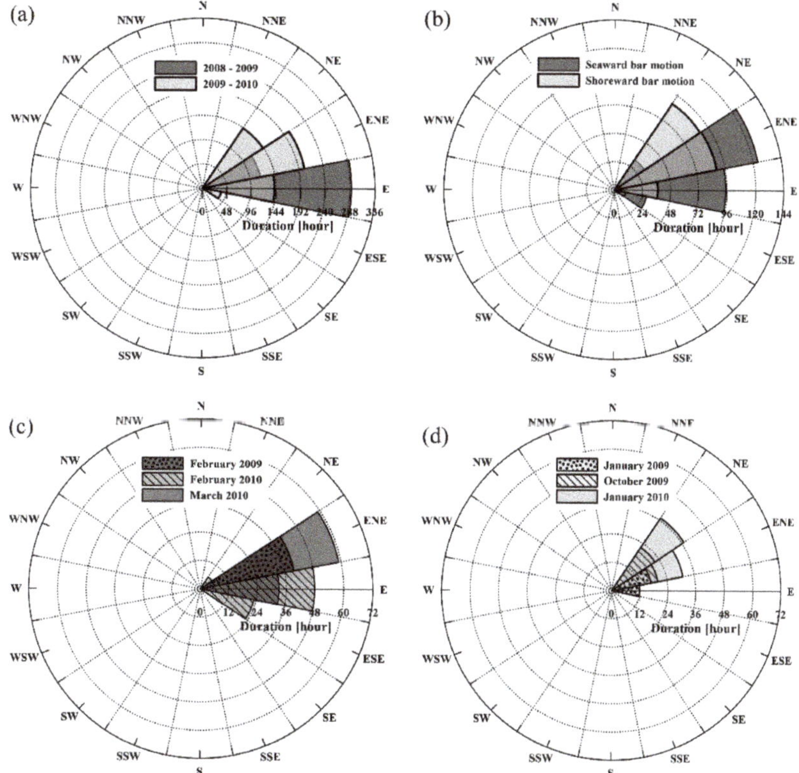

Figure 6. Rose charts of mean direction of wave propagation for: (**a**) storm seasons; (**b**) storm groups SG1&SG2; (**c**) strongest storms moving the outer bar seaward; (**d**) strongest storms moving the outer bar shoreward.

During season 2008–2009 (Figure 6a), waves predominantly were approaching in normal to the shore (E-ENE), while in 2009–2010 oblique incidence dominated (NE-ENE). So, for the present case an assumption was made that waves coming at an angle to the coast contribute to bar being moved shoreward due to the presence of long-shore currents, while those approaching in the normal move the bar seaward, mainly governed by non-linear transformation of waves, when they propagate to the coast. Confirmation was found in results concerning the other two time scales. For Storm groups (Figure 6b) waves displacing the bar shoreward have a more oblique approach (NE-ENE) than those moving it seaward. The assumption was verified at the time scale Monthly storms (Figure 6c,d) because storms causing bar shoreward displacement have a distinct NE approach as opposed to storms moving it seaward (ENE to ESE).

Another factor affecting the outer bar movements is the duration of storms. For both time scales Storm groups and Monthly storms (Figure 6b–d) the storms coming from NE-ENE and moving the bar shoreward have shorter durations than the storms approaching from E-ENE and shifting the bar offshore. These duration differences do not exceed 20–30%, which means that the impact of storms balance the bar's cross-shore migration during the winter season, causing the bar crest to move back-and-forth with respect to the reference distance of 196 m (Figure 3a). As for the time scale Storm seasons results show (Figure 6a) that for season 2009–2010, when the dominant direction of wave approach was NE-ENE the duration is 17.7% longer than the duration in season 2008–2009. These findings were also considered with regard to displacements of the bar crest (Figure 3a). To this end, we determined the maximum cross-shore offsets of the crest from its reference distance. The analysis showed that for storm season 2008–2009 bar's seaward offset is larger (19.87 m) than its offset to the shore (8.24 m). An opposite situation was detected for storm season 2009–2010, when bar's seaward offset equals 11.93 m, while its shoreward displacement is 26.17 m. Therefore, the predominance of oblique wave approach and the longer storm duration in 2009–2010 may have caused the shift of the bar's evolution pattern toward the shore in 2010 (Figure 4d).

Thus far, it may be concluded that intra-annual outer bar evolution follows a seasonal pattern of cross-shore migration, which is mainly governed by scenarios of transformation of highly non-linear waves, and the direction of its off/onshore displacement depends on wave period, duration of wave conditions with steepness >0.04, angle of wave approach and total duration of storms. Annual bar evolution, on the other hand, depends on wave height and storm's parameters as angle of approach and duration.

4.2. Possible Periods of Inter-Annual Outer Bar Evolution due to Variations of Wave Climate

According to analysis in Section 4.1, the location of the outer sand bar predominantly depends on variations of 99th Quantile of the significant wave heights (Q99Hs). As presented in Figure 4 for different years (2009 and 2010) annual variations of bar location depend on the annual variations of Q99Hs, i.e., whether maximum waves were higher or smaller throughout a given year. The fluctuations in wave heights and wave climate depend on wind conditions (wind climate), the variations of which depend on teleconnection patterns determined through the values of the corresponding climatic indices.

Based on long-term data for the Black Sea it was revealed in [68,69] that on a large time scale the indices the Atlantic Multi-decadal Oscillation (AMO), the North Atlantic Oscillation (NAO) and the East Atlantic-Western Russia (EA/WR) have significant influence on the fluctuations of maximum annual wave heights. On a decadal time scale and less, variations of maximum annual wave heights might depend on the NAO, the Arctic Oscillation (AO), the EA/WR, the East Atlantic Oscillation (EA) and the Scandinavian (SCAND) patterns [68,70–72].

In order to determine any periodicity in variations of the maximum annual wave height analysis was conducted to reveal a connection between temporal fluctuations of the Q99Hs and variations of the main climatic indices for the Black Sea study region. The following indices were considered: the NAO, the AMO, the EA, the AO, the EA/WR and the SCAND. Their dimensionless values were taken from the National Oceanic and Atmospheric Administration (NOAA) of the USA [73]. Variations of the Q99Hs for a deep-water point (depth 55 m, Figure 1) offshore the study site and selected climatic indices are shown in Figure 7.

Results show that the maximum annual wave height fluctuates in time around a mean value of 3.04 m. The analysis did not reveal any significant linear trend toward decrease or increase of wave height in time. Minimum wave height was found to be 2.04 m, while the maximum was equal to 4.28 m.

Wavelet analysis by the Morlet wavelet function, representing the temporal evolution of frequency spectrum shows that fluctuations of the maximum annual waves have non-stationary nature (Figure 8, upper panel). Frequency interval 0.08–0.2 [1/year] is of a particular interest, since trends for both increase and decrease were observed therein. Additionally, fluctuations of amplitudes (Figure 8, lower panel) vary in time, as well. Most stable are the temporal variations in frequency interval 0.05–0.03 [1/year] corresponding to periods of 20–30 years. However, there is an insignificant trend toward decrease of fluctuation frequency in time.

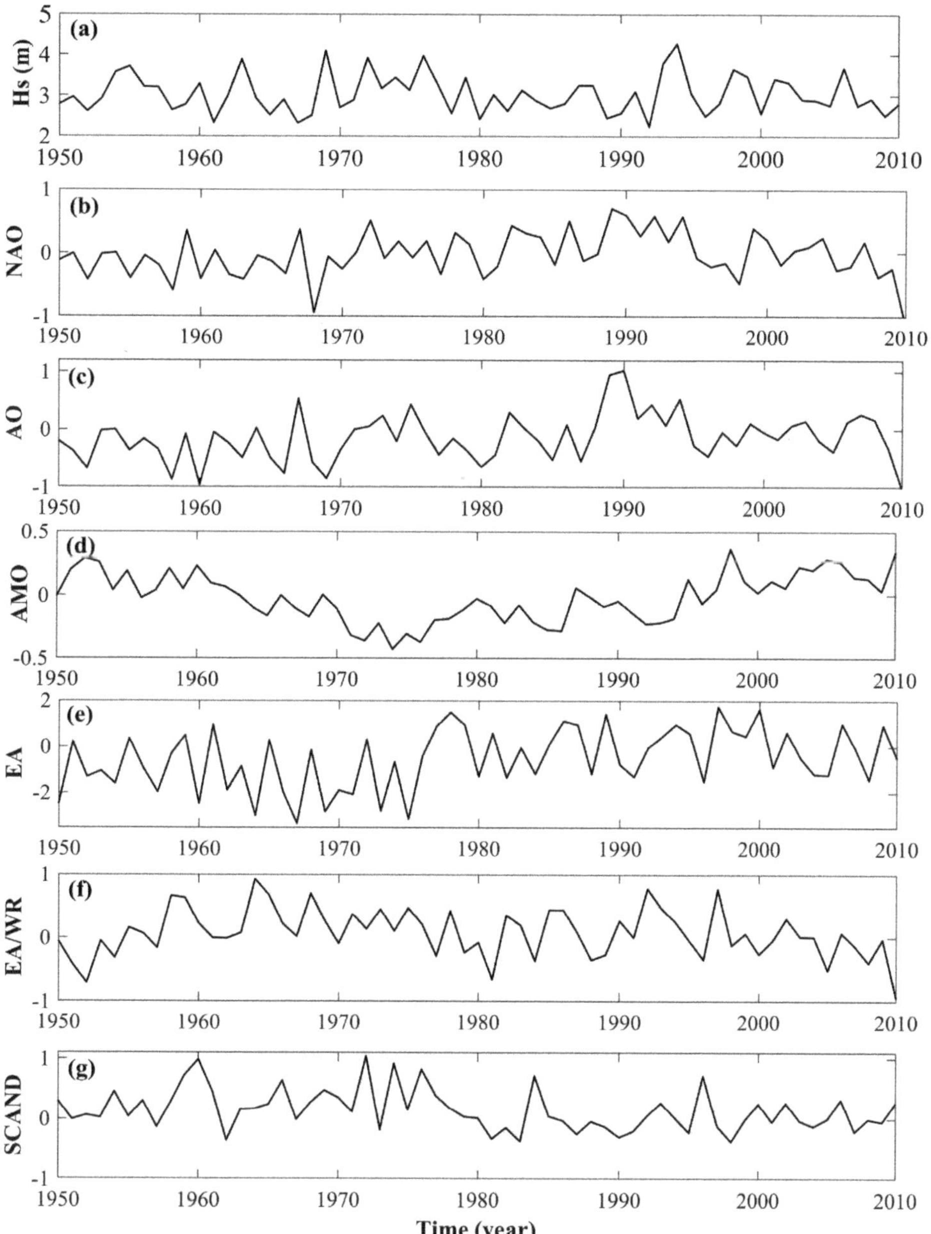

Figure 7. Variations of (**a**) 99th quantile of the significant wave height and (**b**) the North Atlantic Oscillation (NAO), (**c**) the Atlantic Multi-decadal Oscillation (AMO), (**d**) the East Atlantic Oscillation (EA), (**e**) the Arctic Oscillation (AO), (**f**) the East Atlantic-Western Russia (EA/WR) and (**g**) Scandinavian (SCAND) patterns for 1950–2010.

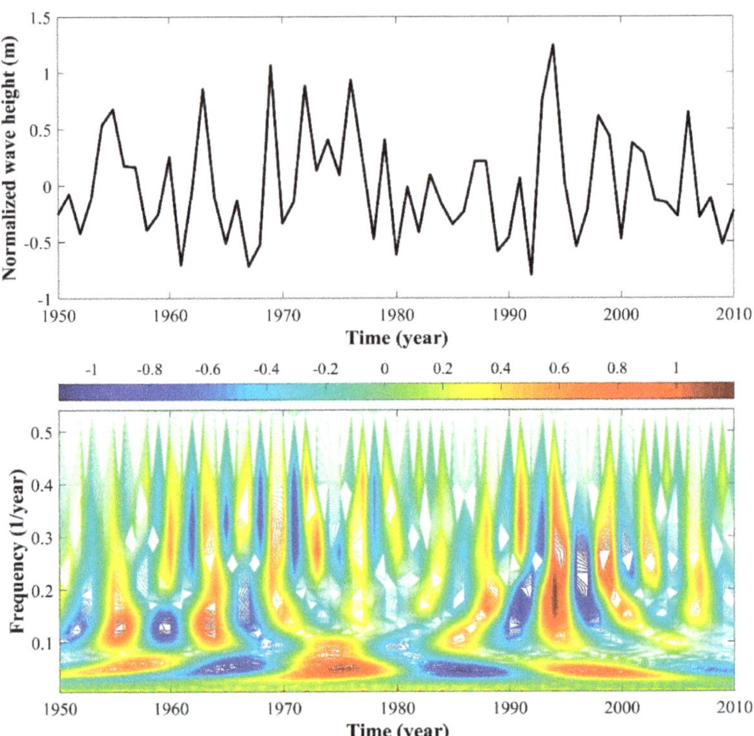

Figure 8. Fluctuations of normalized maximum annual wave time series (mean multi-annual was removed) (upper panel) and wavelet transformation (lower panel).

A previous study by [68] has shown that fluctuations of climate indices are also non-stationary, which makes it impossible to apply the classic correlation analysis to determine the relation between them and the wave height variations. However, a methodology to correlate two non-stationary process based on wavelet analysis was developed in [68,72]. According to this method, mutual correlation functions were calculated between the same frequency scales of two wavelet decompositions, which gives the advantage of having both correlation coefficients and frequencies, where these coefficients have the largest values.

Wavelet correlation coefficients (with time lag = 0) of the maximum wave height variations for selected climate indices are presented in Figure 9. If one considers that correlation coefficient >0.4 represents good correlation between geophysical processes, then for large (multi-decadal) time spans of 20–30 years variations of wave heights depend on indices the EA and the SCAND, respectively. Fluctuations corresponding to 10–15 years depend on the EA/WR (9–13 years), the EA (10–11 years), the AMO (13 years), the NAO (15 years) and the AO (15–16 years). There are also fluctuations of the order of few years that depend on the EA (4–5 years), the EA/WR (4 years), the AMO/AO (3 years), the SCAND and the EA/WR (2 years).

According to the presented analysis, it may be suggested that the Q99 of the wave climate in the coastal region under study at all time spans would be influenced the strongest by the EA and the EA/WR climate indices. Additionally, strong influence might be expected due to variations of indices the AMO, the AO and the SCAND. Significant correlation was established for the NAO index (0.5), but only for variations with time span of 15 years.

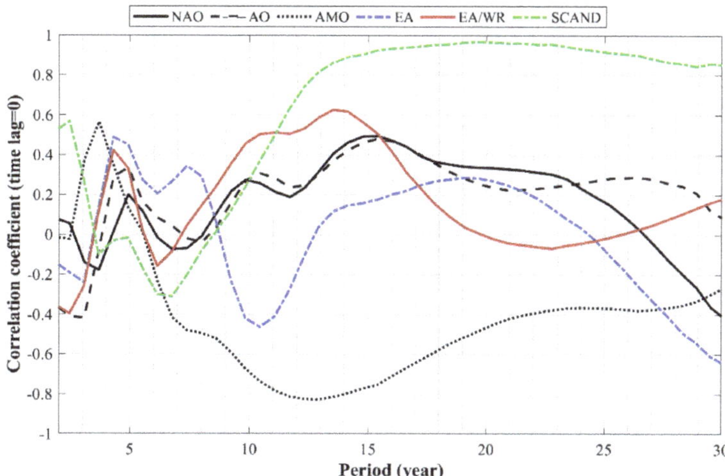

Figure 9. Wavelet correlation coefficients for fluctuations of maximum annual wave height and climate indices: the North Atlantic Oscillation (NAO), the Atlantic Multi-decadal Oscillation (AMO), the East Atlantic Oscillation (EA), the Arctic Oscillation (AO), the East Atlantic Western Russia (EA/WR) and Scandinavian (SCAND).

Thus, it could be expected that mean inter-annual location of the outer bar crest may vary depending on periods of maximum annual wave fluctuations, which in turn predominantly depend on climate indices the EA and the EA/WR. For example, 9 years periodicity of autumn bar location possibly connected with EA/WR can be seen in Figure 2 for years 2007 and 2016.

5. Conclusions

A study has been conducted examining inter- to intra-annual nearshore bar dynamics on the open-coast of Kamchia-Shkorpilovtsi sandy beach at the Bulgarian Black Sea as influenced by regional wave climate, making use of field measurements of seabed morphology, numerically modeled wave data at different time scales and examination of the influence of different telecommunication patterns.

It has been shown that in 2009–2010 on an intra-annual time scale the cross-shore bar (crest) migration followed a certain repetitive seasonal pattern, mainly determined by non-linear transformation of waves in the coastal zone. During summers, the bar retained its stability farthermost off the shore due to weakly non-linear and low-energy wave regimes. At the same time, the most active in the view of crest displacements are autumn-winter and winter-spring periods in both years, as closest to the shore the outer bar was in winter with minor seaward and shoreward shifts near its reference location. It has been revealed that highly non-linear wave regimes and scenarios of waves transformation are responsible for cross-shore bar migration having wave steepness >0.03, significant height over 1 m and peak period more than 6 s. Even though these threshold values were determined based on hindcast monthly average data their validity has been supported by recent scientific studies at the site, as drawing of more precise thresholds could be secured by regular wave and beach profile measurements and analysis. Furthermore, the results on the influence of storms on the bar migration revealed that direction of crest displacement primarily depends on the wave period, the duration of wave conditions with $Hs/Lp > 0.04$, angle of wave incidence and total duration of storms. Thus, among the studied storms having rather similar average significant heights, those moving the outer bar seaward had longer periods (7.2 s), twice as long duration of retained over 0.04 steepness (121 h), lasted by ≈20% longer and were predominantly approaching from E-ENE, as opposed to storms with more oblique angle of incidence NE-ENE displacing the bar shoreward. On annual basis the bar evolution was found the be mainly governed by wave height and storms' parameters as angle of approach and duration, since the predominance of oblique wave incidence and longer storm duration in 2009 could be responsible for shoreward shift of bar evolution pattern and increase of depth above its crest. It is appropriate to comment that according to [74] seasonal fluctuations of individual profiles could be as pronounced as alongshore variations of a rhythmic seabed topography, which recently has been confirmed in [15] based on field observations and numerical modeling.

Results concerning the possible connection between the temporal periodicity of Q99Hs and variability of the climatic indices showed that for all time spans, the regional wave climate would be most influenced by the EA and the EA/WR indices. Thus, it is expected that mean inter-annual outer bar location may vary depending on the periods of maximum annual wave height fluctuations, which in turn predominantly depend on the climate indices the EA (4–5, 10–11, 20–30 years) and the EA/WR (2–4, 9–13 years), but also may be affected by the AMO (3, 13 years), the AO (3, 15 years) and the SCAND (2, 20–30 years). The presented results for the first time demonstrate a connection between the climatic indices and the nearshore bar position for the Bulgarian Black Sea coast.

Author Contributions: Writing of the paper, data processing and analysis, visualization of results, N.A.; Article main idea, analysis of climate change and teleconnection patterns, visualization of results, preparation and writing of one section of the paper, Y.S.; SWAN modelling, hindcast data processing and analysis, N.V.; XBeach modelling, experimental data processing and analysis, visualization of results, P.E.; Analysis of climate change and teleconnection patterns, S.K. All authors contributed to interpretation of results, provided critical feedback and helped shape the research. All authors participated in field experiments on the study site. All authors have read and agreed to the published version of the manuscript.

Funding: This research was partly funded by RFBR, grant number 20-55-46005.

Institutional Review Board Statement: Not applicable.

Informed Consent Statement: Not applicable.

Data Availability Statement: Not applicable.

Acknowledgments: This research was supported by 'National Geo-information Center for monitoring, evaluation and forecasting of natural and anthropogenic risks and disasters' within the Program 'National Roadmap for Scientific Infrastructure (2017–2023)' of Republic of Bulgaria, Contract № ДО1-282/17.12.2019; The research was also performed in the framework of Russian Federation state assignment theme № 0128-2021-0004.

Conflicts of Interest: The authors declare no conflict of interest.

References

1. Plant, N.G.; Holman, R.A.; Freilich, M.H. A simple model for inter-annual sandbar behavior. *J. Geophys. Res.* **1999**, *104*, 15755–15776. [CrossRef]
2. O'Hare, T.J.; Huntley, D.A. Bar formation due to wave groups and associated long waves. *Mar. Geol.* **1994**, *116*, 313–325. [CrossRef]
3. Wijnberg, K.M.; Kroon, A. Barred beaches. *Geomorphology* **2002**, *48*, 103–120. [CrossRef]
4. Dyhr-Nielsen, M.; Sorensen, T. Some sand transport phenomena on coasts with bars. In Proceedings of the 12th International Conference on Coastal Engineering, Washington, DC, USA, 13–18 September 1970; pp. 855–866. [CrossRef]
5. Sallenger, A.H.; Howd, P.A. Nearshore bars and the breakpoint hypothesis. *Coast. Eng.* **1989**, *12*, 301–313. [CrossRef]
6. Roelvink, J.A. Surf Beat and its Effect on Cross-Shore Profiles. Ph.D. Thesis, Delft University of Technology, Delft, The Netherlands, 1993; 150p. Available online: http://resolver.tudelft.nl/uuid:160325cb-e1d2-46ec-855e-4626dc4ab754 (accessed on 7 May 2021).
7. Ruessink, B.G. Infra-Gravity Waves in a Dissipative Multiple Bar System. Ph.D. Thesis, Utrecht University, Utrecht, The Netherlands, 1998; 254p.
8. Boczar-Karakiewicz, B.; Davidson-Arnott, R.G.D. Nearshore bar formation by non-linear wave processes—A comparison of model results and field data. *Mar. Geol.* **1987**, *77*, 287–304. [CrossRef]
9. Falques, A.; Montoto, A.; Iranzo, V. Bed-flow instability of the longshore current. *Cont. Shelf Res.* **1996**, *16*, 1927–1964. [CrossRef]

10. Larson, M.; Kraus, N.C. *Analysis of Cross-Shore Movement of Natural Longshore Bars and Material Placed to Create Longshore Bars*; Technical Report DRP-92-5; U.S. Army Engineer Waterways Experiment Station: Vicksburg, MS, USA, 1992; p. 116. Available online: http://resolver.tudelft.nl/uuid:72cb2f64-5ad7-4e7f-9a15-2a778eab1d9e (accessed on 7 May 2021).
11. Yuhi, M.; Matsuyama, M.; Hayakawa, K. Sandbar migration and shoreline change on the Chirihama Coast, Japan. *J. Mar. Sci. Eng.* **2016**, *4*, 40. [CrossRef]
12. Ruessink, B.G.; Ranasinghe, R. Beaches. In *Coastal Environments & Global Change*, 1st ed.; Masselink, G., Gehrels, R., Eds.; John Wiley & Sons, Ltd.: Chichester, UK, 2015; pp. 149–177.
13. Walstra, D.J.R. On the anatomy of nearshore sandbars: A systematic exposition of inter-annual sandbar dynamics. Doctoral Thesis, Delft University of Technology, Delft, The Netherlands, 2016; 138p. [CrossRef]
14. Ribas, F.; Albert Falqués, A.; Garnier, R. Nearshore sand bars. In *Atlas of Bedforms in the Western Mediterranean*; Guillén, J., Acosta-Yepes, J., Chiocci, F.L., Palanques, A., Eds.; Springer International Publishing: Cham, Switzerland, 2017; Chapter 13, pp. 73–79.
15. Ruggiero, P.; Walstra, D.J.R.; Gelfenbaum, G.; van Ormondt, M. Seasonal-scale nearshore morphological evolution: Field observations and numerical modeling. *Coast. Eng.* **2009**, *56*, 1153–1172. [CrossRef]
16. Shepard, F.P. *Beach Cycles in Southern California. Beach Erosion Board, Technical Memorandum*; Beach Erosion Board Engineer Research and Development Center: Vicksburg, MS, USA, 1950; Volume 20, p. 31. Available online: https://hdl.handle.net/11681/3369 (accessed on 7 May 2021).
17. Dubarbier, B.; Castelle, B.; Ruessink, G.; Marieu, V. Mechanisms controlling the complete accretionary beach state sequence. *Geophys. Res.* **2017**, *44*, 5645–5654. [CrossRef]
18. Ruessink, B.G.; Kuriyama, Y.; Reniers, A.J.H.M.; Roelvink, J.A.; Walstra, J.A. Modeling cross-shore sandbar behavior on the timescales of weeks. *J. Geophys. Res.* **2007**, 112. [CrossRef]
19. Grasso, F.; Michallet, H.; Certain, R.; Barthélemy, E. Experimental flume simulation of sandbar dynamics. *J. Coast. Res.* **2009**, *56*, 54–58. Available online: https://www.jstor.org/stable/25737536 (accessed on 7 May 2020).
20. Walstra, D.J.R.; Ruggiero, P.; Lesser, G.; Gelfenbaum, G. Modeling nearshore morphological evolution at seasonal scale. In Proceedings of the Fifth Coastal Dynamics International Conference, Barcelona, Spain, 4–8 April 2005. [CrossRef]
21. Ruessink, B.G.; Pape, L.; Turner, I.L. Daily to interannual cross-shore sandbar migration: Observations from a multiple sandbar system. *Cont. Shelf Res.* **2009**, *29*, 1663–1677. [CrossRef]
22. Cheng, J. Multiple scales of beach morphodynamic processes: Measurements and modelling. Ph.D. Thesis, University of South Florida, Tampa, FL, USA, 2015; p. 198. Available online: https://scholarcommons.usf.edu/etd/5924 (accessed on 7 May 2021).
23. Vidal-Ruiz, J.A.; de Alegría-Arzaburu, A.R. An annual cycle of sandbar migration on an intermediate meso-tidal beach: Ensenada, Mexico. In Proceedings of the Coastal Dynamics 2017, Helsingør, Denmark, 12–16 June 2017; pp. 575–585, Paper No. 223.
24. Vidal-Ruiz, J.A.; de Alegría-Arzaburu, A.R. Variability of sandbar morphometrics over three seasonal cycles on a single-barred beach. *Geomorphology* **2019**, *333*, 61–72. [CrossRef]
25. Bergsma, E.W.J.; Conley, D.C.; Davidson, M.A.; O'Hare, T.J.; Almar, R. Storm Event to Seasonal Evolution of Nearshore Bathymetry Derived from Shore-Based Video Imagery. *Remote Sens.* **2019**, *11*, 519. [CrossRef]
26. Prusak, Z.; Nikolov, H. Analysis of type-profiles in the conditions of the coastal zones of Poland and Bulgaria. *Proc. IO-BAS* **1992**, *1*, 65–73. (In Bulgarian)
27. Trifonova, E.; Valchev, N.; Andreeva, N.; Eftimova, P. Critical storm thresholds for morphological changes in the western Black Sea coastal zone. *Geomorphology* **2012**, *143–144*, 81–94. [CrossRef]
28. Nikolov, H. Thickness of the eroded layer and tendencies in beach profile changes in front of seaside resorts 'Albena', 'Golden sands' and 'Drujba'. *Oceanology* **1981**, *8*, 54–60. (In Bulgarian)
29. Keremedchiev, S. Seasonal fluctuations of beach pr ofiles. *Oceanology* **1985**, *14*, 63–74. (In Bulgarian)
30. Nikolov, H. Influence of wave steepness and bottom slope on the short-term deformations of the underwater bottom slope. In *Interaction of the Atmosphere, Hydrosphere and Lithosphere in the Nearshore Zone—Results of the International Experiment 'Kamchia 78'*, 1st ed.; Belberov, Z., Antsyferov, S., Zahariev, V., Zats, V., Pykhov, N., Eds.; Publishing House of the Bulgarian Academy of Sciences: Sofia, Bulgaria, 1982; pp. 205–212. (In Bulgarian)
31. Antsyferov, S.; Pykhov, N.; Dachev, V. Dynamics of suspended sediments. In *Dynamical Processes in Coastal Regions*, 1st ed.; Antsyferov, S., Belberov, Z., Massel, S., Eds.; Publishing house of the Bulgarian Academy of Sciences: Sofia, Bulgaria, 1990; pp. 157–159.
32. Trifonova, E.; Valchev, N.; Andreeva, N.; Eftimova, P.; Kotsev, I. Measurements and analysis of storm induced short-term morphological changes in the western Black Sea. *J. Coast. Res.* **2011**, *1*, 149–154. Available online: https://www.jstor.org/stable/26482151 (accessed on 7 May 2021).
33. Eftimova, P.; Trifonova, E.; Valchev, N.; Andreeva, N. Beach erosion caused by storms. Morphological model set up and calibration. In Proceedings of the 11th International Conference on Marine Sciences and technologies—Black Sea'2012, Varna, Bulgaria, 4–6 October 2012; pp. 69–73.
34. Valchev, N.; Andreeva, N.; Eftimova, P.; Trifonova, E. Prototype of early warning system for coastal storm hazard (Bulgarian Black Sea coast). *Compt. Rend. Acad. Bulg. Sci.* **2014**, *67*, 971–978.
35. Valchev, N.; Eftimova, P.; Andreeva, N. Implementation and validation of a multi-domain coastal hazard forecasting system in an open bay. *Coast. Eng.* **2018**, *134*, 212–228. [CrossRef]

36. Walstra, D.-J.R.; Wesselman, D.A.; Van der Deijl, E.C.; Ruessink, G. On the intersite variability in inter-annual nearshore sandbar cycles. *J. Mar. Sci. Eng.* **2016**, *4*, 15. [CrossRef]
37. Shand, R.D.; Bailey, D.G.; Shephard, M.J. An inter-site comparison of net offshore bar migration characteristics and environmental conditions. *J. Coast. Res.* **1999**, *15*, 750–765.
38. Ruessink, B.G.; Kroon, A. The behavior of a multiple bar system in the nearshore zone of Terschelling: 1965–1993. *Mar. Geol.* **1994**, *121*, 187–197. [CrossRef]
39. Wijnberg, K.M.; Terwindt, J.H.J. Extracting decadal morphological behavior from high-resolution, long-term bathymetric surveys along the Holland coast using Eigen function analysis. *Mar. Geol.* **1995**, *126*, 301–330. [CrossRef]
40. Ruessink, B.G.; Terwindt, J.H.J. The behavior of nearshore bars on the time scale of years: A conceptual model. *Mar. Geol.* **2000**, *163*, 289–302. [CrossRef]
41. Masselink, G.; Austin, M.; Scott, T.; Poate, T.; Russell, P. Role of wave forcing, storms and NAO in outer bar dynamics on a high-energy, macro-tidal beach. *Geomorphology* **2014**, *226*, 76–93. [CrossRef]
42. Saprykina, Y.; Kuznetsov, S.; Valchev, N. Multi-decadal fluctuations of storminess of the Black Sea due to teleconnection patterns on the base of modelling and field wave data. In *Lecture Notes in Civil Engineering, Proceedings of the Fourth International Conference in Ocean Engineering (ICOE2018), Chennai, India, 18–21 February 2018*; Murali, K., Sriram, V., Samad, A., Saha, N., Eds.; Springer: Singapore, 2019; Volume 22, pp. 773–781. [CrossRef]
43. D-Maps: Free Maps. Available online: https://d-maps.com (accessed on 7 May 2021).
44. Maxar Technologies—High-Resolution Satellite Imagery, Westminster, CO, USA. Available online: https://www.maxar.com (accessed on 7 May 2021).
45. Ostrowski, R.; Pruszak, Z.; Skaja, M.; Szmytkiewicz, M.; Trifonova, E.; Keremedchiev, S.; Andreeva, N. Hydrodynamics and lithodynamics of dissipative and reflective shores in view of field investigations. *Arch. Hydro-Eng. Environ. Mech.* **2010**, *57*, 219–241.
46. Valchev, N.; Davidan, I.; Belberov, Z.; Palazov, A.; Valcheva, N. Hindcasting and assessment of the western Black sea wind and wave climate. *J. Environ. Prot. Ecol.* **2010**, *11*, 1001–1012.
47. Belberov, Z.; Antsyferov, S. Research station conditions and organization of experiments. In *Dynamical Processes in Coastal Regions: Results of Kamchia International Project*, 1st ed.; Popov, V., Antsyferov, S., Belberov, Z., Massel, S., Eds.; Publishing House of the Bulgarian Academy of Sciences: Sofia, Bulgaria, 1990; pp. 14–22.
48. Davidan, I.; Belberov, Z.; Aubrey, D.; Lavrenov, I. Transformation of wind wave spectral parameters according to the Black sea international experiment. *Proc. IO-BAS* **2005**, *5*, 51–64.
49. Saprykina, Y.; Kuznetsov, S.; Andreeva, N.; Shtremel, M. Scenarios of non-linear wave transformation in the coastal zone. *Oceanology* **2013**, *53*, 422–431. [CrossRef]
50. Kuznetsova, O.; Saprykina, Y.; Shtremel, M.; Kuznetsov, S.; Korzinin, D.; Trifonova, E.; Andreeva, N.; Valchev, N.; Prodanov, B.; Eftimova, P.; et al. Dynamics of sandy beach in dependence on wave parameters. In *Maritime Transportation and Harvesting of Sea Resources, Proceedings of the 17th International Congress of the International Maritime Association of the Mediterranean (IMAM'2017), Lisbon, Portugal, 9–11 October 2017*; Guedes Soares, C., Teixeira, A.P., Eds.; Taylor & Francis Group: London, UK, 2017; Volume 2, pp. 1075–1079.
51. Nikolov, H.; Pykhov, N. Brief deformations along the profile of the underwater coastal slope during storm. In *Interaction of the Atmosphere, Hydrosphere and Lithosphere in the Nearshore Zone. Results of the International Experiment 'Kamchia 77'*, 1st ed.; Belberov, Z., Zahariev, V., Kuznetsov, O., Massel, S., Pykhov, N., Rojdestvensky, A., Filyushkin, B., Eds.; Publishing House of the Bulgarian Academy of Sciences: Sofia, Bulgaria, 1980; pp. 229–237. (In Russian)
52. U.S. Army Corps of Engineers. *Coastal Engineering Manual*, 18th ed.; U.S. Army Corps of Engineers: Washington, DC, USA, 2006; Electronic; ISBN 978-1-60119-026-0. Available online: http://app.knovel.com/hotlink/toc/id:kpCEM0000P/coastal-engineering-manual (accessed on 7 May 2021).
53. The SWAN Team. *SWAN User Manual*; Delft University of Technology: Delft, The Netherlands, 2016.
54. Feser, F.; Weisse, R. Multi-decadal atmospheric modeling for Europe yields multi-purpose data. *EOS Trans. AGU* **2001**, *82*, 305–310. [CrossRef]
55. Yang, F.; Pan, H.; Krueger, S.K.; Moorthi, S.; Lord, S.J. Evaluation of the NCEP global forecast system at the arm SGP site. *Mon. Weather Rev.* **2006**, *134*, 3668–3690. [CrossRef]
56. Valchev, N.; Trifonova, E.; Andreeva, N. Past and recent trends in the western Black Sea storminess. *Nat. Hazards Earth Syst. Sci.* **2012**, *12*, 961–977. [CrossRef]
57. Roelvink, D.; Reniers, A.; van Dongeren, A.; van Thiel de Vries, J.; McCall, R.; Lescinski, J. Modelling storm impacts on beaches, dunes and barrier islands. *Coast. Eng.* **2009**, *56*, 1133–1152. [CrossRef]
58. Vieira, B.F.V.; Pinho, J.L.S.; Barros, J.A.O.; Antunes do Carmo, J.S. Hydrodynamics and Morphodynamics Performance Assessment of Three Coastal Protection Structures. *J. Mar. Sci. Eng.* **2020**, *8*, 175. [CrossRef]
59. Castelle, B.; Bonneton, P.; Dupuis, H.; Sénéchal, N. Double bar beach dynamics on the high-energy meso-macrotidal French Aquitanian Coast: A review. *Mar. Geol.* **2007**, *245*, 141–159. [CrossRef]
60. Saprykina, Y. The influence of wave non-linearity on cross-shore sediment transport in coastal zone: Experimental investigations. *Appl. Sci.* **2020**, *10*, 4087. [CrossRef]

61. Korzinin, D.; Shtremel, M. Deformation of coastal profile during different storm phases. In Proceedings of the 36th International Conference on Coastal Engineering, Baltimore, MD, USA, 30 July–3 August 2018; Volume 1, p. 43. [CrossRef]
62. Saprykina, Y.; Kuznetsova, O. Influence of wave transformation processes on evolution of underwater beach profile. In Proceedings of the 36th International Conference on Coastal Engineering, Baltimore, MD, USA, 30 July–3 August 2018; Volume 1, p. 65. [CrossRef]
63. Saprykina, Y.V.; Shtremel, M.N.; Kuznetsov, S.Y. On the possibility of bi-phase parametrization for wave transformation in the coastal zone. *Oceanology* **2017**, *57*, 253–264. [CrossRef]
64. Saprykina, Y.V.; Kuznetsov, S.Y.; Divinskii, B.V. Influence of processes of nonlinear transformations of waves in the coastal zone on the height of breaking waves. *Oceanology* **2017**, *57*, 383–393. [CrossRef]
65. Longinov, V.V. *Coastal Zone Dynamics in Tideless Seas*; The USSR Academy of Sciences Publisher House: Moscow, Russia, 1963; p. 379. (In Russian)
66. Kuznetsova, O.; Saprykina, Y.; Divinsky, B. Underwater barred beach profile transformation under different wave conditions. In *Coastal erosion and dynamical processes in the nearshore zone, Proceedings of the Materials of XXVI International Coastal Conference Managing Risks to Coastal Regions and Communities in a Changing World, Saint-Petersburg, Russia, 22–27 August 2016*; Academus Publishing: Redwood, CA, USA, 2016. [CrossRef]
67. Kuznetsova, O.; Saprykina, Y. Intra-annual storm deformations on sandy beach by an example of Kamchia-Shkorpilovtsi coast (Black Sea, Bulgaria). *Proc. Geom.* **2017**, *10*, 435–444. (In Russian)
68. Saprykina, Y.V.; Kuznetsov, S.Y. Methods of analysis of nonstationary variability of wave climate of Black Sea. *Phys. Oceanogr.* **2018**, *4*, 156–164. [CrossRef]
69. Polonsky, A.; Evstigneev, V.; Naumova, V.; Voskresenskaya, E. Low-frequency variability of storms in the northern Black Sea and associated processes in the ocean-atmosphere system. *Reg. Environ. Chang.* **2014**, *14*, 1861–1871. [CrossRef]
70. Zăinescu, F.I.; Tătui, F.; Valchev, N.N.; Vespremeanu-Stroe, A. Storm climate on the Danube delta coast: Evidence of recent storminess change and links with large-scale teleconnection patterns. *Nat. Hazards* **2017**, *87*, 599–621. [CrossRef]
71. Saprykina, Y.; Shtremel, M.; Aydoğan, B.; Ayat Aydoğan, B. Variability of the nearshore wave climate in the eastern part of the Black Sea. *Pure Appl. Geoph.* **2019**, *176*, 3757–3768. [CrossRef]
72. Saprykina, Y.; Kuznetsov, S. Analysis of the variability of wave energy due to climate changes on the example of the Black Sea. *Energies* **2018**, *11*, 2020. [CrossRef]
73. National Oceanic and Atmospheric Administration of U.S.A. Available online: https://www.cpc.ncep.noaa.gov (accessed on 23 January 2021).
74. Sonu, C. Three-dimensional beach changes. *J. Geol.* **1973**, *81*, 42–64. Available online: https://www.jstor.org/stable/30060693 (accessed on 7 May 2021). [CrossRef]

geosciences

MDPI

Article

Recent Shoreline Changes Due to High-Angle Wave Instability along the East Coast of Lingayen Gulf in the Philippines

Takaaki Uda [1,*] and Yasuhito Noshi [2]

1 Public Works Research Center, Taito, Tokyo 110-0016, Japan
2 Department of Oceanic Architecture and Engineering, College of Science and Technology, Nihon University, Funabashi, Chiba 274-8501, Japan; noshi.yasuhito@nihon-u.ac.jp
* Correspondence: uda@pwrc.or.jp

Abstract: A small perturbation on the shoreline may develop under high-angle wave conditions, resulting in the formation of sand spits along the shoreline. Serizawa et al. explained the development of sand spits caused by the instability mechanism using the BG model (a model for predicting 3-D beach changes based on Bagnold's concept). However, examples of the development of sand spits caused by this mechanism in the field are limited in number. Lingayen Gulf in the Philippines has a large aspect ratio, so shoreline instability occurs along the coastline, significantly affecting the shore protection along the coast. In this study, the shoreline instability along the river delta coasts around the Balili and Aringay Rivers flowing into Lingayen Gulf and a sand spit were investigated using satellite images together with field observation. The shoreline changes observed south of the Aringay River mouth were compared with those observed in a previous study on the development of a sand spit by San-nami et al. The rate of longshore sand transport to form a sand spit at Santo Tomas in Lingayen Gulf was estimated to be approximately 1.3×10^5 m^3/yr, which is in good agreement with the value measured on the Shimizu coast in Suruga Bay, with a comparable aspect ratio of 1.2 relative to 1.3 in Lingayen Gulf. It was concluded that shoreline undulations have evolved downcoast of two river deltas owing to high-angle wave instability along the east coast of Lingayen Gulf and the formation of a sand spit has occurred. A soft measure, such as sand bypassing, would be better to be adopted along the coasts in Lingayen Gulf instead of hard measures against erosion, to prevent rapid expansion of an artificial, protected coastline.

Keywords: Philippines; Lingayen Gulf; Suruga Bay; high-angle waves; sand spit; shoreline instability; BG model; rate of longshore sand transport

Citation: Uda, T.; Noshi, Y. Recent Shoreline Changes Due to High-Angle Wave Instability along the East Coast of Lingayen Gulf in the Philippines. *Geosciences* **2021**, *11*, 144. https://doi.org/10.3390/geosciences11030144

Academic Editors: Jesus Martinez-Frias and Gianluigi Di Paola

Received: 14 December 2020
Accepted: 9 March 2021
Published: 22 March 2021

Publisher's Note: MDPI stays neutral with regard to jurisdictional claims in published maps and institutional affiliations.

1. Introduction

When waves are obliquely incident to a slender water body at a large angle relative to the direction normal to the shoreline, shoreline undulation may develop owing to high-angle wave instability [1], resulting in the formation of sand spits. Serizawa et al. [2] predicted beach changes under the condition that waves were obliquely incident to the direction normal to the shoreline at an angle over 45° using the BG model (a model for predicting 3-D beach changes based on Bagnold's concept) and showed that sand spits can develop by this instability mechanism. Although real examples of shoreline instability are limited in number because the wave incidence angle is ±20° at most on ordinary coasts, examples can be found in a slender water body. Since Lingayen Gulf in the Philippines is a slender bay, shoreline undulations caused by this instability mechanism occur downcoast of the river deltas, significantly affecting shore protection on the coast. Anthony et al. [3] described the formation of the sand spits associated with the Volta and Senegal River deltas, which show complex patterns of morphodynamic development while also strongly reflecting the recent impacts of human activities, such as the marked reduction in sand supply on the eastern coast of Ghana. In this study, the formation of the shoreline undulations downcoast of the river deltas of the Balili and Aringay Rivers was studied in the same manner using satellite images together with a field observation on 23 February 2018. The evolution of the shoreline undulation downcoast of the Aringay River delta was compared with the results of the numerical simulation using the BG model [4] on the formation of a sand spit at the point where an abrupt change of the shoreline configuration takes place. Furthermore, a sand spit at Santo Tomas located near the bottom of Lingayen Gulf was selected as another study site (Figure 1).

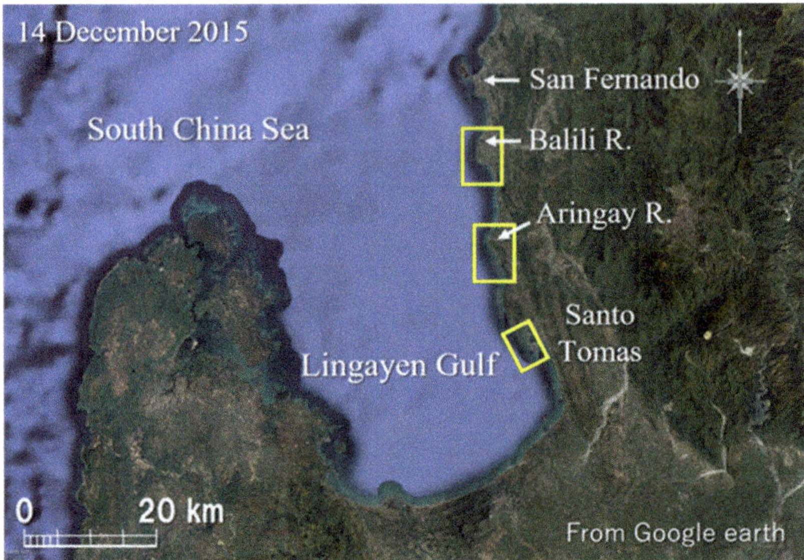

Figure 1. Satellite image of Lingayen Gulf (for location, 16°30′45.27″ N, 120°7′11.62″ E).

Similarly to the studies on the sand spits of the Rhône River delta by Sabatier and Anthony [5], the formation of the sand spits in relation with the evolution and morphology of the Danube mouths and deltaic lobes in the Black Sea basin [6], and the comparison of evolution and dynamics of selected representative deltaic spits [7], the evolution of the shoreline undulations was investigated using satellite images in this study. Furthermore, to predict shoreline changes of a sand spit and consider the measures against beach erosion on a coast around a sand spit, quantitative estimation of the rate of longshore sand transport is important, so the rate of longshore sand transport around this sand spit was estimated from the temporal change in the foreshore area of the sand spit. As mentioned by [5], understanding the functional aspects of spit formation and growth is important for the management of the delta shoreline and its sediment budget by public authorities. In this study, it is pointed out that useful information regarding shore protection of the coasts in Lingayen Gulf can be obtained by the analysis of topographic changes under high-angle wave conditions, given morphological characteristics of the bay shape and arrangement of rivers.

2. Method of Study

First, the morphological characteristics of Lingayen Gulf were briefly described using a satellite image and bathymetric survey data as well as the investigation of wave characteristics using the Global Wave Statistics [8]. Because of the shape of Lingayen Gulf with a large aspect ratio, oblique wave incidence with a large angle to the direction normal to the shoreline prevails in this gulf. Taking these characteristics into account, satellite images of Lingayen Gulf were collected and the shoreline changes along the east coast were determined, selecting the deltas of two rivers (Balili and Aringay Rivers) and a sand spit at

Santo Tomas with a supplementary field observation on 23 February 2018. Then, the wave field in Lingayen Gulf was calculated to explain the mechanism of occurrence of shoreline undulations along the east coast of the gulf due to high-angle wave instability [1] using the energy balance equation [9], based on the seabed contours of the gulf. Along the east coast of this bay, waves are obliquely incident to the direction normal to the shoreline over 45°, causing the shoreline instability. In an area south of the Balili River delta, a sand spit has formed at a location where the shoreline configuration abruptly changes due to high-angle wave instability. Therefore, the results of the numerical simulation regarding the formation of a sand spit on a coast with a sudden change in shoreline configuration were shown and the comparison of the results with the measured data was made. Furthermore, the rate of longshore sand transport along the sand spit at Santo Tomas was estimated from the temporal change in sand volume by integrating the foreshore area of the beach. Finally, the morphological similarity of Lingayen Gulf to that of Suruga Bay in Japan, and the rate of longshore sand transport in both bays were compared to consider the future measures against beach erosion along the coasts in these bays.

3. Morphological Characteristics of Lingayen Gulf

Lingayen Gulf, with a width of 44 km and a length of 58 km, has an aspect ratio of 1.3 and opens to the South China Sea in the direction of N20°W, as shown in Figure 1. This gulf has a semicircular embayment located on the northwestern coast of Luzon [10]. This area has two seasons—the wet season, from April to October, and the dry season, from November to March. Annual average rainfall in this area is 2500 mm with an average peak of 800 mm in August and a low of 1 mm in January. The mountains protect the province from the northeast monsoon and the trade winds but do not spare it from floods during the wet season. The west coast of this gulf has a complicated shoreline with many islands and headlands, whereas alluvial fans have developed along the east coast (Figure 1). A headland protrudes at San Fernando at the north end of the gulf, and the Balili River with a catchment area of 518 km^2 flows into the gulf 9.6 km south of San Fernando. Furthermore, the Aringay River with a catchment area of 405 km^2 flows into the gulf south of the Balili River, forming a delta. In addition, an elongated sand spit forms at Santo Tomas 18 km southeast of the Aringay River. In this gulf, incident waves have a strong directionality; waves are obliquely incident at a large angle relative to the direction normal to the shoreline because of the slender shape, so shoreline undulations due to high-angle wave instability [1] may have occurred.

The wave characteristics of this area can be estimated by the Global Wave Statistics [8]. In this report, the wave statistics in each area, separated into 104 subregions of all the world's oceans, are shown based on the wave observation data and the hindcast of ocean waves. The area offshore of Lingayen Gulf belongs to subregion No. 40, and the results are summarized in Table 1; wave height with a probability of occurrence less than 5% ranges between 4.26 and 5.71 m with the maximum of 5.71 m in the direction of N. This result corresponds to the wave climate in South China Sea offshore of Lingayen Gulf, so the wave height in Lingayen Gulf is reduced, subject to the wave-sheltering effect.

Table 1. Wave climate offshore of Lingayen Gulf [8].

Items	Wave Direction			
	W	NW	N	ALL
Energy mean wave height	2.00	2.04	2.81	2.74
Wave height with a probability of occurrence less than 5% (m)	4.26	4.42	5.71	5.41
Mean wave period (s)	4.9	4.6	6.2	6.3
Most frequent wave period (s)	4.5	4.5	5.5	6.5

Figure 2 shows the seabed topography of Lingayen Gulf with an average depth of 46 m and a maximum charted depth of about 100 m along its northern boundary [10]. Although a 20 m depth contour runs smoothly along the east coast, the contours shallower than 10 m in depth protrude around the Aringay River mouth and offshore of a sand spit at Santo Tomas. The formation of a sand spit on the bottom of the gulf clearly indicates the predominance of southward longshore sand transport under the oblique wave incidence from the north. On this coast with these characteristics, severe beach erosion occurs along the east coast because of the spatial imbalance in southward longshore sand transport.

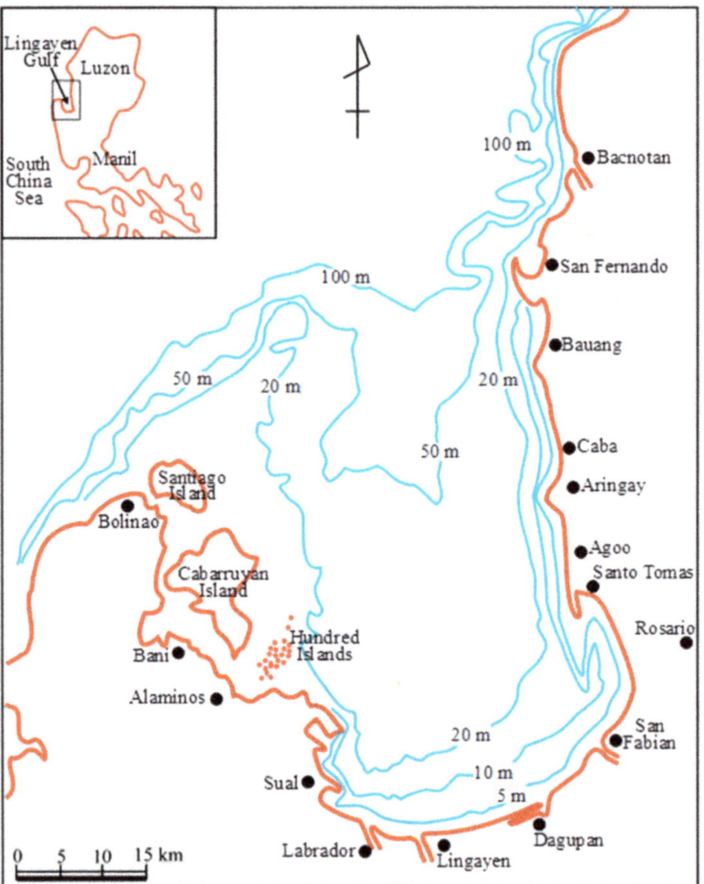

Figure 2. Bathymetric chart of Lingayen Gulf in the Philippines (modified from [10]).

4. Formation of Shoreline Undulations Due to High-Angle Wave Instability

4.1. Balili River Delta

Figure 3 shows a satellite image of a rectangular area around the Balili River mouth, as shown in Figure 1, taken on 15 April 2006. In the lower Balili River, a minor distributary (R_2) is separated from the main distributary (R_1). In Figure 3, the Aringay River mouth was often closed by the elongation of a sandbar from the right bank, which was induced by the deposition of southward longshore sand transport. The shoreline extends straight southward in a 1.9 km stretch between the mouth of the distributary R_2 and point A south of the mouth, and then the shoreline bends around point A at a right angle. The tip of the curved shoreline connects to point B at the jetty of the oil refinery factory with a wide lagoon landward of the barrier island. Thus, a slender wetland separated by a barrier island extended in the west–east direction south of the distributary R_2.

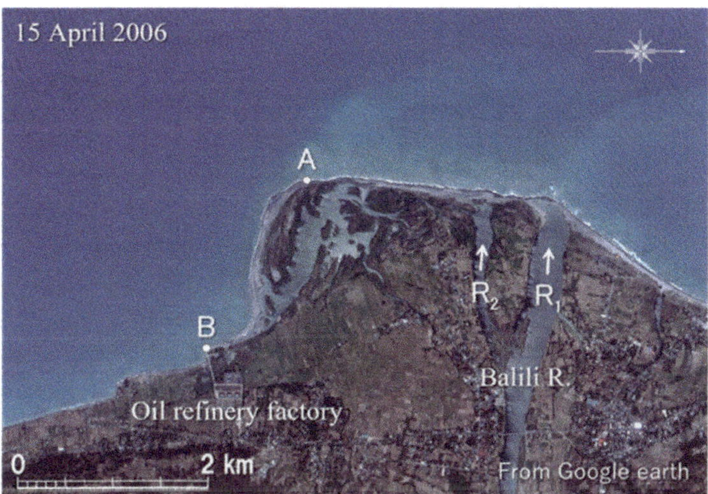

Figure 3. Satellite image of shoreline around Balili River mouth (15 April 2006).

Similarly, Figure 4 shows a satellite image, taken on 14 February 2014, of the same area. Points A, B, P, and Q are indicated in Figure 4 to explain, but in what follows, "point" is omitted for simplicity, and the shoreline configuration measured on 15 April 2006 is shown in each figure. In this case, the shoreline position was directly determined from the satellite images while leaving an occurrence of some error in determining the shoreline position because of the lack of field data, i.e., the tide level and beach slope.

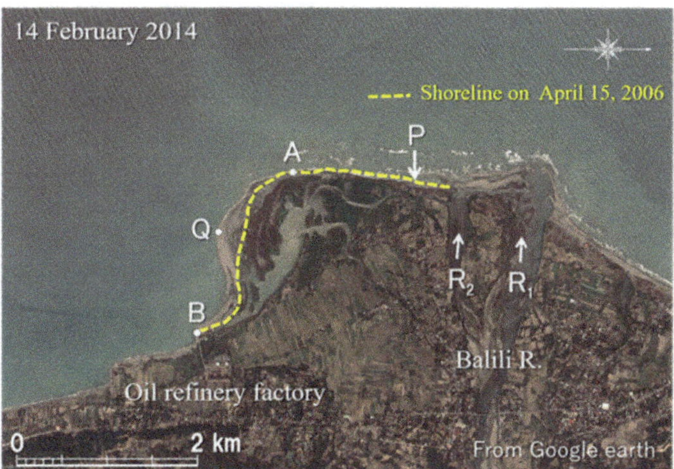

Figure 4. Satellite image of shoreline around Balili River mouth (14 February 2014).

When referring to the tide level measured from January to December 2019 at San Fernando located at the entrance of Lingayen Gulf, the average of the monthly lowest and highest tide levels were −0.09 and +0.90 m, respectively, with a tidal range of 0.99 m. Detailed bathymetric surveys have not been carried out along the coast in Lingayen Gulf, but the berm height and foreshore slope were measured on a beach upcoast of a groin at Agoo, La Union near Narvacan on 23 February 2018, using two measuring staffs. The berm height and the foreshore slope were 2.0 m above mean sea level (MSL) and 1/7.7, respectively [11]. Given the tidal range of 0.99 m and the foreshore slope of 1/7.7, the maximum variance of the shoreline position becomes 7.7 m. This shoreline variance, however, is much smaller than the measured shoreline advance or the shoreline recession.

By 14 February 2014, a river delta had developed owing to the sediment discharge provided by the main channel of R_1. Until this year, no shoreline changes were observed between P and A, although sediment supplied from the Balili River was deposited up to P. In contrast, a semicircular sandbar was newly formed with a shoreline protruding south of A. The formation of the sandbar clearly indicates that sediment supplied from the Balili River was transported southward and deposited as it turned around A. Until this time, the width of the sand spit increased because of the successive sand deposition along the shoreline of the sand spit, and in contrast, the lagoon width behind the sand spit decreased because of the deposition of sand from the distributary R_2, as shown in Figure 4.

In Figure 5 showing an image taken on 11 June 2015, the river delta, which markedly protruded around the mouth, changed its form to a flat shape owing to erosion, and the curved sandbar south of A moved eastward as a whole.

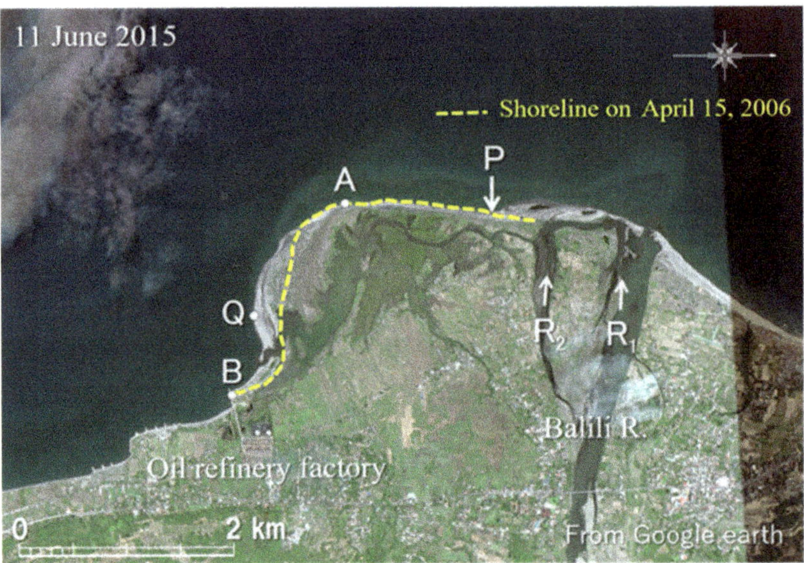

Figure 5. Satellite image of shoreline around Balili River mouth (11 June 2015).

This sandbar further approached B until 20 February 2016 (Figure 6).

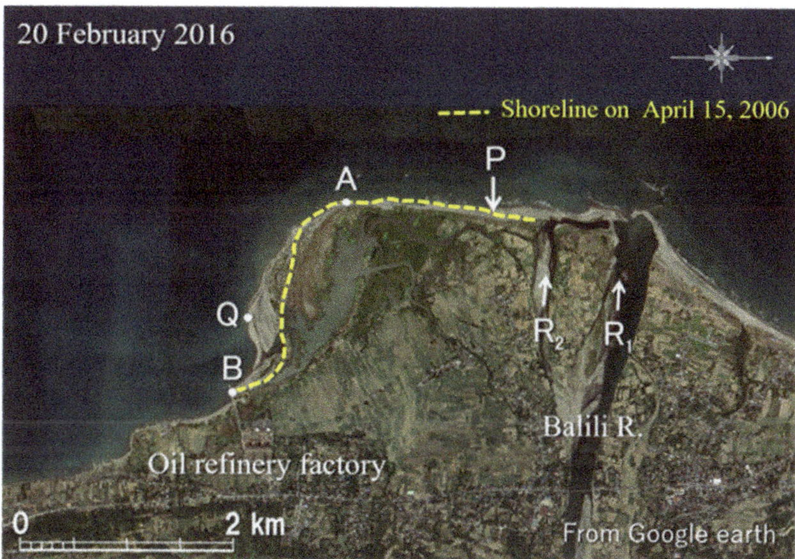

Figure 6. Satellite image of shoreline around Balili River mouth (20 February 2016).

A similar change continued until 1 December 2016, leaving an opening of the lagoon at the east end of the sandbar (Figure 7). In this year, a river delta was formed again around the mouth of R_1 owing to sediment discharge from the river, indicating the occurrence of the intermittent sand supply from the Balili River.

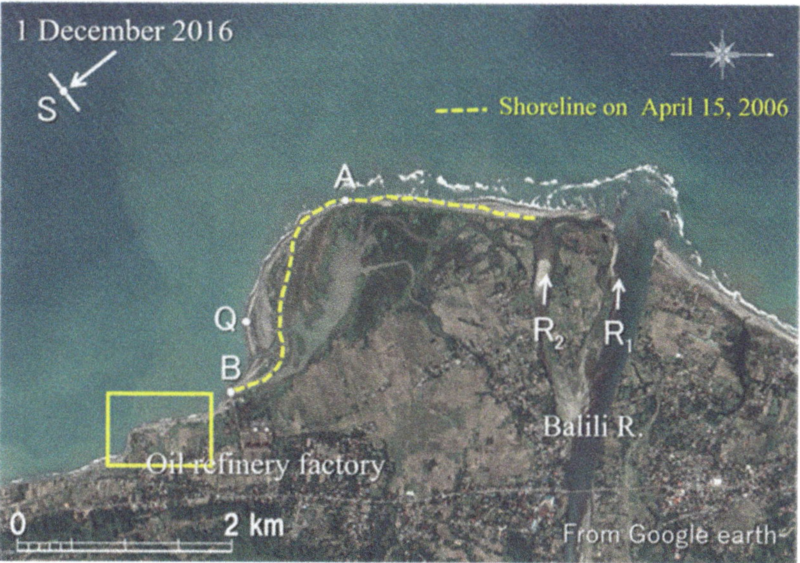

Figure 7. Satellite image of shoreline around Balili River mouth (1 December 2016).

The village of Bauang, La Union south of B, indicated by a rectangular area in Figure 7, is shown in Figure 8.

Figure 8. Enlarged satellite image of rectangular area in Figure 7. Sites 1 and 2 are the locations where site photographs were taken in the field observation.

The center of the village is located 200 m north of the end of the coastal road. A concrete building destroyed by erosion remained in its vicinity, and five groins made of stones were constructed south of this building to maintain the shoreline. Figure 9 shows the coastal condition at Site 1 on the north side of the destroyed building.

Figure 9. North view from the second floor of a damaged building.

A wide sandy beach was formed upcoast of the groins because of the blockage of southward longshore sand transport by the groins. Similarly, Figure 10 was taken at Site 2 on top of groin No. 4, facing south. In this area, a wide sandy beach was formed in front of the village, indicating the effect of the groins that locally blocked longshore sand transport.

Figure 10. South view from groin No. 4.

In Figure 7, it should be noted that waves were obliquely propagating relative to the direction of the mean coastline in the offshore sea south of the Balili River, so the wave direction can be determined as the direction normal to the wave crestline. When setting the offshore point S at a location 3.2 km southeast of B, the wave direction at S was determined to be N37°W. Since the direction normal to the straight shoreline between the mouth of R_2 and A is N84°W, waves were obliquely incident to the direction normal to the shoreline in a clockwise manner at an angle of 47°. In Lingayen Gulf, waves of high directionality are incident from the north because of the shape of the gulf, and the wave direction determined from Figure 7 is in accordance with this general condition of wave direction. Since waves are obliquely incident at an angle over 45° to the direction normal to the shoreline, the shoreline undulation due to high-angle wave instability may develop, resulting in the development of a markedly protruded shoreline south of the Balili River.

When setting Q at a location where the shoreline most protruded in the south between A and B in Figure 7, the eastward distances of Q with reference to the location on 14 February 2014 were 910 m (June 2015), 970 m (February 2016), and 1020 m (December 2016). It was found that Q gradually approached the location by December 2016 as its movement velocity decreased. When the shoreline markedly protruded near A, as shown in Figure 7, the area near B was protected by a significant wave-sheltering effect of the sandbar itself against waves incident from the direction of N37°W, and wave height decreases, resulting in the decrease in the movement velocity of the sandbar. Furthermore, when a sand body moves with the formation of a sand spit, the sand supply downcoast is suspended until the sand body reached downcoast, resulting in beach erosion. Because beach erosion has occurred in this manner in Bauang, La Union, it is considered that the groins have been constructed to protect the village against beach erosion.

4.2. Aringay River Delta

Figure 11 shows a satellite image of the rectangular area enclosing the Aringay River delta, taken on 15 April 2006. Here, a slender sandbar extended southward from the north riverbank, closing the river mouth. A sandy beach continuously extended southward while enclosing a lagoon behind it. Furthermore, a slender sand spit of 308 m length elongated southward from the south end of the river mouth sandbar. The elongation direction of the sand spit was at a large angle of 43° relative to the direction normal to the coastline immediately south of the river mouth. In the following figures, the shoreline configuration measured on 15 April 2006, as shown in Figure 11, is drawn for comparison.

Figure 11. Satellite image of shoreline around Aringay River mouth (15 April 2016).

By 28 October 2009, a slender sand spit elongated to a length of 770 m, while rotating counterclockwise and enclosing a lagoon behind the sand spit, although the shoreline north of the river mouth was stable (Figure 12).

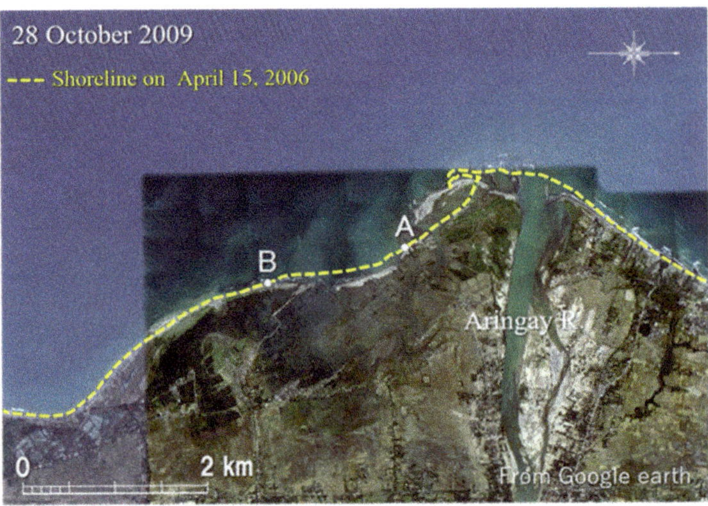

Figure 12. Satellite image of shoreline around Aringay River mouth (28 October 2009).

This southward elongation of a sand spit implies that the entire volume of sand supplied from the Aringay River was transported southward. Because longshore sand has not been supplied to A located on the opposing shore downcoast of the tip of the sand spit during the elongation period of the sand spit, the shoreline of 1.5 km in length between A and B was eroded with a maximum shoreline recession of 115 m at a location 600 m south of A. For comparison, A and B of Figure 12 are shown in the following Figures 13–17.

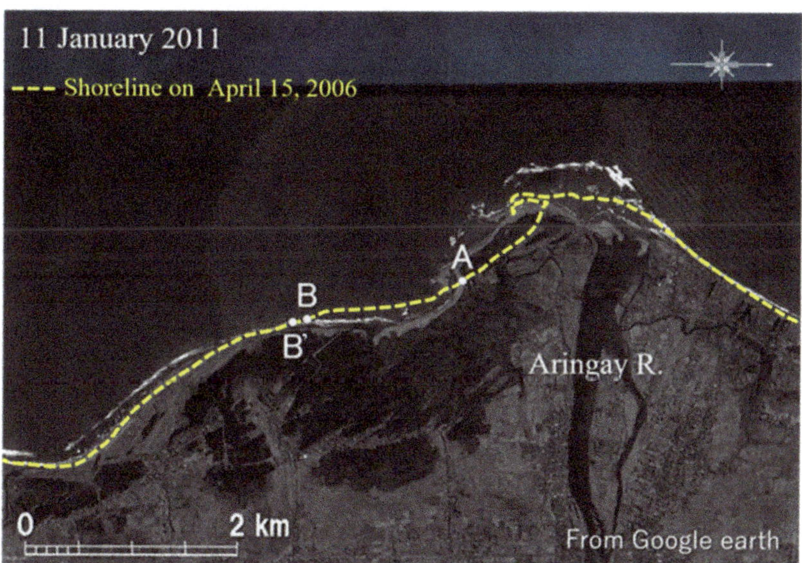

Figure 13. Satellite image of shoreline around Aringay River mouth (11 January 2011).

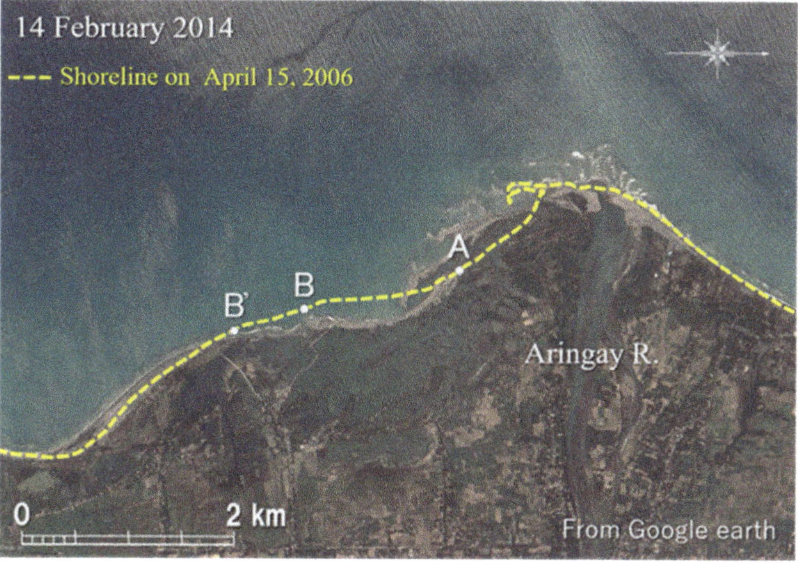

Figure 14. Satellite image of shoreline around Aringay River mouth (14 February 2014).

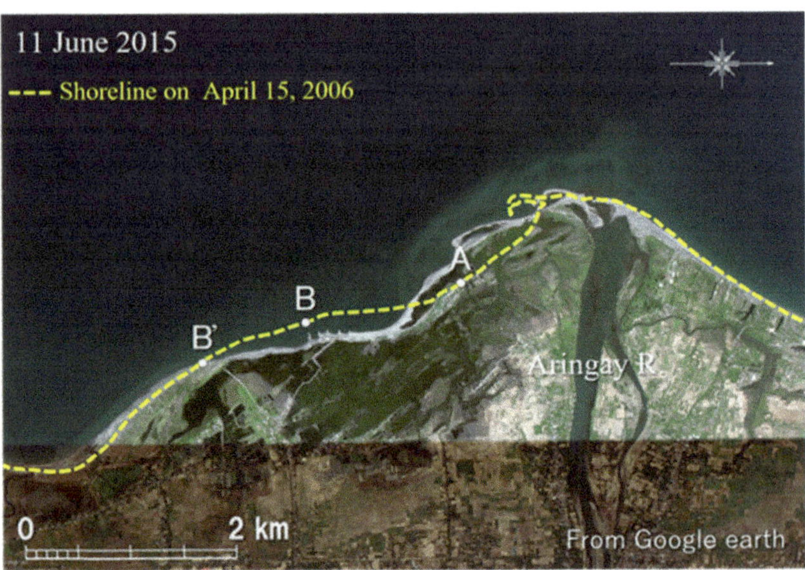

Figure 15. Satellite image of shoreline around Aringay River mouth (11 June 2015).

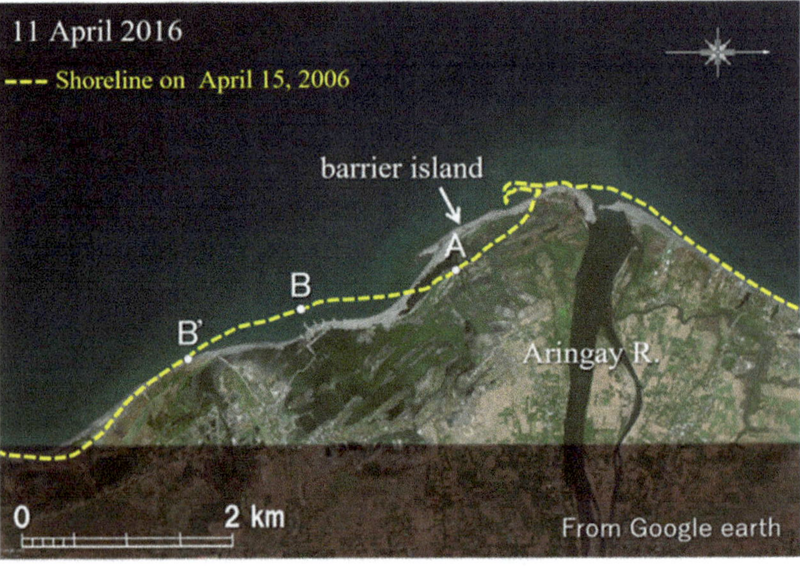

Figure 16. Satellite image of shoreline around Aringay River mouth (11 April 2016).

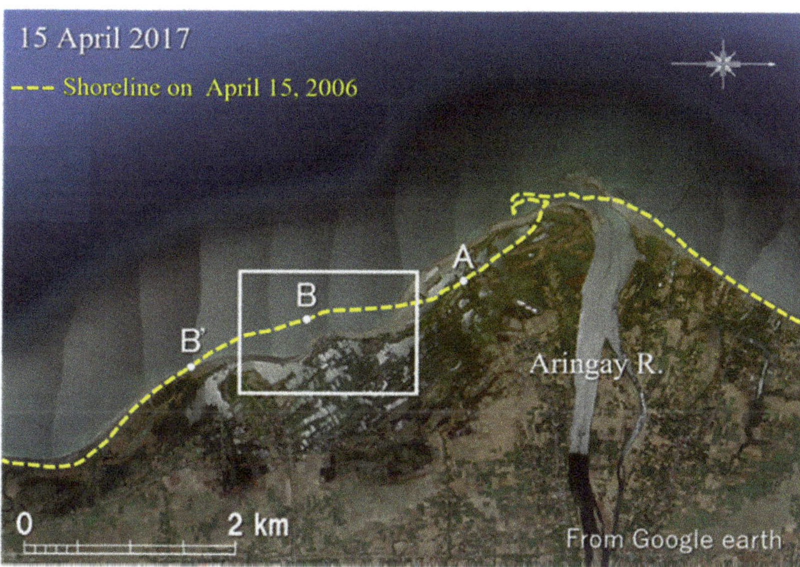

Figure 17. Satellite image of shoreline around Aringay River mouth (15 April 2017).

By 11 January 2011, sand was deposited near A located south of the sand spit (Figure 13). On the other hand, shoreline recession started at B located at the south end of the erosion area, and the erosion zone expanded until B′ 135 m south of it forming an erosion wave. The maximum shoreline recession reached 154 m at a location 580 m south of A. Thus, the shoreline recession increased on the south side of the sand spit with the elongation of the sand spit. In addition, another sand spit with approximately the same size as that in 15 April 2006 started to extend from the left bank of the river.

By 14 February 2014, a barrier island was formed by the extension of a sand spit offshore of A, forming an enclosed lagoon inside (Figure 14). In this period, the tip of the sand spit elongated 730 m south of A, as well as the elongation of another sand spit from the river mouth. With erosion, the eroded area expanded southward, and the south end of the erosion area moved by 670 m south of B. The maximum shoreline recession was 230 m at a location 750 m south of A.

By 11 June 2015, A located at the north end of the erosion zone in 2009 was enclosed by a slender barrier island with the elongation of a sand spit (Figure 15). Simultaneously, the erosion area expanded up to B′ located 1030 m south of B until this year. Even though the erosion expanded over time, the shoreline recession was relatively small near B because of the construction of five stone groins of 80 m in length in its vicinity. These groins prevented the shoreline from receding locally but caused downcoast erosion.

By 11 April 2016, the shoreline changed similarly to that shown in Figure 15, and another sand spit started to extend offshore of the barrier island formed by the extension of a sand spit near A (Figure 16).

Furthermore, the erosion zone extended up to B′ 1180 m south of B, and the shoreline recession downcoast of B reached 256 m. Finally, by 15 April 2017, A was enclosed by double barrier islands, and the distance between A and the shoreline increased until 280 m (Figure 17). Although the south end of the erosion zone is located at the same point of B′ 1180 m south of B, the shoreline recession downcoast of B reached 308 m.

In the vicinity of point B south of the Aringay River, the shoreline upcoast of the groin was locally fixed, whereas the shoreline receded downcoast of the groins because of the blockage of longshore sand transport owing to the groins, as shown in Figure 18; an enlarged satellite image of the rectangular area shown in Figure 17.

Figure 18. Enlarged satellite image of the shoreline in the rectangular area in Figure 17.

In this area, six groins were installed to protect the village, whereas the shoreline markedly retreated south of groin No. 6. For example, the village houses between groins Nos. 4 and 5, made of large boulders with a crown height of 2 m above mean sea level, were protected by a sandy beach between groins, as shown in Figure 19. Similarly, the intertidal beach was left north of groin No. 6.

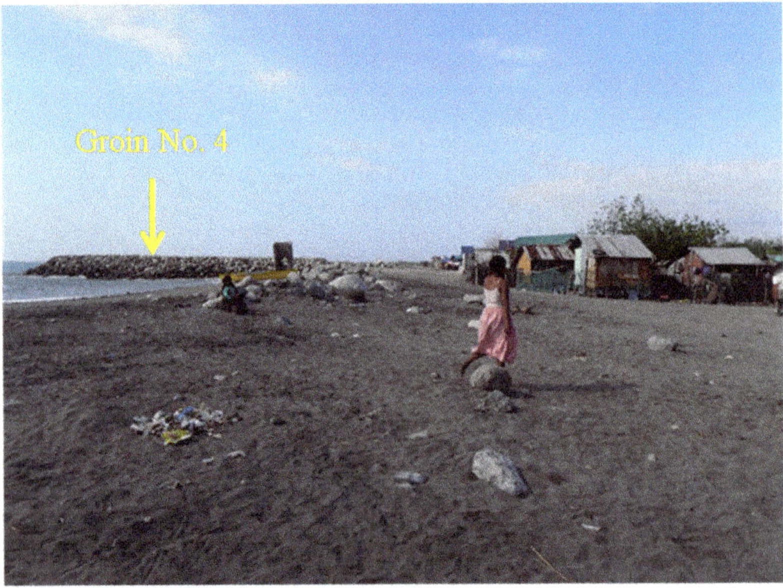

Figure 19. Picture of groin No. 4 taken from behind groin No. 5.

In contrast, the shoreline south of this groin retreated as a water area can be identified landward (left) end of the groin (Figure 20). Thus, the construction of the six groins was successful in locally stabilizing the shoreline, but their impact further expanded downcoast.

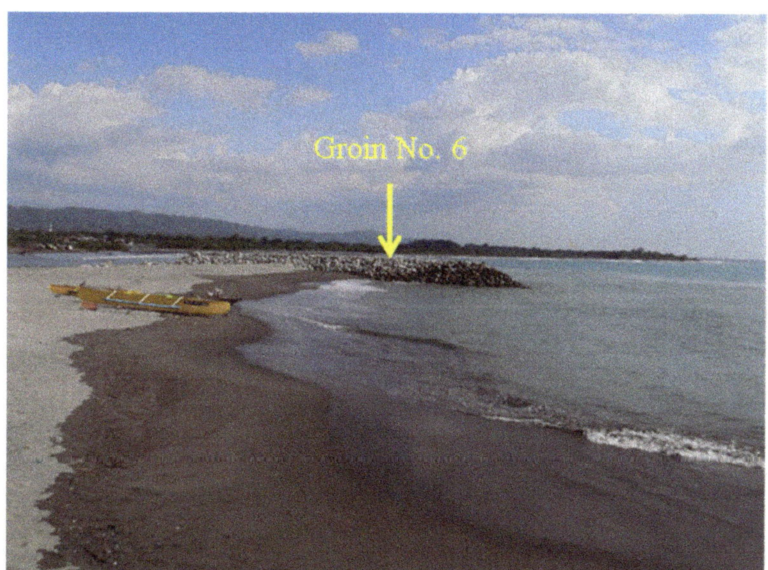

Figure 20. View of groin No. 6 taken from groin No. 5.

4.3. Formation of a Sand Spit at Santo Tomas

The shoreline changes in the rectangular area including a sand spit at Santo Tomas, shown in Figure 1, were investigated. A satellite image taken on 25 October 2003 is shown in Figure 21.

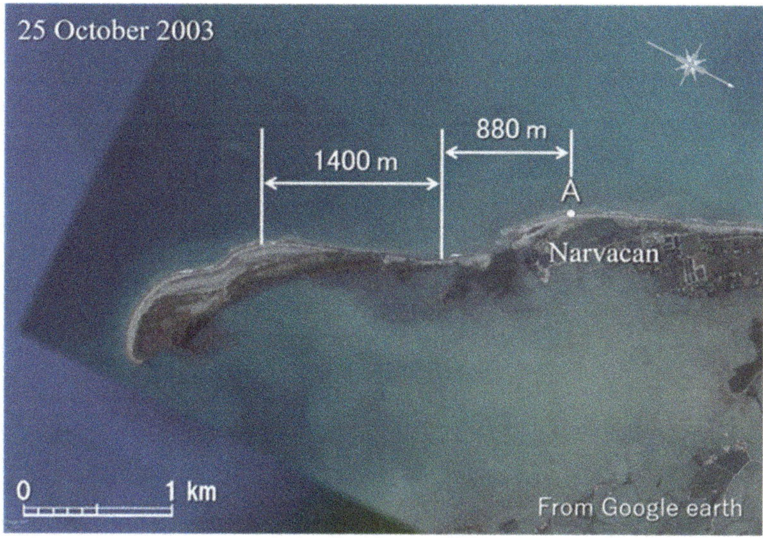

Figure 21. Satellite image around the sand spit at Santo Tomas (25 October 2003).

When setting A at the protruded shoreline in front of Narvacan, a concave shoreline is present near the location 880 m southeast of A, and then the shoreline protrudes again at a location 1400 m southeast of A. Since the shoreline undulation was not so large in 2003, the sand spit was considered to have been simply extending, owing to the continuous sand supply by southward longshore sand transport. However, 19 groins had been constructed until 25 October 2003, north of Narvacan, and continuous sand supply to the tip of the sand spit gradually became difficult.

By 15 April 2006, a significant shoreline change started to occur (Figure 22). By setting points B, C, and D, we found that the shoreline receded between B and C, and advanced between C and D. Until 15 March 2010, dominant shoreline recession occurred between B and C with a maximum shoreline recession of 155 m (Figure 23).

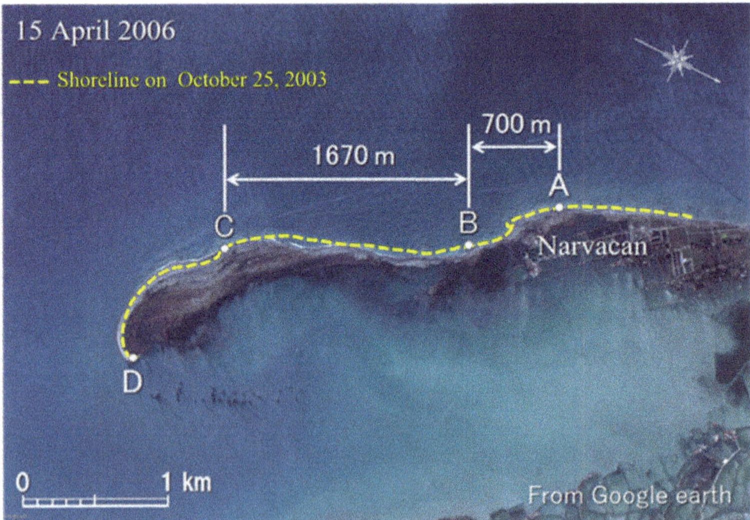

Figure 22. Satellite image around the sand spit at Santo Tomas (15 April 2006).

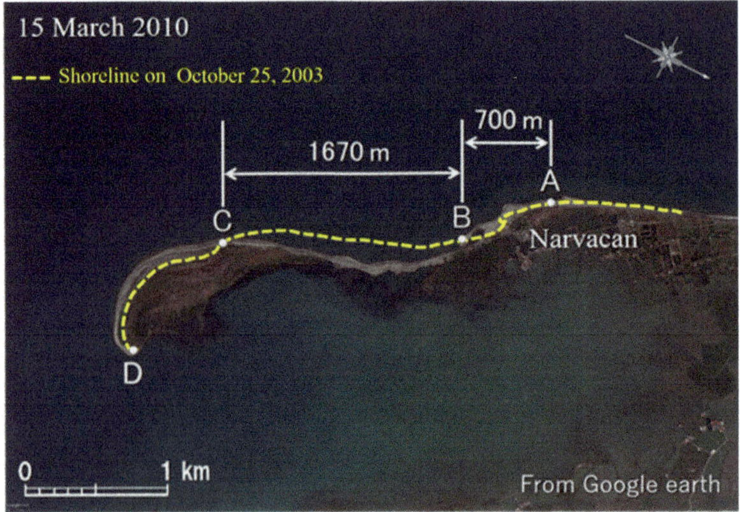

Figure 23. Satellite image around the sand spit at Santo Tomas (15 March 2010).

In contrast, the shoreline advanced by 86 m maximum between C and D. Furthermore, as shown in Figure 23, new sand spits of a small size developed near Narvacan, and simultaneously, the shoreline downcoast of the sand spit retreated because of the discontinuity in longshore sand supply at the tip of the sand spit, causing local shoreline recession.

Figure 24 shows a satellite image taken on 27 October 2013. In this figure, wave crestlines can be clearly observed offshore of the coastline. By setting point P 1.5 km offshore of B, we estimated the incident wave angle at P to be S113°W. Because the angle of the direction normal to the shoreline between B and C is S63°W, the wave incidence angle relative to the direction normal to the mean shoreline becomes 50°, satisfying the occurrence condition of high-angle wave instability. In Figure 24, the critical point where the shoreline recession started was located at B′ 240 m south of B, and the shoreline markedly receded between B′ and C with a maximum shoreline recession of 210 m across transect a–a′. In contrast, the shoreline advanced between C and D with a maximum shoreline advance of 190 m across transect b–b′.

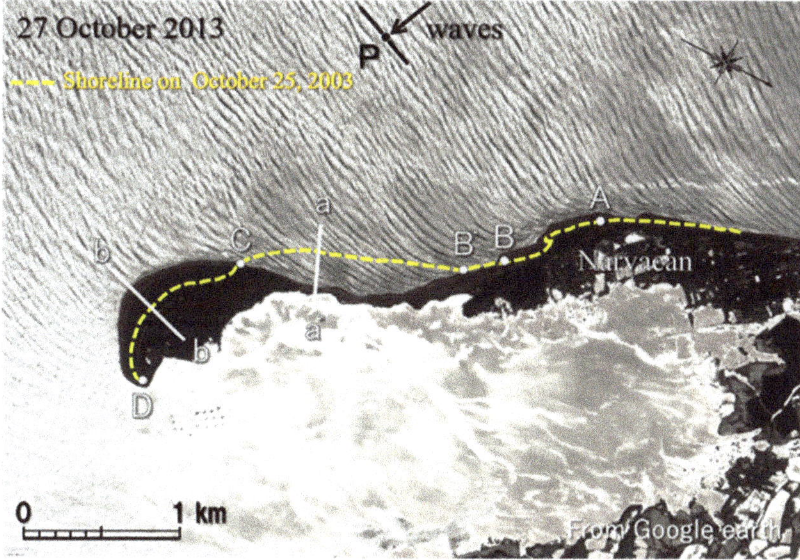

Figure 24. Satellite image around the sand spit at Santo Tomas (27 October 2013).

By 28 February 2015, the shoreline retreated between B′ and C with a maximum shoreline recession of 210 m at transect a–a′, whereas the shoreline advanced between C and D with a maximum shoreline advance of 190 m at transect b–b′ (Figure 25). As mentioned above, accretion occurred in the areas of AB′ and CD with erosion in the area of B′C separating these accretion areas, and the shoreline undulated as a periodic function with increasing amplitude. These shoreline changes are similar to those observed in the area south of the Aringay River mouth, and the amplitude was extremely large at 220 m in the area of B′C.

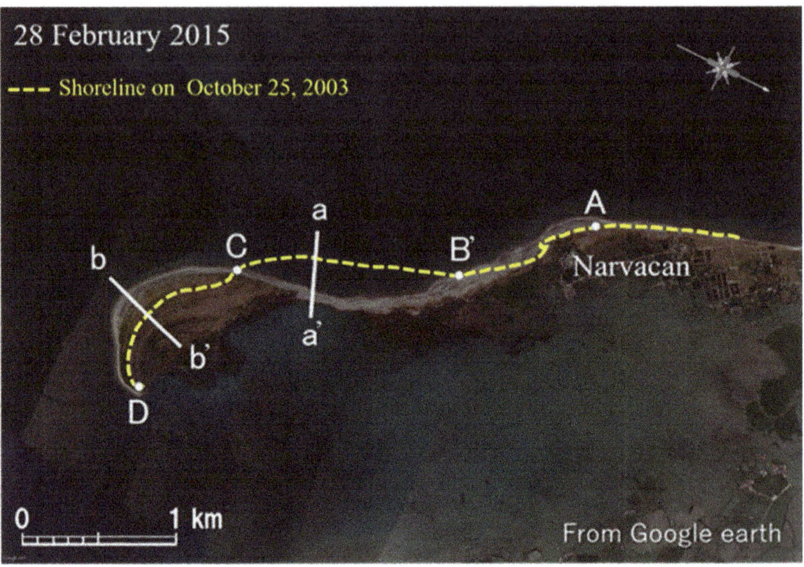

Figure 25. Satellite image around the sand spit at Santo Tomas (28 February 2015).

5. Results of Numerical Calculations

5.1. Wave Field

The formation of shoreline undulations including sand spits along the east coast of Lingayen Gulf may occur due to high-angle wave instability [1] because of oblique wave incidence with a large angle to the shoreline. Thus, the wave field in Lingayen Gulf was calculated using the energy balance equation [9], given the seabed contours of the gulf, as shown in Figure 2. The direction of incident waves was assumed to be N, corresponding to the direction of the most frequent energy mean waves, as shown in Table 1, referring to the Global Wave Statistics [8], with a wave height of 2.81 m and a wave period of 6.2 s. The directional spreading parameter (S_{max}) was selected to be 25 for waves with a large steepness and 75 for ocean waves.

Figure 26 shows the results of the calculation when S_{max} is given as 75. The wave height in the vicinity of the Balili River mouth near the bay mouth is 1.2 m, and the wave height reduces toward the bottom of the gulf, resulting in $H = 0.9$ m near the sand spit at Santo Tomas. The wave direction is shown in Figure 26b, where a positive angle is the value measured counterclockwise with respect to the -Y direction.

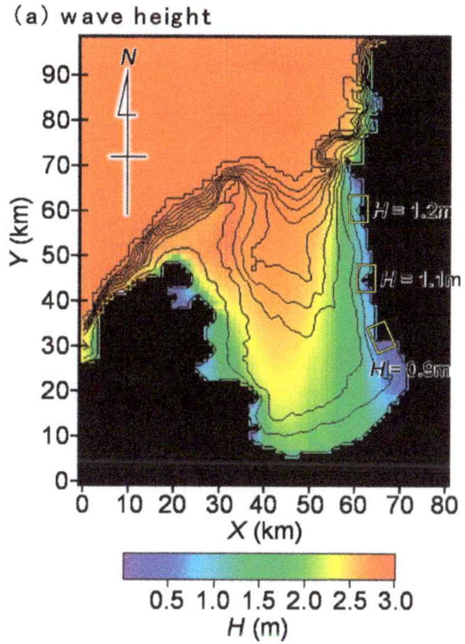

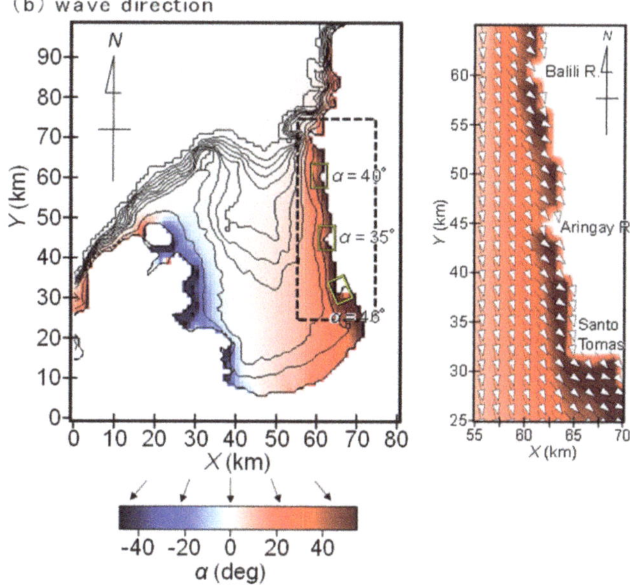

Figure 26. Predicted wave height and wave direction in Lingayen Gulf (S_{max} = 75), (**a**) wave height, (**b**) waive direction.

The wave angle relative to the direction normal to the coastline is over 45° at any point, taking the direction of the coastline into account. Similar results when S_{max} is 25 are shown in Figure 27. The reduction of the wave height from the entrance of the gulf toward the bottom of the gulf is the same as that when S_{max} is 75, with H = 1.37 m around the Balili River and H = 1.1 m near the sand spit at Santo Tomas. Comparing the two results highlights that the wave height along the coastline is greater when S_{max} is 75 than that when S_{max} is 25 because of the concentration of wave propagation. The wave angle relative to the direction normal to the coastline is greater than or close to 45° at any point. When the angle between the direction normal to the shoreline and the wave direction exceeds 45° at a point along the shoreline, the shoreline protrusion occurs at such a point owing to high-angle wave instability [1,2,12]. Thus, shoreline undulations or sand spits are capable of forming along the east coastline in Lingayen Gulf.

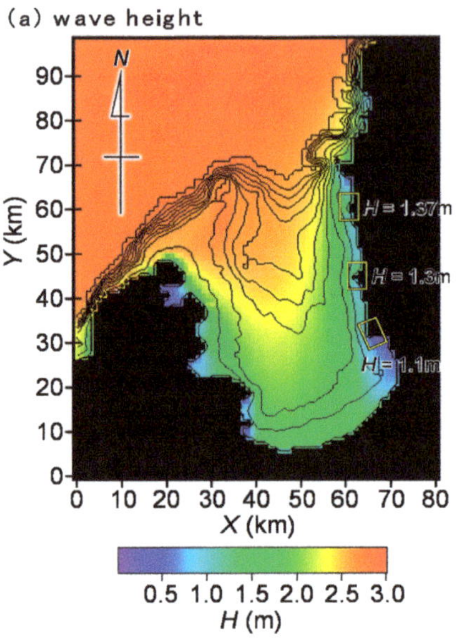

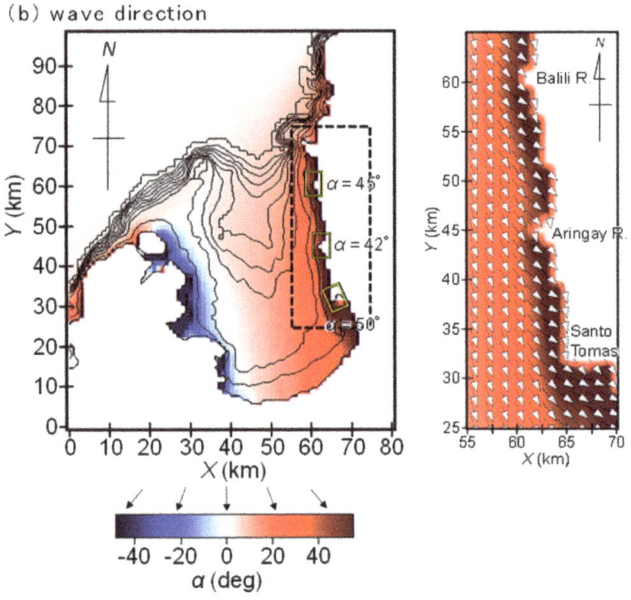

Figure 27. Predicted wave height and wave direction in Lingayen Gulf (S_{max} = 25) (**a**) wave height, (**b**) waive direction.

5.2. Formation of a Sand Spit on a Coast with Abrupt Change in Shoreline Configuration

A sand spit has formed at a location with an abrupt change in shoreline configuration south of the Balili River delta, enclosing a lagoon behind the sand spit, as shown in Figure 3. Such a formation of a sand spit can be numerically predicted using the BG model [4,12]. San-nami et al. [4] carried out a numerical simulation of the elongation of a sand spit with a model case of 1/100 on a seabed with different water depths and slopes around the location where the shoreline configuration abruptly changes. In their study, the water depth where a sand spit is formed were altered to 5, 10, 15, and 20 cm at various seabed slopes of 1/50, 1/40, 1/30, and 1/20. The incident wave height was H_i = 4.6 cm, and the wave period T = 1.27 s. A beach model was produced, as shown in Figure 28a, with incident waves from the Y-axis, so the wave incidence angle normal to the upcoast shoreline was to be 20°. The depth of closure was given as h_c = 2.5 H, where H is the wave height at a point. The berm height and equilibrium slope of sand were assumed as 5 cm and 1/5, respectively, based on the experimental results along with the angle of repose of sand of 1/2. The calculation domain was discretized by meshes of 20 cm, and the 8 h of calculation (8 × 10^4 steps) was carried out using the time intervals of $\Delta t = 10^{-4}$ h. Table 2 shows the calculation conditions.

Table 2. Calculation conditions (numbers in parentheses: experimental conditions).

Wave conditions.	Incident waves: H_I = 4.6 m (4.6 cm), T = 12.7 s (1.27 s), wave direction θ_I = 20° relative to normal to initial shoreline
Berm height	h_R = 5 m (5 cm)
Depth of closure	h_c = 2.5 H (H: wave height)
Equilibrium slope	$\tan\beta_c = 1/5$
Angle of repose slope	$\tan\beta_g = 1/2$
Coefficients of sand transport	Coefficient of longshore sand transport K_s = 0.045 Coefficient of Ozasa and Brampton [13] term K_2 = 1.62 K_s Coefficient of cross-shore sand transport K_n = 0.1 K_s
Mesh size	$\Delta x = \Delta y$ = 20 m (20 cm)
Time intervals	$\Delta t = 10^{-3}$ h (10^{-4} h)
Duration of calculation	80 (8) h (8 × 10^4 steps)
Boundary conditions	Shoreward and landward ends: q_x = 0, right and left boundaries: q_y = 0
Calculation of wave field	Calculation of wave field: energy balance equation [9] Term of wave dissipation due to wave breaking: Dally et al. [14] model Wave spectrum of incident waves: directional wave spectrum density obtained by Goda [15]. Total number of frequency components N_F = 1 and number of directional subdivisions N_θ = 8. Directional spreading parameter S_{max} = 75. Coefficient of wave breaking K = 0.17 and Γ = 0.3. Imaginary depth between minimum depth h_0 and berm height h_R: h_0 = 2 m (2 cm). Wave energy = 0 where $Z \geq h_R$. Lower limit of h in terms of wave decay due to breaking Φ: 0.7 m (0.7 cm)

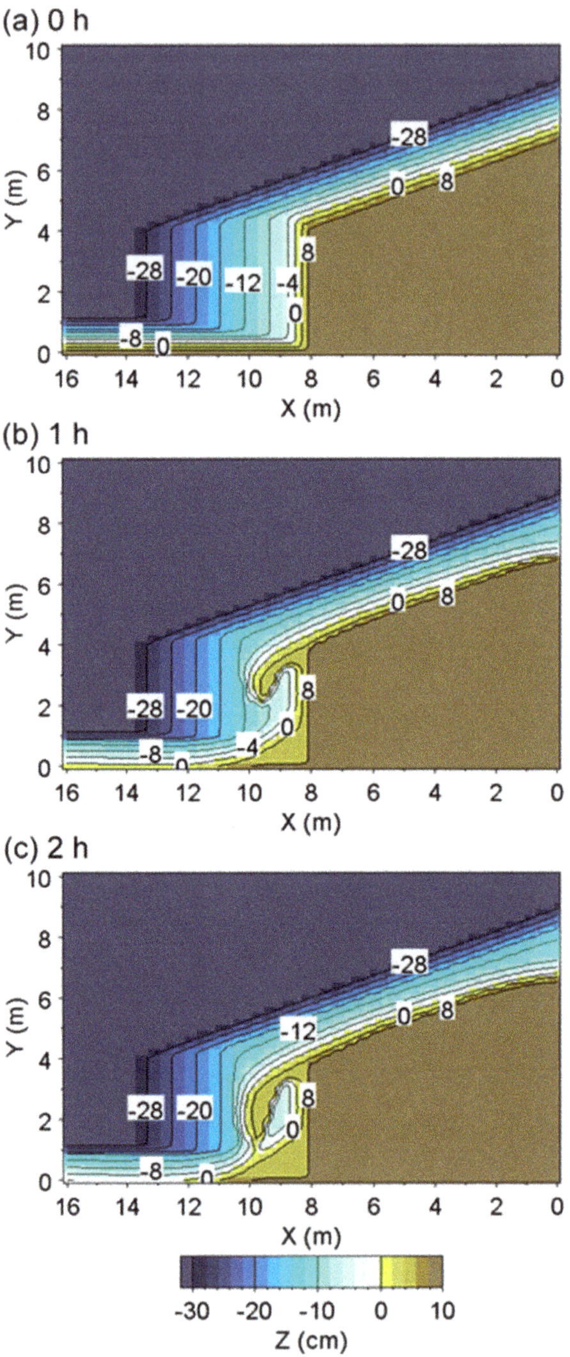

Figure 28. Elongation of sand spit and formation of barrier island [4]. (**a**) Initial topography, (**b**) A sand spit started to elongate from the corner owing to successive deposition of sand supplied by leftward longshore sand transport, (**c**) The tip of the sand spit attached opposite shore, forming a barrier island.

Figure 28b,c shows the results for the condition under which a sand spit is formed on a coast of the seabed slope of 1/20. When waves were incident from the Y-axis where the coastline direction markedly changes, similarly to the area south of the Balili River, leftward longshore sand transport occurred, and a sand spit started to elongate from the location where the coastline abruptly changes. After two hours of wave generation, a sand spit elongated very close to the opposite shore, enclosing a lagoon behind the sand spit. The elongation of a sand spit and the formation of a lagoon behind the sand spit may account for the observation results qualitatively. Thus, the formation of a sand spit south of the Balili River can be explained by the mechanism shown by San-nami et al. [4].

6. Longshore Sand Transport

6.1. Estimation of the Rate of Longshore Sand Transport

Although a sand spit extended south of Narvacan in recent years, many groins have been constructed upcoast of Narvacan, and smooth movement of sand by southward longshore sand transport was blocked by these groins, resulting in the shoreline recession at Narvacan. Here, the change in sand volume can be estimated from the shoreline changes, as shown in Figure 25, assuming that sand supply from upcoast became negligibly small because of the blockage by groins constructed upcoast.

First, the planar area in the shoreline recession zone between B′ and C in Figure 25 between October 2003 and February 2015 is calculated to be 2.1×10^5 m^2. The rate of change in the total volume of sand in the area between B′ and C can be calculated by multiplying this planar area by the characteristic height of beach changes (h), which is a correlation factor when the shoreline change is transformed into the change in cross-sectional area of the beach, and dividing by the period between 25 October 2003 and 28 February 2015. Usually, h can be determined from the bathymetric survey data, but in this case, it was difficult to employ this method because of the lack of bathymetric survey data. Therefore, h was calculated using empirical relationships h = (1.0–1.3) h_c and h_R = 0.31 h_c among h, h_c, and h_R [16]. In this case, the berm height h_R is given as 2.0 m owing to the field observation [11].

Since the mean value of h is determined to be 7.5 m, the rate of change in the total volume of sand in the area between B′ and C became 1.3×10^5 m^3/yr. North of Narvacan, southward longshore sand transport decreased because of the construction of groins. Assuming that the southward longshore sand transport at point B′ is negligible, the rate of longshore sand transport through point C, which is the nodal point between the erosion and accretion areas, can be evaluated to be 1.3×10^5 m^3/yr.

6.2. Interruption of Continuous Longshore Sand Transport at the Sand Spit Tip

In the Balili and Aringay River deltas, the sand supply downcoast was suspended until the sand supplied from the upcoast of the river deltas reaches downcoast, resulting in shoreline recession downcoast. Similarly, the shoreline downcoast of the sand spit retreated owing to the discontinuity in longshore sand supply at the tip of the sand spit at Santo Tomas. This is commonly observed around the tip of the sand spit. Noshi et al. [17] showed an example of the rapid development of a recurved sand spit around the south end of Phan Rang City in Vietnam. In their example, erosion occurred downcoast of the sand spit during the elongation period of the sand spit because of spatial imbalance in longshore sand transport at the tip of the sand spit. After the sand spit further elongated and connected the opposite shore, however, the sand supply by longshore sand transport along the shoreline of the sand spit was possible. In the area first eroded, sand was redeposited. The same phenomena were observed north of Pengambengan fishing port located in Bali Strait [18] and during the development of a barrier island in Nakatsu tidal flat [19].

7. Similarity of Lingayen Gulf and Suruga Bay in Japan

The beach changes along the east coast in Lingayen Gulf could be compared with those measured in another bay with a similar aspect ratio and wave characteristics, such as Suruga Bay in Japan, shown in Figure 29. When a straight line is drawn between Point Irozaki at the tip of Izu Peninsula and Point Omaezaki in Figure 29, the width of the bay mouth is approximately 50 km, and the length of the bay measured from this straight line to the bottom of the bay is 60 km, resulting in an aspect ratio of 1.2. Since the width and length of Lingayen Gulf were 44 and 58 km, respectively, and the aspect ratio is 1.3, both bays are similar to each other in terms of width, length, and aspect ratio. In addition, the west coast of Lingayen Gulf has a complicated coastline with many islands and headlands, similarly to the east coast in Suruga Bay. The development of the Balili and Aringay River deltas along the east coast of Lingayen Gulf is similar to the development of the river deltas around the Ohi and Abe Rivers with the catchment areas of 1280 and 567 km^2, respectively, on the west coast in Suruga Bay. Furthermore, the formation of a sand spit at Santo Tomas is similar to the formation of the Mihono-matsubara sand spit. Thus, the morphologies of both bays resemble each other so that waves are obliquely incident to the direction normal to the shoreline, and a sand spit has been formed near the bay bottom. Therefore, much experience of beach erosion along the west coast in Suruga Bay is useful in considering measures against beach erosion on the east coasts in Lingayen Gulf.

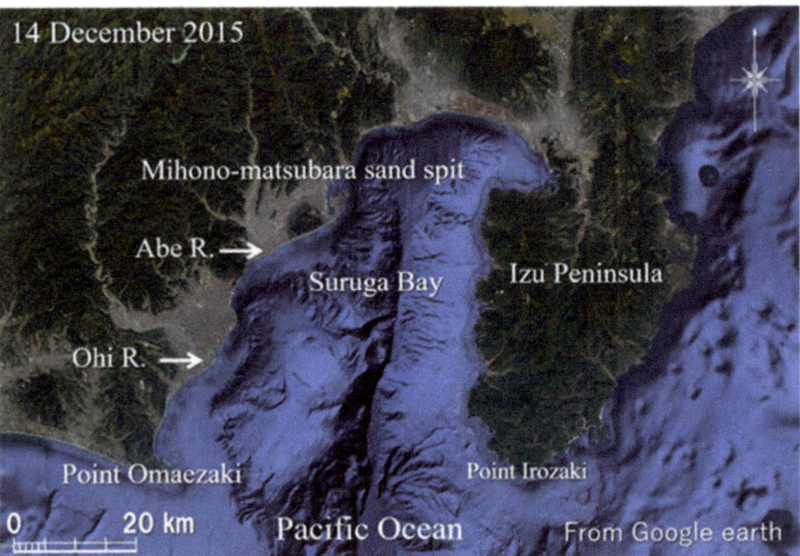

Figure 29. Satellite image of Suruga Bay (for location, 34°36′16.73″ N, 138°16′50.47″ E).

Figure 30 shows the seabed contours of Suruga Bay, which faces the Pacific Ocean to the south [20], and is a deep bay with a maximum water depth of 2000 m compared with that of 100 m in Lingayen Gulf. However, the coastline configuration of the west coast of Suruga Bay resembles the east coast of Lingayen Gulf; the coastline protrudes at the mouths of the Ohi and Abe Rivers, and Mihono-matsubara sand spit develops in the bay. Owing to the wave observations offshore of the Shimizu coast, the wave height with a probability of occurrence less than 5% is 3.0 m with a wave period of 9 s [21]. This wave height is the same order of magnitude in Lingayen Gulf. The rate of longshore sand transport was estimated to be 1.3×10^5 m^3/yr around the sand spit at Santo Tomas, which is in good agreement with the value of 1.3×10^5 m^3/yr estimated on the Shimizu coast surrounding Mihono-matsubara sand spit in Suruga Bay [21]. On the Shimuzu coast, severe beach erosion occurred owing to the decrease in the fluvial supply of sand after extensive riverbed mining in the Abe River, which is the supply source of sand, and the movement of a sand body occurred owing to the blockage of longshore sand transport by a number of detached breakwaters as described in [20]. Thus, the experience of the beach erosion on coasts in Suruga Bay, fully described in detail [20], is useful in explaining the beach changes along the coasts in Lingayen Gulf at least as a first step before detailed numerical simulation.

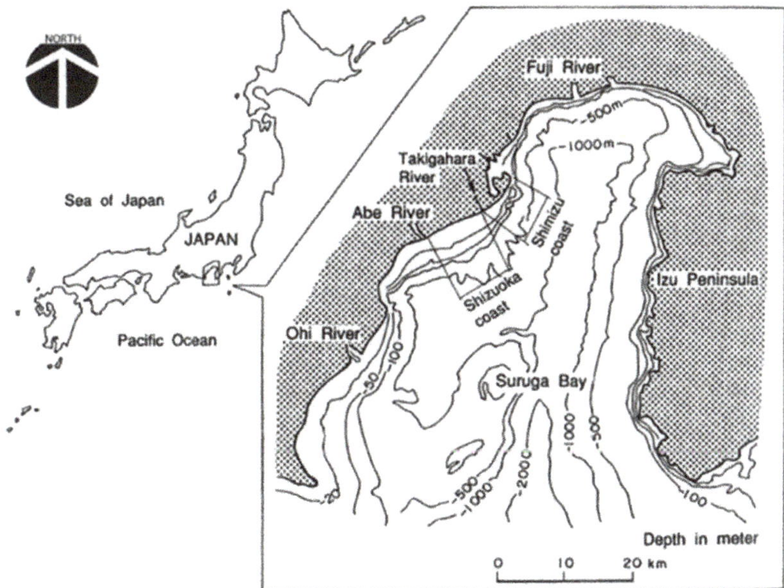

Figure 30. Bathymetric chart of Suruga Bay in Japan [20].

8. Concluding Remarks

The formation of shoreline undulations due to high-angle wave instability was actually observed at two river deltas and around a sand spit at Santo Tomas in Lingayen Gulf. The sand spit at Santo Tomas extended, similarly to a sinusoidal function due to high-angle wave instability, resulting in large shoreline recession and advance. South of the Balili River, the elongation of a sand spit was observed on a coast with an abrupt change in shoreline configuration, because waves were obliquely incident to the shoreline at a large angle. These observation results were explained by the mechanism described by San-nami et al. [4].

The morphologies of the east coast in Lingayen Gulf and the west coast in Suruga Bay are very similar (Figure 1 vs. Figure 29). A protruded shoreline develops north of the Ohi River, similarly to south of the Balili River, and the shoreline orientation abruptly changes. Continuous transport of sand was disrupted in a 20 km stretch between the Aringay River and the sand spit at Santo Tomas because of the construction of many groins to locally protect the coast, causing downcoast erosion. Although the sand supplied from the Abe River was transported northward forming Mihono-matsubara sand spit at a location where the shoreline orientation abruptly changes [21], severe beach erosion occurred on the Shizuoka and Shimizu coasts north of the river mouth owing to the decrease in sediment supply from the river caused by excess riverbed mining in the Abe River [20]. On the Suruga coast north of the Ohi River and the Shizuoka and Shimizu coasts located north of the Abe River, the beaches were severely eroded. As a measure, sand bypassing from the riverbed to the coasts was adopted. Taking into account the fact that the east coasts in Lingayen Gulf have the same condition as those in Suruga Bay, overall management of sand movement is required instead of local optimization using groins. A soft measure, such as sand bypassing, should be preferred along the coasts in Lingayen Gulf instead of hard measures against erosion to prevent an artificial, protected coastline from expanding, similarly to the case of coasts in Suruga Bay in Japan, where sand bypassing has now been adopted as a measure for sediment management.

Author Contributions: Data curation, T.U.; Formal analysis, Y.N. All authors have read and agreed to the published version of the manuscript.

Funding: This research received no external funding.

Institutional Review Board Statement: Not Applicable.

Informed Consent Statement: Not Applicable.

Conflicts of Interest: The authors declare no conflict of interest.

References

1. Ashton, A.; Murray, A.B.; Arnault, O. Formation of coastline features by large-scale instabilities induced by high angle waves. *Nature* **2001**, *414*, 296–300. [CrossRef] [PubMed]
2. Serizawa, M.; Uda, T.; Miyahara, S. Prediction of development of sand spits and cuspate forelands with rhythmic shapes caused by shoreline instability using BG model. In Proceedings of the 33rd ICCE, Santander, Spain, 1–6 July 2012; pp. 1–11. Available online: http://journals.tdl.org/icce/index.php/icce/article/view/6534/pdf_534 (accessed on 20 March 2021).
3. Anthony, E.J. Patterns of sand spit development and their management implications on deltaic, drift-alinged coasts: The causes of the Senegal and Volta River delta sipts, west Africa. In *Sand and Gravel Spits*; Randazzo, G., Jackson, D.W., Cooper, J.A.G., Eds.; Springer International Publishing: Berlin/Heidelberg, Germany, 2015; pp. 21–36.
4. San-nami, T.; Uda, T.; Serizawa, M.; Miyahara, S. Numerical simulation of elongation of sand spit on seabed with different water depths and slopes. In Proceedings of the 34th ICCE, Seoul, Korea, 15–20 June 2014; pp. 1–13. Available online: https://journals.tdl.org/icce/index.php/icce/article/view/7112/pdf_415 (accessed on 9 March 2021).
5. Sabatier, F.; Anthony, E. The sand spits of the Rhone River delta: Formation, dynamics, sediment budgets and management. In *Sand and Gravel Spits*; Splinger International Publishing: Berlin/Heidelberg, Germany, 2015; pp. 259–274.
6. Vespremeanu-Stroe, A.; Preoteasa, L. Morphology and cyclic evolution of Danube Delta spits. In *Sand and Gravel Spits*; Springer International Publishing: Berlin/Heidelberg, Germany, 2015; pp. 327–339.
7. Zainescu, I.F.; Vespremeanu-Stroe, A.; Tatui, F. Comparative spit dynamics. The cause of deltaic river mouth spits. *J. Coast. Res.* **2016**, *75*, 800–804.
8. Hogben, N. *Gloval Wave Statistics*; British Maritime Technology Ltd.: Teddington/Middlesez, UK, 1986.
9. Mase, H. Multidirectional random wave transformation model based on energy balance equation. *Coast. Eng. J. JSCE* **2001**, *43*, 317–337. [CrossRef]
10. MacManus, L.T.; Thia-Eng, C. *The Coastal Environmental Profile of Lingayen Gulf, Philippine, Association of Southeast Asian Nations/United States Coastal Resources Management Project*; Technical Publications Series 5; Association of Southeast Asian Nations: Manila, Philippines, 1990.
11. Uda, T.; Serizawa, M.; Onaka, S.; Ichikawa, S. Development of a sand spit offshore of groin under high-angle wave condition and evaluation of longshore sand transport. *J. Jpn. Soc. Civ. Eng. Ser. B2 Coast. Eng.* **2020**, *76*, 631–636. (In Japanese) [CrossRef]
12. Uda, T.; Serizawa, M.; Miyahara, S. *Morphodynamic Model for Predicting Beach Changes Based on Bagnold's Concept and Its Applications*; INTEC: London, UK, 2018; p. 188. Available online: https://www.intechopen.com/books/morphodynamic-model-for-predicting-beach-changes-based-on-bagnold-s-concept-and-its-applications (accessed on 20 March 2021).
13. Ozasa, H.; Brampton, A.H. Model for predicting the shoreline evolution of beaches backed by seawalls. *Coast. Eng.* **1980**, *4*, 47–64. [CrossRef]
14. Dally, W.R.; Dean, R.G.; Dalrymple, R.A. A model for breaker decay on beaches. In Proceedings of the 19th ICCE, Houston, TX, USA, 6–8 June 1984; pp. 82–97.
15. Goda, Y. *Random Seas and Design of Maritime Structures*; University of Tokyo Press: Tokyo, Japan, 1985; p. 323.
16. Uda, T. *Beach Erosion in Japan*; Sankaido: Tokyo, Japan, 1997; p. 442. (In Japanese)
17. Noshi, Y.; Uda, T.; Kobayashi, A.; Miyahara, S.; Serizawa, M. Beach changes observed in Phan Rang City in Southeast Vietnam. In Proceedings of the 8th International Conference on Asian and Pacific Coasts (APAC 2015), Procedia Engineering, Chennai, India, 7–10 September 2015; Volume 116, pp. 163–170.
18. Uda, T.; Onaka, S.; Serizawa, M. Beach erosion downcoast of Pengambengan fishing port in western part of Bali Island Japan. In Proceedings of the 8th International Conference on Asian and Pacific Coasts (APAC 2015), Procedia Engineering, Chennai, India, 7–10 September 2015; Volume 116, pp. 494–501.
19. Miyahara, S.; Uda, T.; Serizawa, M. Field observation and numerical simulation of barrier island formation as a result of elongation of sand spit and its attachment to opposite shore. In Proceedings of the 35th Conference on Coastal Engineering, Antalya, Turkey, 17–20 November 2016; pp. 1–14.

20. Uda, T. *Japan's Beach Erosion—Reality and Future Measures*, 2nd ed.; World Scientific: Singapore, 2017; p. 530.
21. San-nami, T.; Uda, T.; Ohashi, N.; Iwamoto, H.; Serizawa, M.; Ishikawa, T.; Miyahara, S. Prediction of beach erosion caused by reduction of fluvial sand supply due to excess sand mining and beach recovery after prohibition of mining. In Proceedings of the 33rd ICCE, Santander, Spain, 1–6 July 2012; pp. 1–11.

Article

Beach Gravels as a Potential Lithostatistical Indicator of Marine Coastal Dynamics: The Pogorzelica–Dziwnów (Western Pomerania, Baltic Sea, Poland) Case Study

Cyprian Seul [1,*], Roman Bednarek [1], Tomasz Kozłowski [1] and Łukasz Maciąg [2]

1 Faculty of Civil and Environmental Engineering, West Pomeranian University of Technology in Szczecin, 70-310 Piastów 50, Poland; roman.bednarek@zut.edu.pl (R.B.); tomasz.kozlowski@zut.edu.pl (T.K.)

2 Institute of Marine and Environmental Sciences, University of Szczecin, Mickiewicza 18, 70-383 Szczecin, Poland; lukasz.maciag@usz.edu.pl

* Correspondence: cyprian.seul@zut.edu.pl

Received: 12 May 2020; Accepted: 14 September 2020; Published: 16 September 2020

Abstract: The petrographic composition and grain shape variability of beach gravels in the Pogorzelica–Dziwnów coast section (363.0 to 391.4 km of coastline), southern Baltic Sea, Poland were analyzed herein to characterize the lithodynamics and trends of seashore development. Gravels were sampled at 0.25 km intervals, in the midpart of the berm, following an early-autumn wave storm and before beach nourishment. Individual variations in petrographic groups along the shore were investigated. Gravel data were compared and related to coastal morpholithodynamics, seashore infrastructure, and geology of the study area. The contribution of crystalline rock gravels (igneous and metamorphic) was observed to increase along all coast sections, whereas the amount of less resistant components (limestones, sandstones, and shales) usually declined. This effect is explained by the greater wave crushing resistance of igneous and metamorphic components, compared with sedimentary components. Similarly, the gravel grain shape (mainly elongation or flattening) was observed to change, depending on resistance to mechanical destruction, or due to the increased chemical weathering in mainly the limestones, marbles, and sandstones. Observed increase in contribution of discoid and ellipsoid grains is a potential indicator of depositional trends along the coast sections investigated. On the other hand, increased contents of spheroidal and spindle-shaped grains may be related to erosional trends, where intensive redeposition and mechanical reworking of gravels occurs. However, due to the great number of coastal embankments, the petrographic composition and shape parameters of beach gravels do not always clearly indicate the dominant direction of longshore bedload transport. Increased amount of eroded limestone located east of Pogorzelica indicate increased erosion of glacial tills. These sediments are deposited, building the shallow foreshore, with additional redeposition of morainic material towards the shore.

Keywords: Baltic Sea; cliff; sandy spit; beach; beach gravels; gravel grain shape; lithostatistical analysis; coastal zone dynamics; lithodynamics

1. Introduction

Lithodynamic processes at the land–sea interface result in different morphologies of the coastal zone, dominated by elevated cliff sections or coastal dunes. These forms may reflect the history of development of the coastal zone, where swash hydrodynamics steer the intensity of erosion and accumulation processes [1,2]. The spatial and temporal changes with the coastal sediment budget are usually related to increased cliff and beach erosion and redeposition of sedimentary material [3]. Undercut dunes or cliffs, with visible signs of mass movement, point to domination of erosion and marine ingression, whereas the development of dune ridges on sandy spits, as well as dunes at the forefront of temporarily inactive cliff shores, is indicative of equilibrium or even accumulation occurring in the coastal zone [4–6]. However, the swash-dominated gravel beaches are scarcely mentioned in sediment transport studies, or related to morpholithodynamics [2,7–11].

Gravel beaches are accumulations of shore material formed into distinctive shapes by waves and currents, and contain lithic particles in the gravel size range (2–64 mm). The beach form is a generally seaward-sloping boundary between a water body and mobile sediment, and a flat or landward-sloping surface at the upper limit of the beach. One or more gravel ridges may exist in the subaerial profile. Natural beaches composed of coarse material are common especially in the high latitudes, where coastal sediments indicate glacial transport history and reworking processes. Gravelly beaches also occur in other geographical latitudes, being associated with eroded cliff sections in the backshore and formation of a nearshore sedimentary bench of glacial or postglacial origin [12].

Mixed sand and gravel beaches are often partially composed of coarse material (>2 mm), with supplementary fine fractions, which are the main products of fluvial, glacial, or erosive processes. Gravel beaches are especially important for coastal dynamics, with protection provided by a natural system of shoreline stabilization where its response to sea-level forcing is distinctive and can be preserved, even in the long-term geological record [13].

Distinctive sedimentary features of mixed sand and gravel beaches include (i) a greater amount of coarse sand fraction and relatively high content of fine gravel fraction; (ii) the existence of multimodality within the gravel fraction; (iii) very limited shape sorting; and (iv) high spatial and temporal variability. The beach profile responses occur over semidiurnal, spring-neap, and seasonal time scales [14,15]. The coarse–fine grain sorting processes are usually related to longshore movement of sedimentary material in the swash zone, especially during storms [2,16–19].

Saltation, traction bedload, and sheet flow dominate the nearshore of gravel beaches. The individual clast motion of coarse grains is dictated by a number of micromechanical factors attributable to size and shape variation occurring over a heterogeneous bed. Transport mode affects sediment sorting and morphodynamic feedbacks. Gravels are relatively large, compared to the sandy fraction, so they account for a greater proportion of the sediment volume in the swash zone. Therefore, sheet flow is likely to be most important in gravel beach dynamics, especially when developed as a fluid-thin backwash [2,20].

The different cross-shore size–shape zonation of gravel pebbles, compared to sand, exists on gravel beaches and was postulated by numerous authors, e.g., [21–23]. The discoidal and spheroidal particles usually show a tendency to be transported upslope, acting like a hydrodynamic "wing". The spherical and spindle gravels are usually transported downslope [22,24–26]. Additionally, the flattened particles are prone to longshore sediment transport. However, in some cases it is not clear whether sorting by size and shape, are achieved by these two mechanisms, or what aspects of anisotropy are important (e.g., "shape" or sphericity, aspect ratio, and elongation, and the axially less dominant third dimension, or *c*-axis, which may produce different responses to flow, either individually or as part of mixed beds) [27,28]. The longshore transport of particles cause colliding and rubbing, which may produce different shapes of pebbles that may be used as an indicator of erosion processes [29]. Irregular particles generally indicate a high potential for mechanical reworking, whereas rounded particles are a direct manifestation of the erosion that has occurred. These processes are partially related to lithological variability and textural features of different types of rocks. Soft rocks (limestones, marbles, shales or sandstones) are definitely less resistant, compared to magmatic or metamorphic components of gravel beaches [30]. Coarse sediments may be directly related to changes in lithodynamics, sediment redistribution pathways, or intense material transport. Heavy gravels are less sensitive to wave and wind dynamics, and are applicable for estimating relative changes in sediment transport rates.

Lithological responses to erosion are established in particle size, roundness, sphericity, flatness, and asymmetry. Pebble roundness, flatness, and rate of mechanical destruction depend on wave intensity, initial rock types, and nature of the coast (sheltered or exposed). The intensity of erosion is established by particle morphology, mainly their roundness and flatness. Thus, a greater intensity of erosion is indicated by larger roundness and flatness indices [29,30]. The variability of gravel pebbles may show a different transport pattern under low energy conditions. Spherical grains, resulting from their capability to roll over in the swash zone, usually cover longer distances and indicate more dynamic conditions, when compared to discoidal grains [19]. Compared to sandy beaches, gravel beach sediments are usually spatially differentiated in terms of size and shape to a greater degree, affecting more obvious textural zonation, mainly in the form of mixtures of relatively fine and coarse fractions [21,24]. The step, cusp horns, strands, and berms are composed of larger grains compared with the foreshore, although a number of levels of textural zonation may be discernible as sediments are continuously redistributed [2,24].

Studies on geodynamic processes and lithodynamics of the coastal zone are essential in different branches of the economy (e.g., fisheries, tourism) and in maritime spatial planning [24,31–34]. Increased coastal erosion resulting from a sediment deficit, related to a reduction of river input, dredging processes, and coastal embankments, affects the formation of narrow beaches composed of relatively coarse sedimentary components. Gravel beaches usually form due to erosion of indurated cliff coasts, built mainly of moraine tills with small amounts of loose sediments. Increased volumes of beach rocks exposed due to the coastal erosion can effectively protect sensitive parts of the seacoast; however, during high sea level stands caused by wave storms, water can flow through gaps in coarse material, thereby increasing water saturation and slope instability [3,35].

Recently, erosion has been dominant along nearly the full length of southern Baltic coast. Erosion rates are highly variable and accompanied by simultaneous accumulation processes, either generating or rebuilding accumulative forms. The spatial and temporal distribution of these forms has varied significantly during the last 100 years, related mainly to increases in mean sea level and anthropogenic activity, which affected the intensity and varied the redistribution patterns of sedimentary material. The present rapid rate of dune and cliff erosion is an indirect proof of insufficiency of these sediment resources [36].

Frequent wave storms are the main cause of coast erosion in the southern Baltic. The foreshore in the study area is shallow, with the 10 m isobath located more than one kilometer from the shoreline. The source of redeposited sedimentary material is related to eroded sections of cliff and dune coast; however, the geologic framework of the shallow foreshore also contributes to the total balance of pebble load. During the elevated sea level stands and wave storms, thin layers of sands and gravels developed on moraine tills are mixed, crushed, and redistributed towards the shoreline, and incorporated with the sands and gravels eroded from the cliff sections [1,37].

The Baltic nearshore in the study area is frequently covered by Holocene sands constituting the in-shore bedload [38–40]. The Pleistocene series is lithologically more variable compared to the Holocene sands, and composed of moraine tills, glaciofluvial sands, muds, clays, and organic sediments (mainly peats). Sediments are often laminated and lie in the form of beds or lenses. In some cases, water-bearing horizons are present and promote mass displacements or cliff degradation. Residuals of beach pebbles reflect regional differentiation of glacial tills from which they originate [38]. For example, a greater content of limestone among the other petrographic components of beach gravels located east of Pogorzelica indicates erosion of the bench and morainic material redeposition towards the shore.

The article characterizes the changes in petrographic composition and grain shape variability of beach gravels sampled in the storm berm (or beach face) along the almost 30 km-long shore of the Baltic Sea, extending from Pogorzelica, Poland, (coastline kilometer363.0) and ending at the River Dziwna mouth (391.4 km; Figure 1). Grain size parameters of beach gravels were estimated by application of the traditional Zingg method [41].

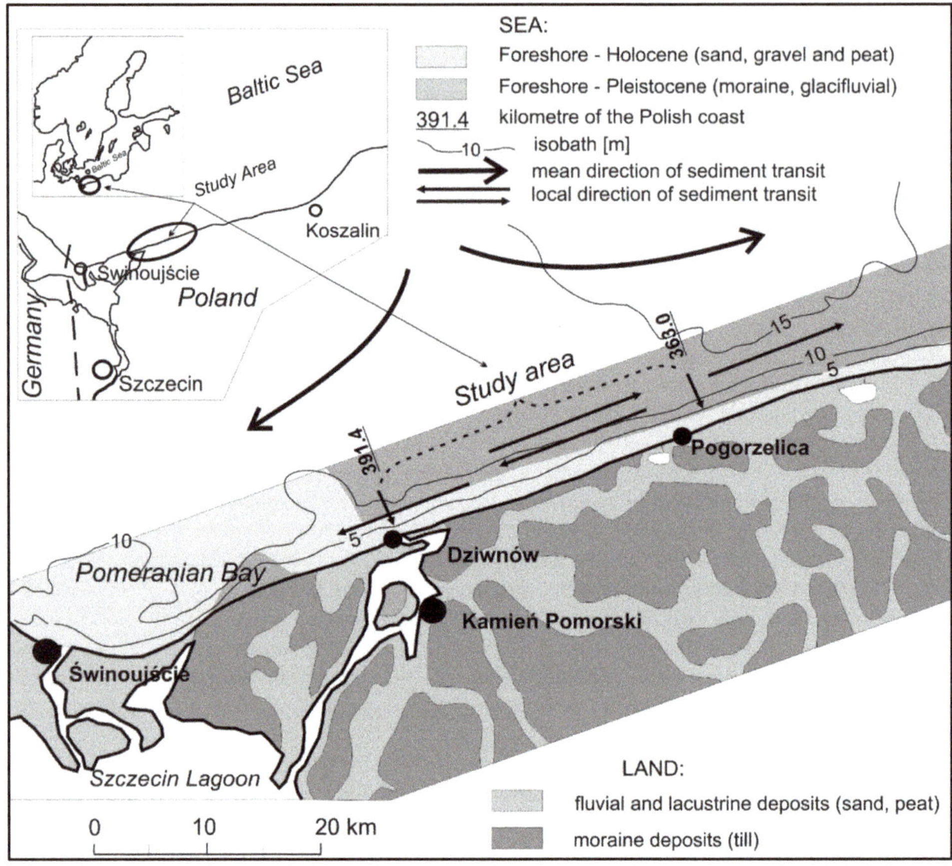

Figure 1. The Western Pomerania coastline of the Baltic Sea, Poland, which is characterized by presence of Pleistocene–Holocene glacial and fluvial deposits, shallow nearshore sediments, and divergence of sedimentary transport [37,42,43].

The major aims of the study are to determine the petrographic composition of beach gravels, compare these with similar initial data for gravels collected from cliff sections, map coast sections affected by erosion and accumulation processes, and then link the results to morphodynamics estimated by applying nongeological methods, accounting for the presence of coastal protection structures. This is also the first detailed study of petrographic composition and measurements of Zingg grain shape parameters of beach gravels conducted along the coastline of Western Pomerania, Baltic Sea, Poland. By assessing of developmental trends of the coast, the petrographic composition and morphometric characteristics of gravels may indicate changes in erosion intensity, and contribute complementary information indicating longshore bedload transport [42,44].

2. Study Area

The section investigated is a part of Trzebiatów Coast, itself a part of the Southern Baltic Coastal Zone, Poland. It features cliff- and dune-type shores [32,45,46]. The study covered the western part of the sandbar–dune Rega Gate area from coastline kilometer 363, the entire Rewal Upland cliff shore section, and the eastern part of the sandbar–dune Dziwna Gate area to the River Dziwna mouth (391.4 km).

Research carried out in the area by various authors [5,32,38,47–53] demonstrated the nature of geodynamic processes and developmental trends. This study builds on research conducted by Jahn [54] on beach pebbles of the Polish Baltic coast, as well as on comprehensive coastal zone research (including beach gravels) in the Niechorze–Trzęsacz section of the coast, carried out in the early 1970s [42].

Owing to the specificity of the sea–land interactions, the Trzebiatów Coast has been divided into the following morphodynamic regions (Figure 2a,b) [4,55]:

1. The sandbar–dune area of the Rega Gate (345.5–366.8 km; Figure 2a,b (profile 1)) exhibits a high and moderate dune shore and a tendency to erode (mainly at 363.0–366.8 km). During the last 25 years, the annual erosion rates have reached 0.25–0.3 m/year [56].
2. The cliff area of the Rewal Upland (366.8–385.7 km; Figure 2a,b (profile 2)) has variable shore types, from the eroded upper foreshore part (e.g., 386.5–371.2 km), to varying developmental phases (e.g., 372.0–374.0 km), and locally with a tendency to erode (e.g., the Pobierowo cliff at 374.0–375.0 km and 378.0–379.0 km). In the area of Rewal, the estimated mean decadal rate of erosion is >1.0 m/year. Between 372–386 km, the erosion rate stabilizes between 0.2 and 0.3 m/year [56].
3. The sandbar–dune area of the Dziwna Gate (385.7–397.4 km; Figure 2a,b (profiles 3 and 4)) exhibits a dominant tendency toward erosion of the upper foreshore part. Varying labile development is seen only in the central part (386.5–388.0 km). The area has a large degree of anthropogenic transformation (groins, revetments, bulkheads) [57] and beach nourishment. The rates of erosion–accumulation are highly variable and vary from 0.3 to +0.4 m/year (dune sections with seawalls) to as much as −0.8 to +0.5 m/year in area of Dziwnów jetty [56].

The entire area of study, including the shallow foreshore, represents a morainic upland built mainly of morainic tills, occasionally separated by glaciofluvial sediments [37,58,59]. The sandy-gravelly deposits of the foreshore are located only in the central part of the Dziwna Gate. The sandbar areas show older valleys of different origins, with distinct horizons of organic sediments, mainly of lake and lacustrine origin [51,60].

The Rega Gate area is a part of a riverine floodplain valley covered by sandy spit and aeolian deposits. The Rewal Upland area is a fragment of a Pleistocene upland with variable denivelation, the depressions of which are frequently covered by aeolian sands [38,61].

The Dziwna Gate area is a fragment of a "bridge" connecting Pomeranian Bay with Kamień Lagoon, remodeled by sandbar deposits and intersected by the River Dziwna mouth [60].

The Pogorzelica–Dziwnów coast section shows a sandbar–dune backshore built by Holocene sands, and cliffs formed by Pleistocene morainic and glaciofluvial deposits. The entire section (except for its westernmost part with the old River Dziwna bed) features an erosional platform built mostly by morainic tills or locally by glaciofluvial sediments, and frequently covered by thin layers of marine sands. During storms, these sands (or gravelly sands) work as an active layer and their movement erodes the platform [58,62].

The estimated rate of cliff sections erosion during last 100 years erosion has varied from 0.5 to 2 m/year. Due to the presence of embankments along Rewal cliff, the accumulation along the dunes slowed to <2 m/year, sometimes indicating even erosion (~0.7 m/year), especially around the Dziwna spit, where intensive erosion dominates [56].

According to meteorological conditions, the study area is dominated by southwestern and western winds annually, with mean wind speeds of 2 to 4 m/s, and frequency reaching 20 to 25% (light to gentle breeze). The frequency of 6–8 m/s winds varies 15 to 20% (gentle to fresh breeze). Frequency of strong winds and storms (>8 m/s) is only about 1–2%. The highest wind speeds occur during autumn and winter, usually 0.6–1.4 m/s and more intense compared to summer and spring [6].

The frequency and intensity of storm events in the study area increased during last 70 years. Storms induce an increase of the wave splash zone, which usually reach approximately 2–2.5 m above the seasonal mean high water. The frequency of waves with heights <0.5 m, depending on the season, exceeds 90%. The annual mean sea level increase induced by global climate forcing varies from 0.04 to 0.08 cm/y [63,64].

The dune and cliff foot positions in the study area are positively correlated, especially in the area between Rega and Rewal (correlation coefficient >0.75; significance level $\alpha = 0.05$). In the western part of the study area, the correlation is less (<0.25; $\alpha = 0.1$). During the last 40 years, the correlations decreased, mainly due to more embankments and the increasing mean sea level, which heavily disturb circulation of pebbles in the study area [36]. In the western part, longshore currents distribute particles towards Wolin Island. Meanwhile, in the eastern part, bedload transport towards Kołobrzeg dominates. In the middle part, the divergence zone was observed, with no visible processes of gravel redeposition (temporal erosion–accumulation) [43].

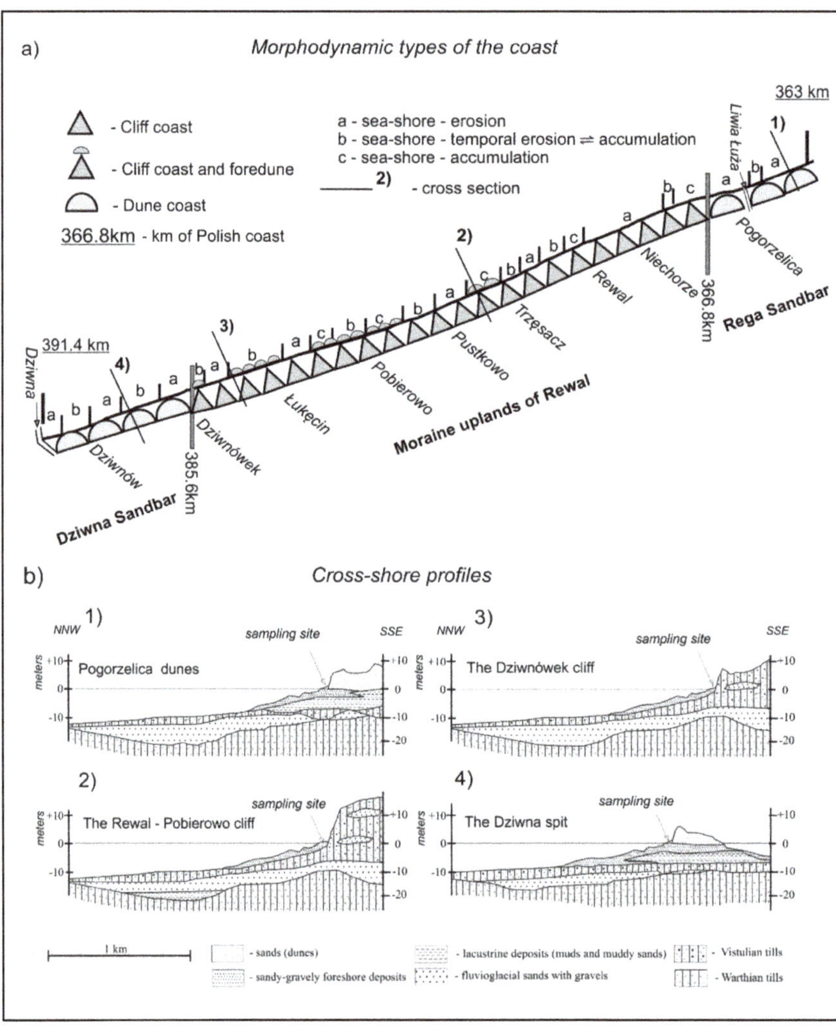

Figure 2. (**a**) Morphodynamic types of the coast in the study area, (**b**) representative geological cross-shore profiles 1–4 (adapted from [65]).

Lithodynamic processes in the study area are strongly related to seasonal variability of meteorological and hydrodynamic conditions. The grain size composition along sandy coast sections can be characterized by individual groups of so-called "lithodynamic belts". The transformation of initial grain size depends on stormy conditions and intensity of aeolian processes. These differences are related to (i) grain size variability in the surf zone and backshore, (ii) intensity of wave storms, (iii) spatiotemporal changes of winds and air streams, (iv) different rates of seashore drying or wetting by rainfall, (v) influxes of the water from the surf zone, and (vi) marine aerosols or dew [66]. Deposition and redeposition of coarse sedimentary material may be attributed to three stages:

1. During the storm conditions, the greatest effect of particle dispersion is usually observed in the backshore. The weakest dispersion is noted in the foreshore. Additionally, the balanced liability between scattered and drifting particles is visible.
2. In poststorm conditions, the selective processes of granular differentiation in dunes material and backshore are observed. The coarse layer becomes thicker, and predominance of sediment movement and accumulation is obvious, rather than redeposition (erosion and secondary accumulation). Accumulation in the foreshore is negligible. The processes of redeposition and movement of sedimentary material slowly appear to develop on poorly dampened surfaces.
3. The prestorm conditions are dominated by aeolian processes, which are mainly represented by redeposition (wind erosion and sediment mass movement). The thinnest and finest fractions are removed and gravels are exposed. The fine fractions are flushed from the cliff sections and redeposited into the deeper parts of the Odrzana and Pomorska Bays [66].

The beach sections in the study area are composed mainly of fine and medium sands. The amount of gravel is generally low, especially during periods between storms. Among the gravels, the 2–5 cm subfraction dominates. During the wave storms the relative amount of coarse grains increases, mainly due to the removal of fine fractions, or by redeposition of foreshore gravels [42,44].

3. Methodology of Field and Laboratory Studies

Field work was conducted following stormy weather and in the second half of September 2004. At of the end of 2004 and the start of 2005, parts of the coast section studied were subjected to beach nourishment with dredged material. Beach sampling was carried out for two days, during field classes given to students in the Construction Program of the former Technical University of Szczecin (currently the West Pomeranian Technological University in Szczecin, Poland). Gravels with a dominant grain size diameter of 2–5 cm were collected from the berm (or beach face) of dune and cliff sections of the Polish seacoast.

Owing to the low water temperature, the underwater part of the beach was not sampled. During storms and strong winds, the lower and middle parts of the beach were flooded; therefore it was assumed that gravel samples collected in the beach just after the storm cessation are representative for the assessment of morpholithodynamic processes at the land–sea interface. The weather during sampling was rainy. The sea state was 2–3 Beaufort scale, and the wind was westerly. Immediately before the sampling, the weather was windy (moderate breeze), with SW and W winds dominating (4–7 m/s), and 4–5 Beaufort scale. The atmospheric pressure was decreasing from 1020 to 998 hPa [67].

During the sampling, the beach width varied from 10 to 40 m. The beach width in the cliff sections varied 10–15 m, whereas in cliff sections with developed foredune, the width was 20–40 m.

The beach gravels were collected at a radius of about 50 m, along 250 m intervals of the coastline. Individual samples weighed from 2 to 5 kg and were collected in the area of 10 × 10 m squares. If the amount of collected rock material was above 2 kg, the sampling procedure was finished; if <2 kg, then the sampling area was expanded to 50 × 10 m rectangles. Gravel samples were collected simultaneously by two groups of eight individuals. One group started sampling in Pogorzelica (363.0 km), the other starting from the Dziwnów jetty (391.4 km). The sampling operation took two days, because—once the windy conditions were over—the gravels were rapidly covered by sand blown in by the wind, and the beach became sandy. The gravels deposited during the storm lay beneath a thin aeolian layer [68].

Next, the gravel samples were examined for their petrographic composition. Additionally, the grain shape among individual petrographic groups was determined.

The basic petrographic groups included (i) high resistant igneous and metamorphic crystalline rocks with quartz; (ii) low resistant limestone; and (iii) low resistant sandstone, dolomite, and others (shales, flints, brick and concrete fragments). The samples consisted of 200–400 gravel grains. The grain axes were measured in each petrographic group to arrive at the maximum length, maximum width, and maximum height. Gravel morphometry was determined using the flatness index (*Ws*) of Wentworth and the gravel shape as described by Zingg [41]. The *Ws* was estimated using equation $Ws = (A + B)/C$ [69]. Zingg's method describes the grain shape based on (i) the ratio between the intermediate (*B*) and longest (*A*) axis, and (ii) the ratio between the shortest (*C*) to intermediate (*B*) axis. The gravel grains were classified by their shape as discoid, ellipsoid, spheroid, and spindle-shaped (Figure 3).

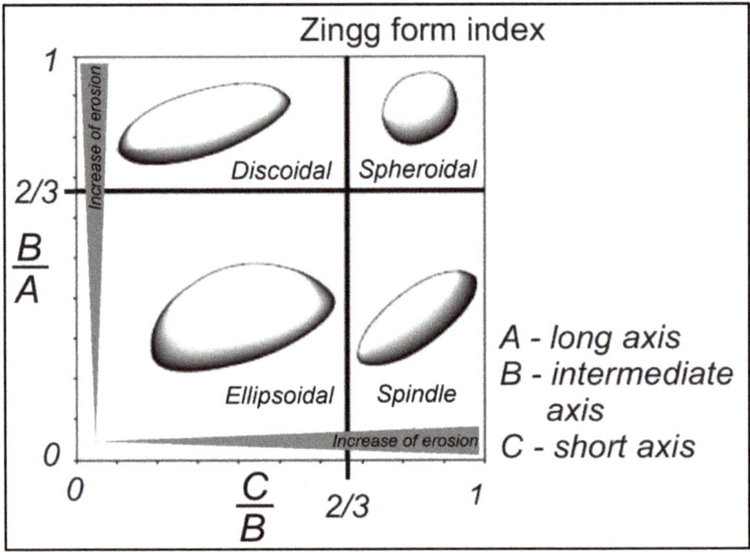

Figure 3. The Zingg form index diagram [41]. Arrows indicate the tendency of shapes related to dominance of erosion and mechanical reworking of sedimentary material (after [44]).

The basic statistical parameters (arithmetic mean and standard deviation) of grain shape indices and petrographic group classification were calculated for the set of data consisting of 113 sampling stations and total number of 17,882 gravel pebbles. The mean number of gravel grains collected from single stations was 172. At nine stations gravels were not found, so only 104 stations were used in the further analysis of lithodynamics.

The petrographic composition of beach gravels may point to the source of the initial material and directions of transport in the in-shore zone. A comparison of the proportion of gravel in individual petrographic groups with the initial material allows a qualitative assessment of the intensity of the gravel-destructing processes in those groups. Also, the gravel shape may useful for interpreting the coastal dynamics. When assessing the intensity of shore erosion, lithodynamics of the coastal zone may be subject to rank-based evaluation [51,70]. Assuming the general knowledge of individual sections of the Polish coast (erosive or accumulative), and simple mechanical properties of selected group of rocks, such as hardness or presence of slaty cleavage (e.g., greater hardness of crystalline rocks, compared to sandstones, limestones, or shales), the zones of increased mechanical reworking of rocky material can be distinguished. Proportions between different types of rocks, compiled with the basic Zingg measurement statistics indicate primary depositional patterns (instantaneous deposition with no mechanical reworking traces, accumulation of initial sedimentary material) or redeposition (erosion and accumulation, reworking, or sorting traces). Presence of cusps and embayments located along the study area affects coastal dynamics, and may decrease movement of sedimentary debris (i.e., greater amount of discoidal and ellipsoidal gravels that indicate shorter mechanical reworking) [57,71].

Finally, the data obtained were subjected to lithostatistical analysis. Considering the large number of collected and measured gravel samples (n = 17,882), the basic statistics (means, standard deviations, and percentages) were calculated with 90% confidence intervals. Calculated parameters (amount of gravels in selected petrographic groups, shape parameters, and *Ws*) were compared with coastal dynamics data (1986–2004) and presence of groins, wavebreakers, and seawalls [56,57].

The mean values of each parameter were estimated and subsets between half the standard deviation (0.5 SD) above and below the mean were considered as unimportant for the lithodynamics interpretation. The subsets above half of the standard deviation (+0.5 SD) were treated as positive anomalies, whereas the subsets below the half the standard deviation (–0.5 SD) were negative anomalies. The positive anomalies in the group of discoidal and ellipsoidal gravels are indicators of accumulation (direct deposition without secondary reworking of sedimentary material). The positive anomalies in the group of spheroidal and spindle-shaped gravels may indicate the dominance of erosion, increased mechanical reworking, and longshore movement of sedimentary debris [70]. In this particular case study, the statistical application of 0.5 SD may be justified by robust sampling data and a flattening of distribution. Additionally, the shorter subsets expose properly obtained anomalies.

Similarly as above, the positive anomalies of the gravel flattening index (*Ws*) are indicators of accumulation, whereas negative anomalies are related to erosion and redeposition processes [44,70]. The *Ws* index positively correlates with contents of ellipsoidal and discoidal gravels.

4. Variability of Petrographic Composition

The petrographic composition of morainic structures making up the cliff shore and the shallow foreshore is dominated by fragments of crystalline rocks (about 40–50%), limestone gravels (30–50%) and sandstones with dolomites (10–15%). The remaining gravels contribute less than 10% [59,72–74]. The gravels occurring on the beach eroded from morainic deposits and were rapidly sorted in both petrographic composition and shape. The contribution of mechanically resistant components (mainly crystalline-rock gravels) increases and contents of less resistant gravels (shales, sandstones, limestones) decrease with distance from the source of the initial material.

Table 1 summarizes contributions of initial gravels originating in morainic tills, which make up the cliffs and the gravels collected on the beach. The percentages of beach gravels are shown in relation to the total gravel weight and to the number of gravel grains collected. In individual petrographic groups, there are small differences between the percentages associated with the gravel weight and number of grains. Because the grain shape was measured on each grain, the contribution of individual petrographic groups to the total number of gravels examined is given (Table 1).

Table 1. The mean contribution of individual petrographic gravel groups in the study area (n = 17,882; confidence interval 90%). Contents of initial material are from archival data [72,74].

Petrographic Gravel Group	Initial Material (Glacial Pleistocene Sediments)	Beach Gravels	Contribution by			
			Weight		Number of Fragments	
	range (%)	interval (%)	mean (%) $\bar{x}$	standard deviation σ	mean (%) $\bar{x}$	standard deviation σ
crystalline	39–52	50–90	75.8	8.7	72.4	8.4
limestone	30–62	5–50	14.8	6.9	17.2	7.6
sandstone	10–15	3–15	7.7	3.1	7.8	2.8
other	5–10	1–5	1.6	1.0	2.5	2.0

The petrographic composition along the shore section showed considerable variability (Figure 4a–c). The mean weight content of crystalline rock gravels was 75.8%, and ranged within 50–90%. The highest contribution of crystalline gravels (over 75%) was found in the Rewal Upland cliff, where glacial tills dominate in the shallow nearshore and are intensively mechanically reworked during stormy weather, creating an additional source of crystalline components. These processes are especially visible between 375.0 and 383.0 km. The lowest proportion of those gravels (below 70%) was revealed in a zone of contact between the cliff area and the sandbar section of the Dziwna Gate (384.0–387.0 km), as well as on the Rewa Gate dune shore sections in the vicinity of the Rewal Upland dune coast (364.5–368.0 km), and the Dziwna Gate (390.0–391.4 km). The contribution of limestone gravels averaged about 15%. The highest proportions of limestone gravels (above 25%) were recorded in the area of active cliffs from Niechorze to Trzęsacz (368–373 km) and (about 20%) between Łukęcin and Dziwnówek (384.0–386.0 km), as well as in the vicinity of the Dziwnów jetty. The lowest contribution (about 10%) of limestone gravels occurred in the cliff area between Pustkowo and Łukęcin (375.0–384.0 km) and east of the Liwia Łuża mouth (365.5 km). The average percentages of sandstone gravels ranged within 5–15%; elevated contributions (10–15%) of sandstones were found only in the dune shore section in Pogorzelica (364.5–365.5 km) and in the zone of contact between the cliff and dunes at Dziwnówek (386.0–366.5 km). The remaining shore sections showed a similar content of sandstone gravels (5–10%).

The ratio between gravels of less resistant rocks (limestones and sandstones) and more resistant rocks (crystalline rocks with quartz) was at its highest (above 0.4) in the area of active cliff and the Liwia Łuża mouth between Niechorze and Rewal, somewhat lower ratios (0.3–0.4) were recorded between Łukęcin (384.0 km) to the cliff–dune contact zone (386.5 km) to the River Dziwna mouth (391.4 km). The lowest ratios, about 0.2 and below 0.2 were found in sandbar areas east of the Liwia Łuża mouth towards Mrzeżyno, and at a temporally active cliff at Pobierowo, respectively, as well as at a cliff in a labile developmental phase (371–377 km), except for clearly eroded sections (Figure 4a). The greatest domination of crystalline rock gravels over limestones and sandstones was revealed east of Pogorzelica and between the Pobierowo cliff to Łukęcin (375–383 km). Similar were the relationships between sandstones and limestones. At the Pobierowo cliff, the sandstone to limestone ratio was the highest, declining east of Rewal (371–364 km) and west of Łukęcin (381–391 km). Increased ratios were also related to the presence coastal protection structures, which trap less-resistant gravel components and decrease mechanical reworking of the material.

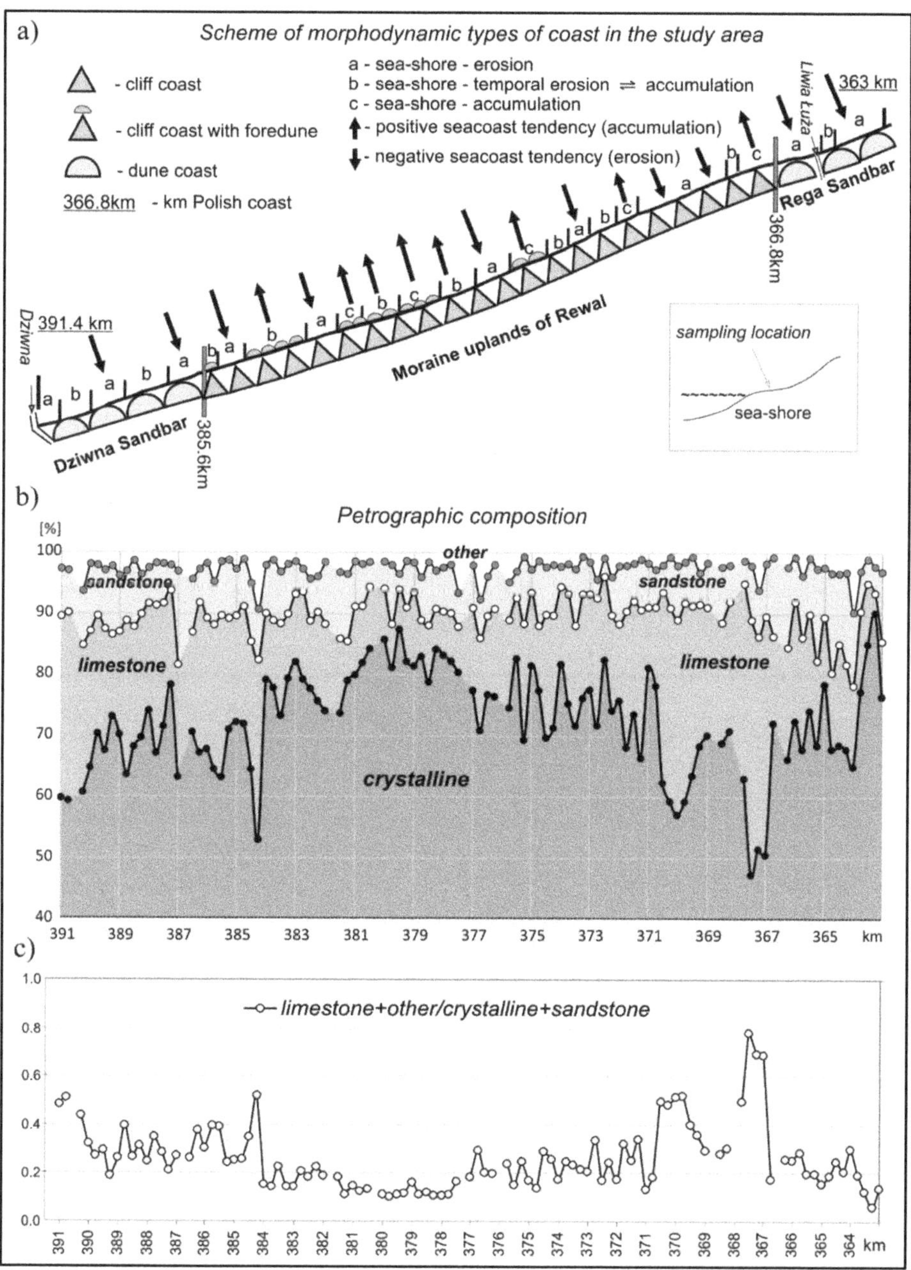

Figure 4. Coastal morphodynamics and petrographic composition of gravels between Pogorzelica and Dziwnów (363.0–391.4): (**a**) the seashore morphodynamics with erosion–accumulation tendencies, (**b**) the petrographic composition of beach gravels, and (**c**) the ratio between mechanically less resistant sedimentary rocks and more resistant gravels. The broken lines indicate a lack of gravel along the coast section investigated.

5. Shape Variability

The mean values (and standard deviations) of the Zingg gravel shape index and flatness index are shown in the Table 2. The table also contains contributions of remaining gravels, but their contributions were very low (below 5%) and are disregarded in the subsequent discussion, as they are not important for insights into the beach lithodynamics.

Table 2. The shape measurement data of beach gravels in selected petrographic groups (n = 17,882; confidence interval 90%).

Gravel Grain Shape and Type	Crystalline		Limestone		Sandstone		Other	
	mean (%) $\overline{x}$	standard deviation σ	mean (%) $\overline{x}$	standard deviation σ	mean (%) $\overline{x}$	standard deviation σ	mean (%) $\overline{x}$	standard deviation σ
Discoidal	72.1	10.1	75.1	10.9	78.6	13.8	70.8	29.7
Ellipsoidal	9.3	4.5	16.3	10.4	10.8	9.8	22.4	26.2
Spheroidal	13.7	7.8	5.9	5.5	8.1	10.0	4.0	9.8
Spindle	4.6	3.0	2.5	3.5	2.3	4.9	3.2	10.5
Flatness index (Ws)	2.3	0.3	3.2	0.9	2.9	0.8	2.3	1.1

The beach was generally dominated by discoidal gravels, which contributed greatly (78.6%) to the sandstones, as well as the limestones (75.1%), crystalline rocks (72.1%), and other gravel (70.8%). A high content of discoidal pebbles is related to processing intensity during storm events or strong winds, when less resistant gravels are flushed out from the shallow foreshore and partially reworked in the surf zone. The lowest standard deviation (10.1%) was found in the mean contribution of crystalline gravels, and the highest in other gravel (29.7%). Ellipsoid gravels contributed most (22.4%) to the other gravel group, but with the highest standard deviation (26.2%). This is one of the reasons why this group is disregarded in the further discussion on lithodynamics. Ellipsoid gravels contributed 16.3, 10.8, and 9.3% to limestone, sandstone, and crystalline gravels, respectively. The lowest standard deviation was found for the crystalline gravels. Spheroid gravels contributed most (13.7%) to the crystalline rock gravels and least (5.9%) to the limestone gravels. Spindle-shaped gravel grains accounted for 4.6 and 2.3% of the crystalline and sandstone gravels, respectively. Figures 5, 6 and 7a,b illustrate the variability of gravel grain shape in the petrographic groups.

As the discoid gravels contributed most to all of the petrographic groups, they should form the basis for evaluating the sedimentary environment dynamics in the coastal zone between Pogorzelica and Dziwnów. The remaining data should be treated as auxiliary.

The preponderance of subsets containing discoid and ellipsoid gravels, in relation to the values within the standard deviation from the mean for the entire coastal section, is indicative of the domination of accumulation over erosion and redeposition in the individual shore sections. On the other hand, the preponderance of subsets containing spheroid and spindle-shaped grains, compared to the values within the standard deviation from the mean along the section analyzed, points to domination of erosion and redeposition over accumulation (instantaneous deposition).

The contributions of the discoid crystalline gravels above half the standard deviation occur primarily in the western part of the Rega Sandbar (364.5–366.5 km), in the area of a stabilized cliff (379.5–381.5 km), and locally in the area of the Dziwna Sandbar shore armoring (386–391 km). A similar distribution is characteristic of the crystalline ellipsoid gravels. It is only in the Rega Sandbar that higher values occur within 363.0–364.5 km, whereas a very similar distribution is found in the Dziwna Sandbar (Figure 5a,b).

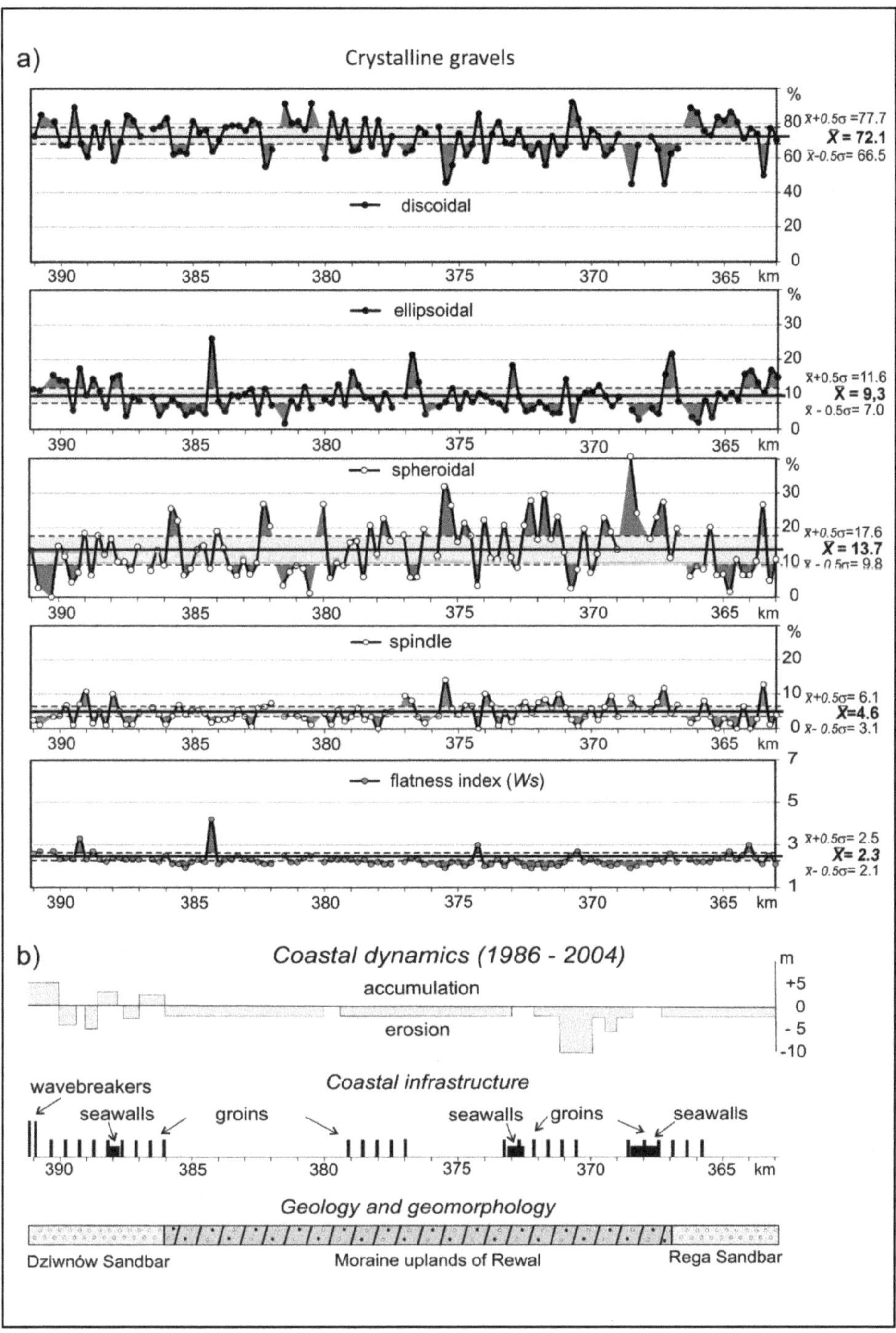

Figure 5. (**a**) Variability of morphometric indices of mechanically high resistant crystalline gravels along the investigated coast section, (**b**) comparison with coastal dynamics and coastal infrastructure data. The broken lines indicate lack of gravels along the beach (n = 17,882; confidence interval 90%).

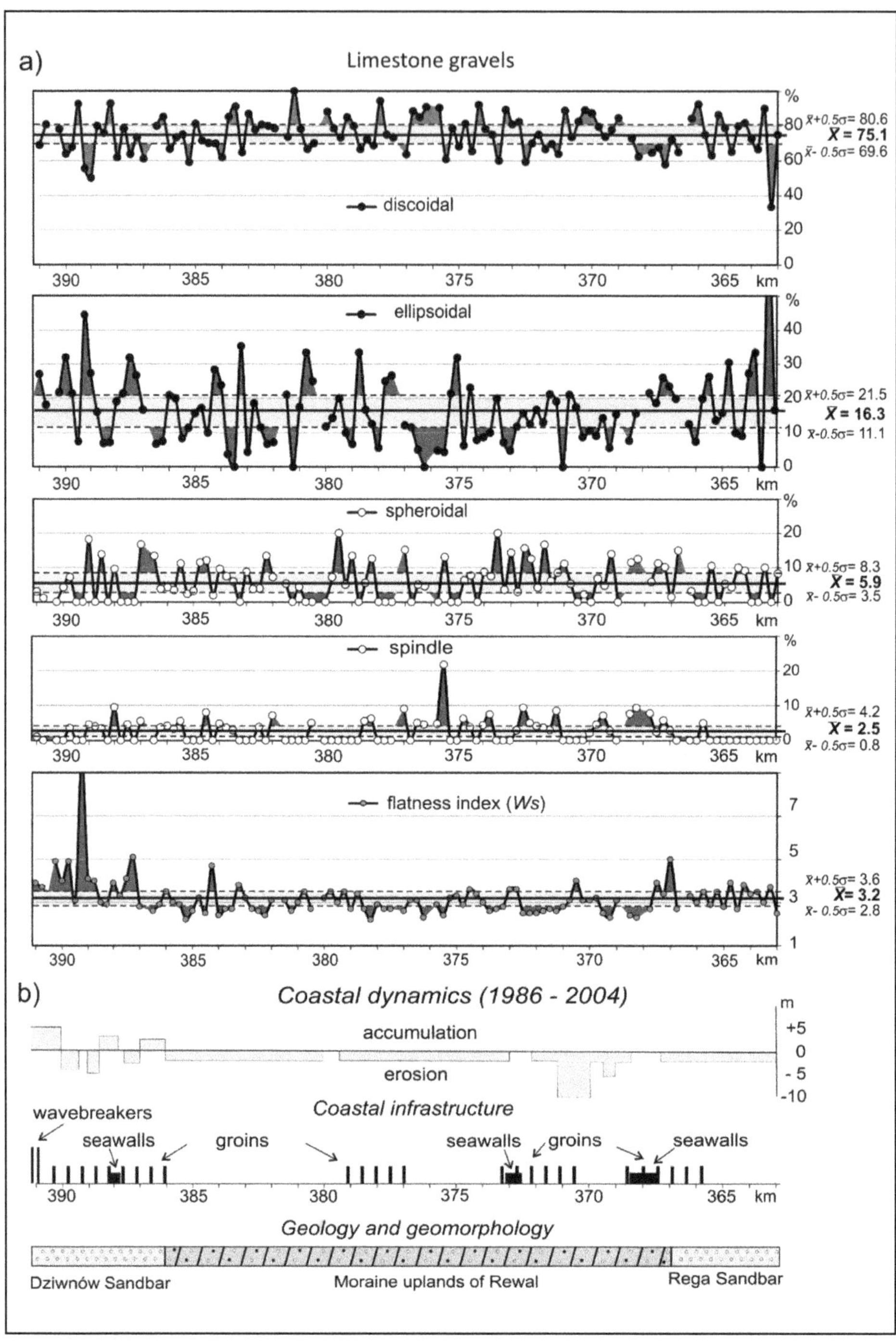

Figure 6. (**a**) Variability of morphometric indices of mechanically low resistant limestone gravels along the investigated coast section, (**b**) comparison with coastal dynamics and coastal infrastructure data. The broken lines indicate lack of gravels along the beach (n = 17,882; confidence interval 90%).

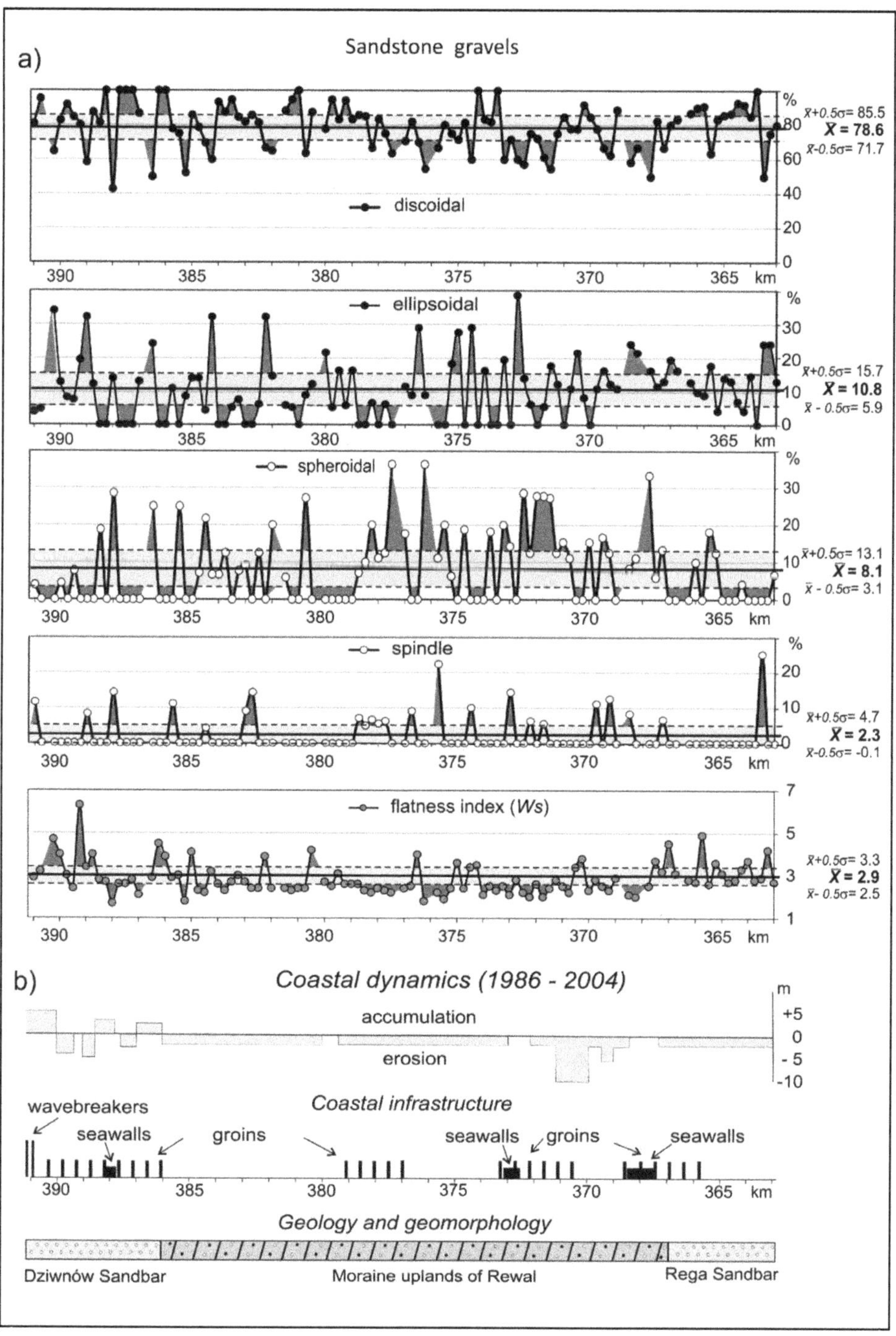

Figure 7. (**a**) Variability of morphometric indices of mechanically low resistant sandstone gravels along the investigated coast section, (**b**) comparison with coastal dynamics and coastal infrastructure data. The broken lines indicate lack of gravels along the beach (n = 17,882; confidence interval 90%).

Contributions of discoid crystalline gravels below half the standard deviation occur mainly in the eroded cliff sections, except for those parts of the cliffs characterized by labile development, and even by dune accretion in front of the cliff. Contributions of ellipsoid gravels less than half of the standard deviation are similar to those of the discoid gravels, except for the areas of the Rega Sandbar and the Dziwna Sandbar (Figure 5a,b).

Contributions of discoid limestones and sandstones accentuate the preponderance of deposition on sandy spit areas, with certain exceptions, as well as domination of deposition in the cliff section (376.0–382.0 km). The distributions of limestone and sandstone gravels are similar to those of the crystalline gravels, but more visible is their increase in the Rega and Dziwna Sandbars, as well as in the western part of the cliff, less susceptible to erosion processes (Figures 6 and 7).

Contributions of spheroid and spindle-shaped gravels above half the standard deviation in all the petrographic groups point to domination of the cliff coast erosion within 367–378 km, and to a labile equilibrium in the western part of the cliff (378–385 km), with a tendency toward stabilization. Contributions below half the standard deviation were found in the Rega Sandbar and, partly, in the Dziwna Sandbar, indicating domination of accumulation over erosion. However, local variation in development trends is visible (Figures 5–7).

Increased values of the gravel flatness index (*Ws*) provide information on the degree of gravel roundness. The values obtained indicate increased mechanical reworking of gravel grains. The *Ws* is related to domination of discoidal and ellipsoidal gravels, and additionally with increased contents of mechanically less resistant rocks. If the *Ws* value is low, it means that the gravels are dominated by rounded (spheroidal or ellipsoidal) pebbles; if it is high or extremely high, it means domination by flat nonrounded grains. The flatness index distributions along the shore are shown for different petrographic groups (Figures 5a, 6 and 7a). The crystalline rock gravels are less rounded than the less resistant limestones and sandstones. In the sandbar areas (363–368 km) and in the eastern part of the Dziwna Sandbar (387–391 km), the limestone and sandstone gravels show higher flatness index values (above the mean and half the standard deviation). In the remaining areas, the *Ws* were very low (except for a few samples). Other gravels, due to their low abundance, are disregarded, although their *Ws* values were similar to those of sandstones and limestones, with a somewhat larger standard deviation from the mean.

The *Ws* values of limestone and sandstone gravels above half the standard deviation occur on the sandbars and locally in the cliff areas, but only in those at the labile development phase or with domination of sediment accumulation. The lowest values are characteristic of erosion-dominated cliff areas. The flatness index values of crystalline gravels show no particular variability along the coastal section under study. It is only the most eroded cliff section (368–378 km) that shows lower values of the index.

6. Discussion

The analyses showed that the beach gravels collected from Pogorzelica to Dziwnów differed widely in their petrographic composition. The cliff sections of the beach, in labile equilibrium, showed the highest proportion of crystalline rock gravels. Their contribution declined in sections affected by erosion and in the sandbar sections. Limestone gravels dominated in active cliff areas (Niechorze–Pustkowo and Łukęcin–Dziwnówek) and in a part of the Dziwnów Sandbar, indicating increased erosion. Occasionally, their contribution was twice that found in the stable or nonactive cliff sections (e.g., cliff shore in Pobierowo). Attention must be drawn, however, to the eroded foreshore that nourishes the beach, too. Increased reworking of gravel material leaves only the most resistant gravels and their distribution along the shore may be indicative of the direction of the longshore bedload transport.

The direction of the longshore bedload transport in the shore section under study is still unclear. In the 1950s, Zenkowicz [75] pointed to west-to-east bedload transport. Subsequently, Bączyk [43] noticed a divergence of the sandy bedload transport off the Pobierowo cliff, eastwards towards the

Koszalin Bay and westwards towards the Pomeranian Bay. In his studies of the shallow foreshore, Cieślak [76] demonstrated the direction and intensity of the sandy bedload transport to be both eastward (measured at Niechorze, 368 km) and westward (measured at the River Dziwna mouth, 391.4 km). In the Niechorze–Dziwnów section, Furmańczyk [71] showed the locally varying directions of the bedload transport along the shallow foreshore, and associated the variation with cellular currents occasionally producing embayments, as well as larger or smaller cusps. Between Pogorzelica and Dziwnów, channel-like locations termed "underwater circulation gates" were identified [77]. They distort the bedload transport and serve as outlets allowing export of substantial amounts of material during storms.

In general, the coast sections investigated are affected by long term erosion, which can be seasonally less intensive, or limited by coastal infrastructure (so-called "unstable equilibrium"). Gravel circulation is strongly related to erosion and accumulation (direct deposition), considered as a primary removal of sedimentary material, e.g., from morainic cliff sections, and secondary accumulation directly within the cliff base or shallow nearshore, with no significant mechanical reworking patterns. Redeposition occurs due to secondary movement of this material, mainly resulting from strong wave storms. The same process may be related to erosion of shallow seashore, which remobilizes sedimentary material towards the coastline and partially rounds the gravels.

The contribution of individual petrographic groups does not allow for strictly pinpointing the dominant direction of the bedload transport. The increased proportions of crystalline gravels in the cliff area between Rewal and Łukęcin would be indicative of washing-out of this part of the erosional platform and the deposition of "fresh" gravel material in sandbar–dune areas. This is evidenced by an increased contribution of gravels representing rocks less resistant to destruction, both in the Rega and Dziwna Sandbars.

The longshore distribution of less resistant and more resistant gravels may indicate the absence of any distinct direction of bedload transport along the shore, e.g., from the southwest to the northeast or back. Shoreline armoring (breakwaters, revetments, bulkheads, groins, etc.) significantly limits the unimpeded bedload transport along the shore. Most likely, the most intensive bedload movement proceeds seaward from the shore and back, and it is then that the material is most exposed to destruction. Zones with dominance of crystalline magmatic and metamorphic pebbles may indicate increased redeposition. On the other hand, the increased amount of soft limestones or sandstones may suggest longshore bedload transport (high rate of accumulation). The greater contents of crystalline pebbles found along the cliff bench sections with developed foredunes may indicate a long period of mechanical reworking. Thus, the soft pebbles were mechanically destroyed or chemically dissolved. Additionally, the greater contents of crystalline pebbles may be related to redeposition from cliff sections or shallow foreland.

The increased proportions of discoid and ellipsoid gravel grains and higher flattening indices are indicative of the dominance of material accumulation, and were recorded mainly in the sandbar–dune areas, or in the western part of the cliff shore (378.0–386.0 km). These areas include shore sections with lower proportions of the discoid and ellipsoid gravels, indicative of intensified erosion of the beach. Most likely, this is associated with earlier gravel deposition during periods of strong storms. Shore sections with evidence of accumulation (elevated contributions of discoid and ellipsoid gravels) are separated by sections with lower proportions of these gravels. This is most likely associated with the presence of 100–500 m long cusps and embayments in the shore area discussed. Those cusps and embayments were shown by Furmańczyk to be structures affecting the shore dynamics [71]. Therefore, areas of increased accumulation may feature sections indicating erosion characterized by erosion–redeposition processes (instantaneous redeposition).

Increased proportions of spheroid and spindle-shaped gravels were found mainly in the eastern part of the cliff shore (368.0–378.0 km), indicating domination of erosion processes and further mechanical reworking of the sedimentary material.

Generally, the lithostatistical data provided indicate increased erosion and redeposition processes occurring in the middle part of coast section studied, which are significantly reduced in the area of Dziwna and Rega Sandbars. The variability of *Ws* index in the group of limestones indicates low rates of erosion in areas of seawalls and groins, where bedload transport is also significantly reduced, especially between 367–369 km and 387–391 km. The large number of flat and relatively mechanically low resistant gravels, such as limestones and marbles, indicate tendencies of dominating accumulation. These gravels do not experience sufficient time and energy to be rounded. Discoidal or ellipsoidal grains may indicate lower energy and shorter bedload transport.

Elevated contents of spheroidal gravels in the crystalline and limestone groups, visible especially between 370 and 375 km, may be related to increased erosion and material reworking, and disturbed by the presence of seawalls and groins. A similar situation occurs between 374 and 375 km, or in the eastern part Dziwnów embankments, where low number of spheroidal gravels and slightly increased discoidal and ellipsoidal shape indicate low accumulation. The variability of sandstones is comparable, with highest contents of spheroidal pebbles in low erosion sections, and discoidal or ellipsoidal pebbles in areas of increased accumulation.

It must be borne in mind that the coastal zone dynamics pattern presented here, as concluded from the petrographic composition and gravel grain shapes, reflects the post-early-autumn storm conditions. Generally, characteristics indicative of material deposition and redeposition contribute to the developmental trends in the part of the Polish coast studied. Additionally, the results can inform comprehensive coastal protection programs.

Author Contributions: Conceptualization, C.S. and T.K.; methodology, C.S.; software, C.S. and R.B.; formal analysis, C.S., R.B. and Ł.M.; investigation, C.S., R.B. and T.K.; resources, C.S. and T.K.; writing—original draft preparation, C.S.; writing—review and editing, C.S., Ł.M. and R.B.; visualization, C.S.; funding acquisition, C.S., R.B. and T.K. All authors have read and agreed to the published version of the manuscript.

Funding: The publication was cofinanced by Faculty of Civil Engineering and Architecture, West Pomeranian University of Technology in Szczecin, Poland.

Conflicts of Interest: The authors declare no conflict of interest.

References

1. Rudowski, S. Sedimentary environment at the barred coast of a tideless sea; Southern Baltic, Poland. *Studia Geol. Pol.* **1986**, *87*, 1–86. (In Polish with English summary)
2. Buscombe, D.; Masselink, G. Concepts in gravel beach dynamics. *Earth-Sci. Rev.* **2006**, *79*, 33–52. [CrossRef]
3. Pranzini, E.; Cinelli, I.; Cipriani, L.E.; Anfuso, G. An Integrated Coastal Sediment Management Plan: The Example of the Tuscany Region (Italy). *J. Mar. Sci. Eng.* **2020**, *8*, 33. [CrossRef]
4. Bohdziewicz, L. Przegląd budowy geologicznej i typów polskich wybrzeży. In *Materiały do Monografii Polskiego Brzegu Morskiego*; Mielczarski, A., Ed.; IBW PAN Gdańsk-Poznań: Gdańsk, Poland, 1963; Volume 5, pp. 11–41.
5. Musielak, S.; Furmańczyk, K. Fizyczno—Geograficzna charakterystyka odcinka Niechorze—Międzyzdroje. In *ZZOP w Polsce—Stan Obecny i Perspektywy. Problemy Erozji Brzegu*; Furmańczyk, K., Ed.; INoM US: Szczecin, Poland, 2005; pp. 73–84.
6. Sydor, P.; Łabuz, T.A. Development of beach and coastal dune relief in the Pogorzelica area as based on studies of sedimentary structures. *Przegląd Geogr.* **2012**, *84*, 123–137. (In Polish with English summary) [CrossRef]
7. Van Wellen, E.; Chadwick, A.J.; Mason, T. A review and assessment of longshore sediment transport equations for coarse grained beaches. *Coast. Eng.* **2000**, *40*, 243–275. [CrossRef]
8. Elfrink, B.; Baldock, T. Hydrodynamics and sediment transport in the swash zone: A review and perspectives. *Coast. Eng.* **2002**, *45*, 149–167. [CrossRef]
9. Jennings, R.; Shulmeister, J. A field based classification scheme for gravel beaches. *Mar. Geol.* **2002**, *186*, 211–228. [CrossRef]
10. Butt, T.; Russell, P.; Puleo, J.; Miles, J.; Masselink, G. The influence of bore turbulence on sediment transport in the swash zone and inner surf zone. *Cont. Shelf Res.* **2004**, *24*, 757–771. [CrossRef]

11. Bartholomae, A.; Ibbeken, H.; Schleyer, S. Modification of gravel during longshore transport (Bianco Beach, Calabria, southern Italy). *J. Sediment. Res.* **1998**, *68*, 138–147. [CrossRef]
12. Carter, R.W. *Coastal Environments. An Introduction to the Physical, Ecological, and Cultural Systems of Coastlines*; Academic Press Limited: London, UK, 1988; pp. 1–617.
13. Schulmeister, J.; Jennings, R. Morphology and morphodynamics of gravel beaches. In *Coastal Zones and Estuaries, Encyclopedia of Life Support Systems (EOLSS)*; Ignacio Isla, F., Iribarne, O., Eds.; Eolss Publishers/UNESCO: Oxford, UK; pp. 244–261.
14. Pontee, N.I.; Pye, K.; Blott, S.J. Morphodynamic Behaviour and Sedimentary Variation of Mixed Sand and Gravel Beaches, Suffolk, UK. *J. Coast. Res.* **2004**, *20*, 256–276. [CrossRef]
15. Stark, N.; Hay, A.E.; Cheel, R.; Lake, C.B. The impact of particle shape on the angle of internal friction and the implications for sediment dynamics at a steep, mixed sand–gravel beach. *Earth Surf. Dyn.* **2014**, *2*, 469–480. [CrossRef]
16. Sherman, D.J.; Nordstrom, K.F. Beach scarps. *Z. Geomorphol.* **1985**, *29*, 139–152.
17. Chadwick, A.J.; Karunarathna, H.; Gehrels, W.R.; Massey, A.C.; O'Brien, D.; Dales, D. A new analysis of the Slapton barrier beach system. In Proceedings of the Institution of Civil Engineers-Maritime Engineering; Thomas Telford Ltd: London, UK, 2005; Volume 158, pp. 147–161.
18. Ciavola, P.; Castiglione, E. Sediment dynamics of mixed sand and gravel beaches at short timescales. *J. Coast. Res.* **2009**, *SI 56*, 1751–1755.
19. Grottoli, E.; Bertoni, D.; Pozzebon, A.; Ciavola, P. Influence of particle shape on pebble transport in a mixed sand and gravel beach during low energy conditions: Implications for nourishment projects. *Ocean Coast. Manag.* **2019**, *169*, 171–181. [CrossRef]
20. Savage, S.B. The mechanics of rapid granular flow. *Adv. Appl. Mech.* **1984**, *24*, 289–366.
21. Orford, J.D. Discrimination of particle zonation on a pebble beach. *Sedimentology* **1975**, *22*, 441–463. [CrossRef]
22. Williams, A.T.; Caldwell, N. Particle size and shape in pebble beach sedimentation. *Mar. Geol.* **1988**, *82*, 199–215. [CrossRef]
23. Isla, F.I. Overpassing and armouring phenomena on gravel beaches. *Mar. Geol.* **1993**, *110*, 369–376. [CrossRef]
24. Bluck, B. Sedimentation of beach gravels: Examples from South Wales. *J. Sediment. Petrol.* **1967**, *37*, 128–156.
25. Bluck, B. Clast assembling, bed forms and structure in gravel beaches. *Trans. R. Soc. Edinb. Earth Sci.* **1999**, *89*, 291–323. [CrossRef]
26. Petrov, V.A. The differentiation of material on gravel beaches. *Okeanologiya* **1989**, *29*, 279–284.
27. Illenberger, W.K. Pebble shape (and size!). *J. Sediment. Petrol.* **1991**, *61*, 756–767.
28. Le Roux, J.P. Shape entropy and settling velocity of natural grains. *J. Sediment. Res.* **2002**, *72*, 363–366. [CrossRef]
29. Dei, L. Geomorphological significance of pebble shapes along the littoral zone of Ghana. *J. Trop. Geogr.* **1975**, *40*, 19–30.
30. Anim, M.; Kofi Nyarko, B. Lithological responses to sea erosion along selected coastlines between Komenda and Saltpond, Ghana. *Ghana J. Geogr.* **2017**, *9*, 109–126.
31. Dudzińska-Nowak, J.; Furmańczyk, K.; Łęcka, A. Ochrona brzegu na odcinku Międzyzdroje—Niechorze. In *ZZOP w Polsce—Stan Obecny i Perspektywy. Problemy Erozji Brzegu*; Furmańczyk, K., Ed.; INoM US: Szczecin, Poland, 2005; pp. 96–105, INoM US: Szczecin, Poland, 2005.
32. Łabuz, T.A. Morphodynamics and rate of Cliff erosion in Trzęsacz (1997–2017). *Landf. Anal.* **2017**, *34*, 29–50. [CrossRef]
33. Pranzini, E.; Wetzel, L.; Williams, A.T. Aspects of coastal erosion and protection in Europe. *J. Coast. Conserv.* **2015**, *19*, 445–459. [CrossRef]
34. Kubowicz-Grajewska, A. Morpholithodynamical changes of the beach and the nearshore zone under the impact of submerged breakwaters—A case study (Orłowo Cliff, the Southern Baltic). *Oceanologia* **2015**, *57*, 144–158. [CrossRef]
35. Coastal Erosion and Protection in Europe. Pranzini, E.; Williams, A. (Eds.) Taylor & Francis: London, UK; Routledge: London, UK, 2013; pp. 1–488. [CrossRef]
36. Zawadzka-Kahlau, E. Trends in South Baltic Coast Development During the Last Hundred Years. *Peribalticum* **1999**, *VII*, 115–136.
37. Uścinowicz, S. Relative sea level changes, glacio-isostatic rebound and shoreline displacement in the southern Baltic. *Polish Geol. Inst. Spec. Pap.* **2003**, *10*, 1–79.

38. Tomczak, A. Geological structure and Holocene evolution of the Polish zone. *J. Coast. Res. Spec. Issue* **1995**, *22*, 15–31.
39. Emelyanov, E.; Lemke, W.; Harff, J.; Kramarska, R.; Uścinowicz, S. Sediments maps of the Western Baltic. Prace Państwowego Instytutu Geologicznego CXLIX. In Proceedings of the Third Marine Geological Conference "The Baltic", Sopot, Poland, 21–24 September 1993; pp. 133–137.
40. Uścinowicz, S. *Geological Map of the Baltic Sea Bottom, 1:200000, Sheet Kołobrzeg*; Państwowy Instytut Geologiczny: Warszawa, Poland, 1989.
41. Zingg, T. Beiträge zur Schottenanalyse. Schweitz. *Min. Petr. Mitt.* **1935**, *15*, 39–140.
42. Racinowski, R. *Zróżnicowanie Petrograficzne i Morfometryczne Żwirów w Górnej Części Strefy Potoku Przyboju Między Trzęsaczem a Pogorzelicą. PNPS IIW*; No. 45; Geotechnika i Gruntoznawstwo: Szczecin, Poland, 1975; pp. 25–45.
43. Bączyk, J. Les masses d'eaux de la Mer Baltique mevidionale et l'infulence de leur mouvements sur la zone litorale Polonais. *Pr. Geogr. PAN* **1968**, *65*, 1–120, In Polish with French summary.
44. Dobrzyński, S. *Contemporary Developmentof a Land Part of a Tideless Seashore Zone in the Light of Lithological Studies (on the Distance Jarosławiec-Czołpino)*; Wydawnictwo Uczelniane WSP: Słupsk, Poland, 1998; pp. 1–172, (In Polish with French summary).
45. Łabuz, T.A. Environmental Impacts-Coastal Erosion and Coastline Changes. In *Second Assessment of Climate Change for the Baltic Sea Basin*; The BACC II Author Team, Ed.; Regional Climate Studies; Springer: Cham, Switzerland, 2015; pp. 383–396. [CrossRef]
46. Musielak, S. Shoreline dynamics between Niechorze and Świnoujście. *J. Coast. Res. Spec. Issue* **1995**, *22*, 289–291.
47. Subotowicz, W. *Litodynamika Brzegów Klifowych Wybrzeża Polski*; GTN—Ossolineum: Gdańsk, Wrocław, Poland, 1982; pp. 1–150.
48. *Fotointerpretacyjny Atlas Dynamiki Strefy Brzegu Morskiego Świnoujście—Pogorzelica w skali 1:5000*; Musielak, S. (Ed.) Urząd Morski w Szczecinie, INoM US, OPGK: Szczecin, Poland, 1991.
49. Musielak, S.; Łabuz, T.A.; Wochna, S. Współczesne procesy brzegowe na Wybrzeżu Trzebiatowskim. In *Środowisko Przyrodnicze Wybrzeży Zatoki Pomorskiej i Zalewu Szczecińskiego*; Borówka, R.K., Musielak, S., Eds.; Oficyna in Plus: Szczecin, Poland, 2005; pp. 61–71.
50. Dudzińska-Nowak, J.; Furmańczyk, K. Zmiany położenia linii brzegowej Zatoki Pomorskiej (w latach 1938–1996). In *Środowisko Przyrodnicze Wybrzeży Zatoki Pomorskiej i Zalewu Szczecińskiego*; Borówka, R.K., Musielak, S., Eds.; INoM US, PTG: Szczecin, Poland, 2005; pp. 71–78.
51. Racinowski, R.; Baraniecki, J. Przydatność litologicznych wskaźników dla charakteryzowania wzdłuż brzegowego potoku rumowiska na polskim wybrzeżu Bałtyku. *Rozpr. Hydrotech.* **1990**, *51*, 159–201.
52. Musielak, S.; Furmańczyk, K.; Bugajny, N. Factors and Processes Forming the Polish Southern Baltic Sea Coast on Various Temporal and Spatial Scales. In *Coastline Changes on the Baltic Sea from South to East*; Harff, J., Furmańczyk, K., von Storch, H., Eds.; Springer: Cham, Switzerland, 2017; Volume 19, pp. 69–85. [CrossRef]
53. Zawadzka–Kahlau, E. *Tendencje Rozwojowe Polskich Brzegów Bałtyku Południowego*; GTN: Gdańsk, Poland, 1999; pp. 1–147.
54. Jahn, M. Beach pebbles of the Pomerania Bay. *Czasopismo Geogr.* **1962**, *23*, 129–135. (In Polish with English summary)
55. Racinowski, R.; Seul, C. Actual morphodynamic attributes of Szczecin shore. In *Lithodynamics of Sea Shore*; Meyer, Z., Ed.; PAS, Technical University of Szczecin: Szczecin, Poland, 1996; pp. 107–115.
56. Łęcka, A.; Furmańczyk, K. Application of geovisualisation to the assessment of the magnitude of coastal erosion. In *Człowiek i Środowisko Przyrodnicze Pomorza Zachodniego, III. Środowisko Przyrodnicze i Problemy Społeczno-Ekonomiczne*; Koźmiński, C., Dutkowski, M., Radziejewska, T., Eds.; Uniwersytet Szczeciński: Szczecin, Poland, 2006; pp. 99–106.
57. Dudzińska-Nowak, J. Effects of coastal protection methods on changes in the position of dune base line, as shown in a selected example. In *Człowiek i Środowisko Przyrodnicze Pomorza Zachodniego, III. Środowisko Przyrodnicze i Problemy Społeczno-Ekonomiczne*; Koźmiński, C., Dutkowski, M., Radziejewska, T., Eds.; Uniwersytet Szczeciński: Szczecin, Poland, 2006; pp. 91–98.
58. Tomczak, A. Budowa Geologiczna Strefy Brzegowej (I). In *Atlas Geologiczny Południowego Bałtyku*; Mojski, J.E., Ed.; PIG Sopot: Warszawa, Poland, 1995; Tab. XXXIII.

59. Krzyszkowski, D.; Dobracka, E.; Dobracki, R.; Czerwonka, J.A.; Kuszell, T. Stratigraphy of Weichselian deposits in the Cliff section Baltic Coast, northwestern Poland. *Quat. Stud. Pol.* **1999**, *16*, 27–45.
60. Seul, C. Changeability of lithological features of deposits of the shallow basement shore of Dziwnowska Sandbar. In *Seas and Oceans. Proceedings of the 1st International Congress of Seas and Oceans*; Guziewicz, J., Ed.; Szczecin: Międzyzdroje, Poland; Volume 1, pp. 491–496.
61. Dobracka, E.; Ruszała, M. Geological and geomorphological characteristics of the coastal zone between Międzyzdroje, Trzęsacz and Niechorze. In *Prace Naukowe Politechniki Szczecińskiej*; Wydawnictwo Naukowe Politechniki Szczecińskiej: Szczecin, Poland, 1988; Volume 378, pp. 17–52.
62. Uścinowicz, S. Baltic sea continental shelf. *2014; Geol. Soc. Mem.* **2014**, *41*, 69–89. [CrossRef]
63. Furmańczyk, K.; Dudzińska-Nowak, J. Extreme Storm Impact to the coastline changes—South Baltic example. *J. Coast. Res.* **2009**, *SI56*, 1637–1640.
64. Wiśniewski, B.; Wolski, T. Physical aspects of extreme storm surges and falls on the Polish coast. *Oceanologia* **2011**, *53*, 373–390. [CrossRef]
65. Dobracki, R.; Zachowicz, J. 1: 10,000 Geodynamic map of the Polish coastal zone—Results, utilisation and further plans. *Quat. Stud. Pol. Spec. Issue* **1999**, *16*, 85–89.
66. Racinowski, R. Eolic Differentiation in the Granulation of Rubble on the Marine Dune Shore Basing on Dziwnów Sand-bar. *Peribalticum* **1999**, *VII*, 65–87.
67. IMGW Institute of Meteorology and Water Management. Available online: https://dane.imgw.pl/data/dane_pomiarowo_obserwacyjne/dane_meteorologiczne/terminowe/synop (accessed on 12 June 2020).
68. Seul, C.; Kozłowski, T. Petrografia żwirów plażowych na odcinku Pogorzelica—DZIWNÓW (Pomorze Zachodnie). In *XVII Seminarium Naukowe z Cyklu Regionalne Problemy Ochrony Środowiska*; Wyd. Zachodniopomorskiego Uniwersytetu Technologicznego w Szczecinie: Szczecin, Poland, 2006; pp. 87–95.
69. Wentworth, C.K. The shapes of beach pebbles. United States Geological Survey. *Prof. Pap.* **1922**, *131*, 75–83.
70. Baraniecki, J.; Racinowski, R. The application of graining parameters of the rubble from the lower part of the back-swash of the shore stream zone to the determination of evolution tendencies of the Wolin Island coast. In *Lithodynamics of Sea Shore*; Meyer, Z., Ed.; PAS, Technical University of Szczecin: Szczecin, Poland, 1996; pp. 27–38.
71. Furmańczyk, K. *Współczesny Rozwój Strefy Brzegowej Morza Bezpływowego w Świetle Badań Teledetekcyjnych Południowego Bałtyku. Rozprawy i Studia*; Wydawnictwa Uniwersytetu Szczecińskiego: Szczecin, Poland, 1994; Volume 161, pp. 1–149.
72. Lagerund, E.; Malmberg-Perrson, K.; Krzyszkowski, D.; Johansson, P.; Dobracka, E.; Dobracki, R.; Panzing, W.A. Unexpected ice flow directions during the Late Weichselian deglaciations of the South Baltic area indicated by a new lithostratigraphy in NW Poland and NE Germany. *Quat. Int.* **1995**, *28*, 127–144. [CrossRef]
73. Olszak, I.; Florek, W.; Seul, C.; Majewski, M. Lithology of sediments and stratigraphy of some cliff section of central and western Polish Coast. *Geologija* **2011**, *53*, 1–9. [CrossRef]
74. Racinowski, R.; Sochan, A. *Lithostratigraphical Characteristics of Pleistocene Deposits on the Northern Part of Szczecin Shoreland*; Prace Naukowe Politechniki Szczecińskiej: Szczecin, Poland, 1981; Volume 128, pp. 1–110. (In Polish (English summary))
75. Zenkowicz, W.P. Niektóre zagadnienia brzegów polskiego Bałtyku. *Tech. Gospod. Morska* **1955**, *9*, 222–226.
76. Cieślak, A. Ruch rumowiska wzdłuż wybrzeża Polski—Hipotezy i propozycje. *Kwart. Geol.* **1986**, *30*, 426–427.
77. Furmańczyk, K.; Musielak, S. Important Features of Coast Line Dynamics in Poland "Nodal Points" and "Gates". In *Baltic Coastal Ecosystems—Structure, Function and Coastal Zone Management, Central and Eastern European Development Studies*; Schernewski, G., Schiewer, U., Eds.; Springer: Berlin/Heidelberg, Germany, 2002; pp. 191–194.

Article

Predicting Crenulate Bay Profiles from Wave Fronts: Numerical Experiments and Empirical Formulae

Mariano Buccino, Sara Tuozzo *, Margherita C. Ciccaglione and Mario Calabrese

Department of Civil and Environmental Engineering, University of Naples Federico II, Via Claudio 21, 80125 Naples, Italy; buccino@unina.it (M.B.); margheritacarmen.ciccaglione@unina.it (M.C.C.); mario.calabrese@unina.it (M.C.)

* Correspondence: sara.tuozzo@unina.it

Abstract: For crenulate-shaped bays, the coastal outline assumes a specific shape related to the predominant waves in the area: it generally consists of a tangential zone downcoast and a curved portion upcoast. Many coastal engineers have attempted to derive an expression of the headland bay shapes that emerge when a full equilibrium is reached (stable or dynamic). However, even though models for static equilibrium bays exist, they are merely of an empirical kind, lacking further insight on relationships between incident wave characteristics and beach shape. In addition, it is commonly believed that shoreline profiles tend to follow wave fronts, but this has been never fully verified. In this paper, we investigate a possible correlation between static equilibrium profiles and wave front shapes. Numerical experiments have been performed using the MIKE 21 Boussinesq Wave module, and the generated wave fronts have been compared to the hyperbolic-tangent equilibrium profile. A thoughtful analysis of results revealed that a single-headland equilibrium profile is merely the wave front translated perpendicularly to the wave direction at the headland tip, without any influence of wave period or in wave direction. A new function called the *"wave-front-bay-shape equation"* has been obtained, and the application and validation of this formula to the case-study bay of the Bagnoli coast (south-west of Italy) is described in the paper.

Keywords: crenulated-bay beach; hyperbolic-tangent shape; shoreline response; Boussinesq Wave model; coastal defences

Citation: Buccino, M.; Tuozzo, S.; Ciccaglione, M.C.; Calabrese, M. Predicting Crenulate Bay Profiles from Wave Fronts: Numerical Experiments and Empirical Formulae. *Geosciences* **2021**, *11*, 208. https://doi.org/10.3390/geosciences11050208

Academic Editors: Jesus Martinez-Frias, Gianluigi Di Paola, Germán Rodríguez and Carmen M. Rosskopf

Received: 19 February 2021
Accepted: 6 May 2021
Published: 10 May 2021

1. Introduction

Crenulate-shaped or headland bays are quite common on exposed sedimentary coasts containing headlands, and represent about 50% of the world's coastline [1,2]. The term headland-bay beach has been used to define a shoreline bounded by rocky outcrops or headlands, either natural or man-made, which lead to diffraction of incoming waves. Particularly, the predominant waves are diffracted in such a way as to break simultaneously around the periphery of the bay once an equilibrium plan-shape has been established (static equilibrium condition, [3]). The fame of the headland bay beaches, in fact, lies in their equilibrium condition, static or dynamic, which ensures that they are considered a way to achieve coastline stabilization [4]. Static equilibrium is a condition characterized by the absence of littoral drift, without the need for sediment supply to preserve its long-term stability; on the other hand, a dynamic equilibrium condition requires sediment supply, from updrift and/or from another kind of source, to maintain its stability and not retreat towards the limit defined by static equilibrium position [3].

Typically, the plan-shape of a single-headland bay is characterized by an upcoast curved zone (diffraction zone), a gentle transition zone and a relatively straight tangential segment on the downdrift end of the bay (illuminated zone), which is largely orthogonal to the dominant wave direction; this equilibrium plan-form is that assumed by the bay at a relatively long-term scale (e.g., annual to decadal) as a response to the predominant wave direction. Short-term fluctuations arising from beach-storm interactions, which could cause severe beach berm retreat, can be neglected due to their reversibility.

Devising a headland bay beach system in static equilibrium in such a way that the reorientation of the shoreline cancels the longshore sediment transport is called "*headland control*", and can be considered as a valuable option for coastal stabilization, which is frequently addressed via traditional detached low-crested breakwaters [5,6] or artificial reefs [7,8].

The headland control concept has been suggested for engineering use by [9,10] as a naturally functioning and preferable means of shore protection. In this regard, the formation of a crenulate-shaped bay on a sedimentary coastline, under oblique attack of persistent swell, is the most stable beach generated by nature [11]. The headland control approach appears to be of the utmost importance in the field of coastal stabilization against beach erosion, which increasingly affects many parts of the world's coastline. Hence, the employment of a crenulate-shaped bay to stabilization of a shoreline may be a powerful tool for engineering purposes.

Several studies and researches have been carried out in order to develop functional models describing static equilibrium bays' shapes: the logarithmic-spiral model [12], hyperbolic-tangent model [13,14] and parabolic model [15–17]. However, they are merely of an empirical nature, lacking further insight on relationships between incident wave characteristics and beach shape. As a matter of fact, none of these shapes are derived directly from the acting physical processes that developed the shoreline; rather, they are observational. Consequently, despite the strength in the assessment of stability of existing beaches, they are affected by uncertainties in the design of new artificial beaches through the headland control approach, which instead requires a deep knowledge of the dominant physical processes that govern the plan-shape of the bay. Without that, the project design of a new beach could result extremely challenging.

With the aim of overcoming such drawbacks and, broadly, to establish a relationship between wave forcings (diffraction and refraction) and bay shape response, the main task of this paper is founding a possible correspondence between static equilibrium profiles and wave characteristics, accounting for the relationship between equilibrium shape profiles and wave fronts. In fact, it is commonly believed that equilibrium beach profiles follow the wave front trend, but this has not been proved in literature so far, and no research has clarified in depth how wave characteristics, and particularly wave fronts, could shape a crenulated stretch of coast. For these reasons, this research represents a first step towards a development of a guidance which could help in engineering and morphological practice.

The analysis has been carried out via numerical modeling, which has long been proved to be a powerful tool, suited to even complex hydrodynamic phenomena [18–20].

Numerical experiments have been performed using the MIKE 21 Boussinesq Wave Module (BW) [21], where wave fronts have been compared to the hyperbolic-tangent equilibrium profile, analysing the influence of wave direction, wave period and refraction phenomenon. A correspondence function, called the "wave-front-bay-shape equation" has been established, offering an easy application to engineering uses due to the simple geometric interpretation of its controlling parameters. It is noteworthy that, being based on wave front analysis, the employed approach is essentially linear (e.g., [22]), although Boussinesq wave models account for weak non-linearity. As such, the parametric study discussed in Section 3 treats the wave height as an invariant of the problem.

The simulation's outcomes seem to indicate that equilibrium beach profiles of a single headland bay correspond to a simple translation of wave fronts normal to the propagation line at the headland tip. Moreover, the application of the "wave-front-bay-shape equation" to the case-study bay of the Bagnoli coast (south-west of Italy) is described in the paper.

2. Background

Since the beginning of nineteenth century, many coastal geologists, geographers and engineers have been trying to predict the shoreline plan-shape of headland-bay beaches.

Despite the complexity involved in coastal processes, simple empirical expressions have been derived since the 1940s in order to fit part or whole of the bay periphery. Among these approaches, the logarithmic spiral model [12], hyperbolic tangent model [13,14] and parabolic model [15–17] have turned out to be the most acceptable expressions for practical applications to headland bay beaches in static equilibrium. However, none of these models are derived directly from the acting physical processes that developed the shape; rather, they are based on the observation of the shoreline plan-shape, and so are lacking in a correlation between shoreline response and wave forcings (refraction and diffraction).

The logarithmic-spiral model was introduced by [12] observing an early 1940 imagery of Half-Moon Bay in California, USA. With field observations, the author realized that the bay adopted an equilibrium shape that is similar to a log-spiral. A definition sketch is given in Figure 1. The log-spiral equation, in polar coordinates, is given by:

$$R_1 = R_2 \exp(\theta cot\alpha) \tag{1}$$

where:

- R = Radius from origin
- θ = Angle between origin and R
- α = Constant angle of the tangent to the curve with radii R_1 and R_2

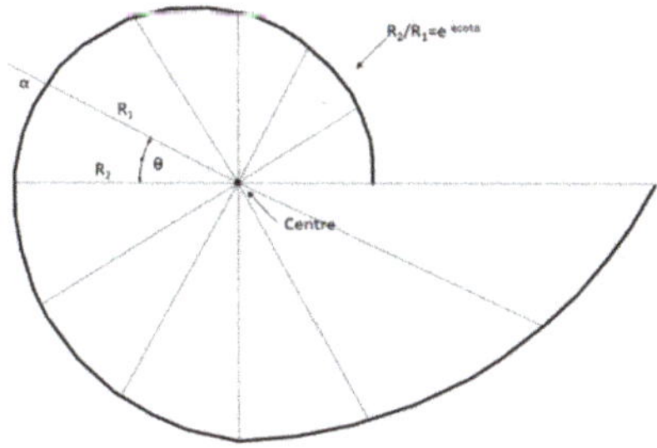

Figure 1. Definition sketch for the log spiral.

Despite of being the very first model proposed, representing a milestone in the field of crenulated bays research, the log-spiral model has long been criticized [3,23] because of several drawbacks which make it tricky to use. First, due to its constant curvature, the equation does not fit the relatively straight section of the headland bay beach; secondly, the pole of the log-spiral curve must be found by trial and error, because it does not coincide with the diffraction point, and could even deviate from it by a distance ranging from centimetres to kilometres (see [24]); therefore, it is not possible to predict the effect of relocating the headland (e.g., designing a headland bay beach introducing a coastal structure); and finally it does not take into account wave direction. In light of these disadvantages, two other empirical equations were developed: the hyperbolic-tangent equation and the parabolic equation.

The parabolic bay shape equation is a second-order polynomial equation developed by [15] and [16,17] in two separate works, from fitting the plan-shape of 27 mixed cases of prototype and model bays believed to be in static equilibrium:

$$\frac{R}{R_\beta} = C_0 + C_1\left(\frac{\beta}{\theta}\right) + C_2\left(\frac{\beta}{\theta}\right)^2 \tag{2}$$

where:

- R_β = Control line length
- R = Radius to a point along the curve at an angle θ

- β = Wave obliquity
- C = Constants generated by regression analysis
- θ = Angle between wave crest and radius to any point on the bay periphery

The two basic parameters are the reference wave obliquity β and control line length R_β (Figure 2). The variable β is a reference angle of wave obliquity, or the angle between the incident wave crest (assumed linear) and the control line, joining the upcoast diffraction point to a point on the near straight beach, namely the downcoast control point. The radius R to any point on the bay periphery in static equilibrium is angled θ from the same wave crest line radiating from the point of wave diffraction upcoast. The three C constants, generated by regression analysis to fit the peripheries of the 27 prototype and model bays, differ with reference angle β. Their analytical expressions have been provided by [14]:

$$C_0 = 0.0000000479\beta^4 - 0.00000879\beta^3 + 0.000352\beta^2 - 0.00479\beta + 0.0715 \tag{3}$$

$$C_1 = -0.000000128\beta^4 + 0.0000182\beta^3 - 0.000487\beta^2 + 0.00771\beta + 0.955 \tag{4}$$

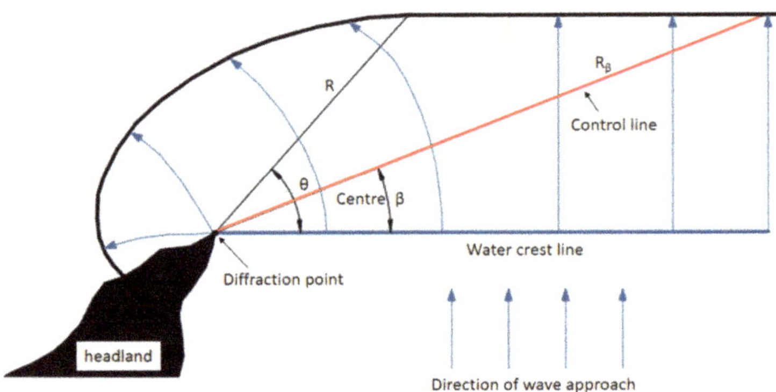

Figure 2. Definition sketch for the parabolic equations.

$$C_2 = 0.0000000944\beta^4 - 0.000012\beta^3 + 0.000316\beta^2 - 0.00828\beta + 0.0265 \tag{5}$$

In contrast to the log-spiral, the parabolic equation origin coincides with the diffraction point and therefore the equation is directly related to wave direction. However, it is affected by a drawback: the uncertainty of locating the downdrift control point. Despite that there are several interpretations of the downcoast control point of the parabolic bay shape equation [14], it is a considerable limitation which inhibits the application of the parabolic model in designing new beaches.

The hyperbolic-tangent shape model was derived by [13] through the analysis of 46 beaches around Spain and North America. Its equation is defined in a relative Cartesian coordinate system (Figure 3) as:

$$y = \pm a tanh^m(bx) \tag{6}$$

where:

- y = Distance across shore [m]
- a = Empirically determined dimensional coefficient [m]
- b = Empirically determined dimensional coefficient [1/m]
- m = Empirically determined coefficient [dimensionless]
- x = Distance alongshore [m]

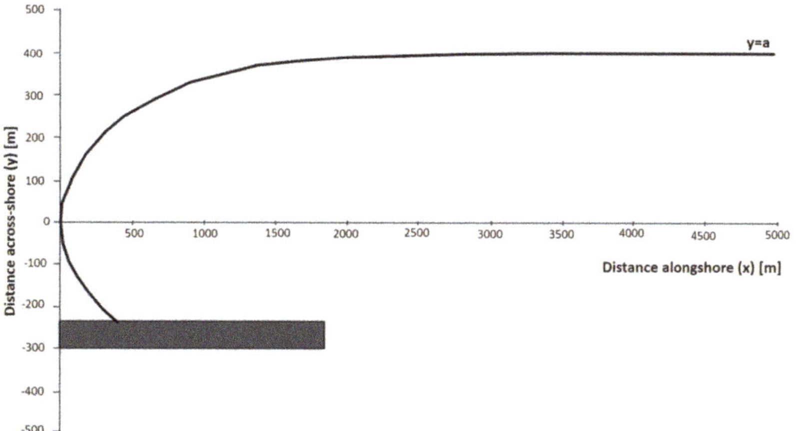

Figure 3. Definition sketch for the hyperbolic-tangent shape equation.

The x-axis is parallel to the general trend of the shoreline while the y-axis points shoreward; the origin of the coordinates is placed at the point where the local tangent to the beach is perpendicular to the general trend of the shoreline. The hyperbolic tangent curve is symmetric with the x-axis and produces two asymptotes, found at $y = \pm a$. The line $y = +a$ indicates the location on the shoreline which is no longer under the influence of the headland. The parameter a controls the magnitude of the asymptote (distance between the relative origin of coordinates and the location of the straight shoreline), b is a scaling factor controlling the approach to the asymptotic limit and m controls the curvature of the shape. These unknowns were found by using trial and error and an optimisation procedure that minimises rms errors. The authors of [13] found the following relationships: $ab \cong 1.2$; $m \cong 0.5$.

The model indirectly considers wave direction and, initially, it does not correlate the diffraction point with the origin of the hyperbolic tangent.

In 2018, [14] presented a development of the hyperbolic-tangent equation, establishing a relationship between the existing hyperbolic-tangent shape equation with the wave diffraction point. The authors used a database of case studies comprised of 46 beaches in Spain, Southern France and North-Africa, determining the coefficients a, b and m. The value of m is about 0.496, supporting the original findings of [13]. Moreover, the authors estimated a correlation between a and b, and the hyperbolic tangent shape equation became:

$$y = \pm a tanh^{0.496}\left(1.794a^{-1.097}x\right) \tag{7}$$

Overall, the research's relevance has been constituted by two relationships which link the hyperbolic tangent profile and location of diffraction point. This makes the model easy to use, both in order to verify the bay's stability and design a new beach. In the former case, starting from the distance between the location of diffraction point and asymptote, it is instantaneous to locate the origin of the equation and obtain the hyperbolic tangent profile; in the latter case, based on the shoreline advancement established in the illuminated zone, the location of the new diffraction point (i.e., new headland's tip) can be directly carried out. The relationships, which correlate parameters c, d and a shown in Figure 4, are:

$$\frac{c}{a} = 1.256 \tag{8}$$

$$\frac{d}{a} = 0.517 \tag{9}$$

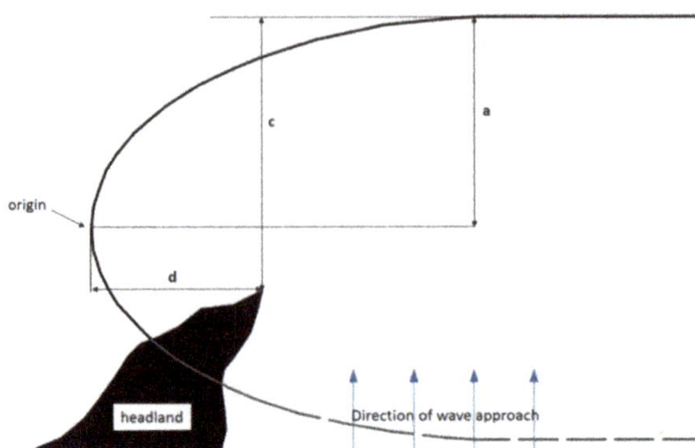

Figure 4. Definition sketch of parameters for the modified hyperbolic-tangent equation.

It is worth noting that the hyperbolic-tangent model could be applied not only to bay in static equilibrium conditions, but also to fit a bay under non-equilibrium conditions [3]. Among the three aforementioned models, the parabolic model prevailed over the others since it was the only one that used the wave diffraction point as the origin of the co-ordinate system, ensuring that the effect of relocating the diffraction point can be assessed [14]. Despite that, in comparing the origin of the three coordinates system in studying a headland bay (using a computer program based on a trial and error approach), [25] found that the parabolic equation's origin was located in the middle of the ocean (Figure 5). This outcome weakens the strong point of parabolic model, revealing an uncertainty about its alleged robustness.

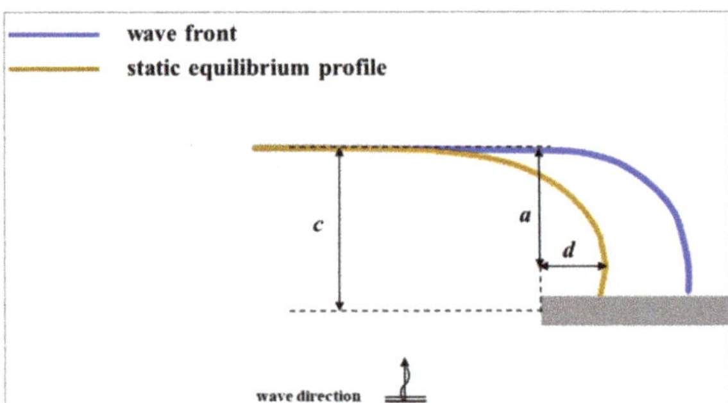

Figure 5. Definition sketch to obtain the hyperbolic tangent profile starting from a wave front.

On the whole, regardless of models' possible limitations, their description here presented shows how none of them model the relationship between wave characteristics and beach shape, but instead simply characterize the geometry of the headland bay profile. Nevertheless, over the last few decades, further investigations have been carried out towards the comprehension of the correlation between crenulate-shaped bays and wave characteristics. Particularly, several works focused on wave directional spreading, demonstrating that the crenulate-beach plan-form suffers under its influence. The authors of [26], searching for a new criterion to locate the downdrift control point of the parabolic shape equation, observed that the broader the directional spreading the farther the position of the downdrift control point. This behaviour indicates that, with a wider directional spreading of waves, a greater area could be influenced by wave diffraction behind the coastal structure (i.e., the transition zone increases).

During their numerical experimentation, [27] observed that a narrow directional spreading produces a more curved and asymmetric planform, confirming that the crenulated-beach behaviour is related to the modal directional spreading of incident waves [26,28]. Furthermore, [28] evaluated how the degree of directional spreading influences diffraction's effect on the headland shape, observing that diffraction appears less important with a high directional spreading degree. The latter outcome is in accordance with [29], which demonstrated that diffraction shapes bay morphology only for wave climates with restricted approach angles.

Although important progress has been made, the effective correlation between wave forcing (i.e., diffraction and refraction) and static equilibrium profile has not been fully examined and understood. In this work we attempt to understand this issue, starting from the commonly engineering assumption that an equilibrium profile follows the wave front trend, with the aim of verifying the validity of this statement.

3. Numerical Experiments

The primary aim of this research is to investigate the relationship between the headland-bay static equilibrium profile and wave propagation characteristics; to do so, numerical modelling of a single headland case study has been implemented. Moreover, the possible influence of wave direction and wave period has been verified on this researched correlation. Several experiments have been carried out using the wave driver BW of MIKE 21 (Danish Hydraulic Institute) [21], which is based on the numerical solution of the time domain formulations of Boussinesq-type equations [30–33]. The BW module is able to reproduce the combined effects of all important wave phenomena, among them diffraction and refraction, which play an important role in headland-bay equilibrium profile shaping.

3.1. BW Model of MIKE 21 (DHI)

The Boussinesq model is capable of reproducing the combined effects of important wave phenomena, such as diffraction and refraction. Therefore, it has been used in numerical experimentation to obtain diffracted wave fronts generated by headland.

MIKE 21 BW includes two modules, the 2DH wave model and 1DH wave model, both based on the numerical solution of time domain formulations of Boussinesq-type equations, which are solved using a flux-formulation with improved linear dispersion characteristics. The enhanced Boussinesq equations were originally derived by [30,31], making the modules suitable for simulation of the propagation of directional wave trains travelling from deep to shallow water. Moreover, it contains wave breaking and moving shorelines, as described in [32–34]. The 2DH BW model has been used in this work.

The enhanced Boussinesq equations are expressed in terms of the free surface elevation, ξ, and the depth-integrated velocity-components, P and Q. The continuity and the momentum-conservation equations read:

$$n\frac{\partial \xi}{\partial t} + \frac{\partial P}{\partial x} + \frac{\partial Q}{\partial y} = 0 \tag{10}$$

$$n\frac{\partial P}{\partial t} + \frac{\partial}{\partial x}\left(\frac{P^2}{h}\right) + \frac{\partial}{\partial y}\left(\frac{PQ}{h}\right) + \frac{\partial R_{xx}}{\partial x} + \frac{\partial R_{xy}}{\partial y} + n^2 gh\frac{\partial \xi}{\partial x} + n^2 P\left[\alpha + \beta\frac{\sqrt{P^2+Q^2}}{h}\right] + \frac{gP\sqrt{P^2+Q^2}}{h^2C^2} + n\psi_1 = 0 \tag{11}$$

$$n\frac{\partial Q}{\partial t} + \frac{\partial}{\partial y}\left(\frac{Q^2}{h}\right) + \frac{\partial}{\partial x}\left(\frac{QP}{h}\right) + \frac{\partial R_{yy}}{\partial y} + \frac{\partial R_{yx}}{\partial x} + n^2 gh\frac{\partial \xi}{\partial y} + n^2 Q\left[\alpha + \beta\frac{\sqrt{P^2+Q^2}}{h}\right] + \frac{gQ\sqrt{P^2+Q^2}}{h^2C^2} + n\psi_2 = 0 \tag{12}$$

with the dispersive Boussinesq terms ψ_1 and ψ_2, defined by:

$$\psi_1 = -\left(B+\tfrac{1}{3}\right)d^2(P_{xxt}+Q_{xyt}) - nBgd^3(\xi_{xxx}+\xi_{xyy}) - dd_x\left(\tfrac{1}{3}P_{xt}+\tfrac{1}{6}Q_{xt}+ngBd(2\xi_{xx}+\xi_{yy})\right) - dd_x\left(\tfrac{1}{6}Q_{xt}+ngBd\xi_{xy}\right) \tag{13}$$

$$\psi_2 = -\left(B+\tfrac{1}{3}\right)d^2(P_{xyt}+Q_{yyt}) - nBgd^3(\xi_{xxy}+\xi_{yyy}) - dd_x\left(\tfrac{1}{6}P_{xt}+\tfrac{1}{3}Q_{xt}+ngBd(2\xi_{yy}+\xi_{xx})\right) - dd_x\left(\tfrac{1}{6}P_{yt}+ngBd\xi_{xy}\right) \tag{14}$$

where d is the still water depth; g is the gravitational acceleration; n is the porosity; C is the Chezy resistence number; α is the resistance coefficient for laminar flow in porous media; β is the resistance coefficient for turbulent flow in porous media; B is the Boussinesq dispersion coefficient; the terms R_{xx}, R_{xy} and R_{yy} indicate the incorporation of the wave breaking by means of the surface roller model [35].

In MIKE 21 BW, the waves may either be specified along open boundaries or be generated internally within the model through the generation line; the latter must be placed in front of a sponge layer absorbing all outgoing waves. Moreover, porosity (e.g., to model partial transmission through porous structures) and sponge layers can be used on an ad hoc basis. At open boundaries, either a level boundary, namely wave energy given as time series of surface elevation, or flux boundary, where flux density is perpendicular to the boundary, can be set.

3.2. Methodology

The method followed involves comparing the diffracted wave fronts generated by the BW simulations with static equilibrium profiles sketched out by the hyperbolic-tangent model [13,14]. The procedure obeys to the following steps:

1. generating the diffracted wave front through the BW module;
2. extrapolating the wave front from the model;
3. starting with the illuminated zone of the front, drawing out the hyperbolic-tangent profile.

The third step is directly related to one of the major findings in the field of static equilibrium bays, namely that equilibrium crenulated beaches tend to align transverse to the direction of dominant waves. The main assumptions in applying the methodology here proposed are that (1) the equilibrium headland-bay planform follows the wave front; and (2) that the predominant wave direction is perpendicular to the straight area of the bay. In this way, in order to obtain the hyperbolic tangent plan-form of every model configuration, we supposed that the asymptote of the hyperbolic tangent matches the illuminated zone of the wave front, out of the shadow zone where the influence of diffraction is negligible. Therefore, once the distance between asymptote and diffraction point is measured (distance c in Figure 5), the origin of the hyperbolic tangent is automatically obtained through Equations (8) and (9), and the x,y coordinates of hyperbolic tangent profile are achieved through Equation (7) (Figure 5). It is worth pointing out that, in order to compare equilibrium static profile and wave front, the latter must be able to expand without any kind of physical interference (e.g., presence of additional obstacles).

Additionally, in order to investigate the influence of wave characteristics on the relationship between wave fronts and equilibrium profiles, we performed different numerical simulation scenarios by varying wave direction, wave period and refraction conditions. For each scenario, more wave fronts were extrapolated from the BW module, each progressively further away from the headland tip, in order to examine the influence of dimensionless distance (c/L) on the researched correlation.

Finally, before describing the model set up, a clarification regarding the choice of the headland bay shape model is necessary. The two possible models to be implemented to sketch out the static equilibrium profile were the hyperbolic-tangent model [13,14] and the parabolic model [15–18] (the logarithmic spiral model [12] has been rejected a priori given its difficulty in practical application). Thus, the same methodology has been implemented to both the hyperbolic model and parabolic model. The results proved that there was no difference: there is no change, regardless of the model adopted. Nevertheless, given the uncertainty due to the determination of the downdrift control point of the parabolic profile, the hyperbolic tangent model [13,14] turned out to be the best static equilibrium model to be compared to the BW simulations wave fronts.

3.3. Boussinesq Wave Module Set Up

The primary aim of this research is to investigate the relationship between headland-bay static equilibrium profile and wave propagation characteristics (diffraction and refraction). Therefore, we created different numerical simulations scenarios, in order to generate different wave fronts, by varying wave direction, wave period and refraction conditions.

First, to evaluate refraction conditions, two bathymetry configurations were been investigated. The first one (called "*gentle slope*") is characterised by a linearly varying cylindrical bathymetry, with a gentle bottom slope of 1/100 and a diffraction point, modelled through a breakwater, located at a water depth of 3 m. The water depth at the offshore boundary is 20 m (Figure 6a). A gentle slope has been adopted to allow an expansion of the wave fronts without the influence of bottom abrupt raising. The second configuration (called "*flat bottom*") exhibits a bottom with a constant water depth equal to 10 m (Figure 6b).

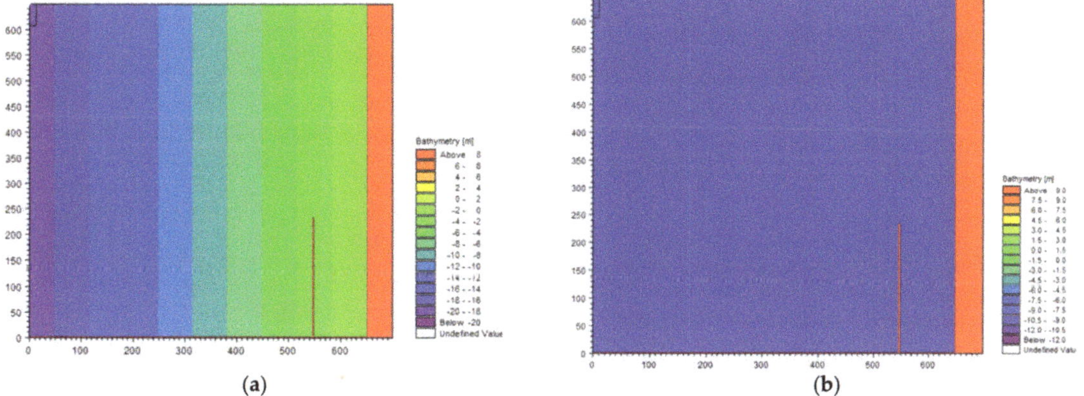

Figure 6. (**a**) Gentle slope configuration bathymetry; (**b**) flat bottom configuration bathymetry.

Secondly, numerical experiments have been conducted considering the possible effect of wave period and wave direction; conversely, the influence of wave height was not explored. This is because the correlation function describing the relationship between static equilibrium profiles and wave characteristics, at this first stage of the research, is obtained through a "linear approach" based on simple geometric consideration. Therefore, the only parameters that could geometrically affect the aforementioned correlation are period, wave direction and bottom inhomogeneity. Conversely, wave height does not represent a variable of the problem. Three different wave directions have been investigated; one is normal to the breakwater, while the other two are angled of 25° and −35° with respect to being perpendicular to the structure. For each direction, three wave periods have been tested: 5.9 s, a typical wave period of the Mediterranean wave climate; and 10 s and 15 s, which simulate swell conditions. The wave height used was 0.8 m for each numerical scenario. Since the predominant wave shapes the crenulated beach, for the numerical experimentation the value of wave height has been chosen equal to the LDR equivalent wave height value for the case study of Bagnoli bay (see Section 5.2). Simulations have been implemented using regular waves. It is important to underline that the effect of breaking waves was not considered in the trials. For each bathymetry configuration ("gentle slope" and "flat bottom"), the scenarios analysed are summarized in Table 1.

Table 1. Scenarios analysed in the numerical experimentation.

	Gentle Slope	Flat Bottom
Wave Direction	0°; 25°; −35°	0°; 25°; −35°
Wave Period	5.9 s; 10 s; 15 s	5.9 s

Models are made up of a fine grid (square cells with grid spacing 3 m), upon which orientation coincides with wave direction; the time step used is 0.1 s. The wave generation line has been used and wave absorbing sponge layers have been applied at the model boundaries. Near the land, one sponge layer has been applied in order to avoid the occurance of wave reflection which could influence the expansion of the wave fronts. Geometric characteristics are summarized in Table 2, and are valid for both bottom configurations.

Table 2. Grid geometric characteristics.

Grid Dimension [m^2]	Grid Spacing [m]	Grid Orientation [°]	Headland Extension [m]
2100 × 1950	3	0	690
2700 × 1500	3	25	330
2700 × 1500	3	−35	630

4. Results

At first sight, comparisons suggest that wave front does not represent a static equilibrium profile: within the shadow zone, the wave front is located backward from the static planform position, according to the wave propagation direction. In fact, as seen in Figure 7, the hyperbolic tangent profile (red line) crosses various fronts until it tends to the asymptote; the farther the asymptote from the breakwater tip, the more the profile is placed on additional wave fronts. On the other hand, wave fronts distant less than a wave length from the diffraction tip represent an exception, since they coincide with their corresponding hyperbolic tangent profile (Figure 8).

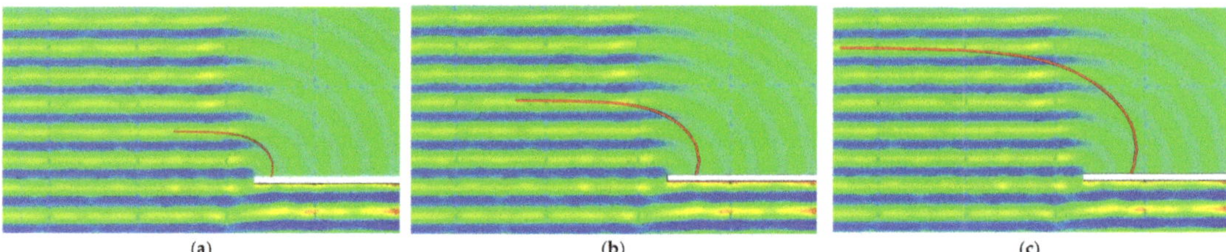

Figure 7. Comparison between wave front (obtained with flat bottom configuration) and hyperbolic tangent profile (red line), images from (**a–c**) show profiles sketched out, starting with fronts gradually further from the headland.

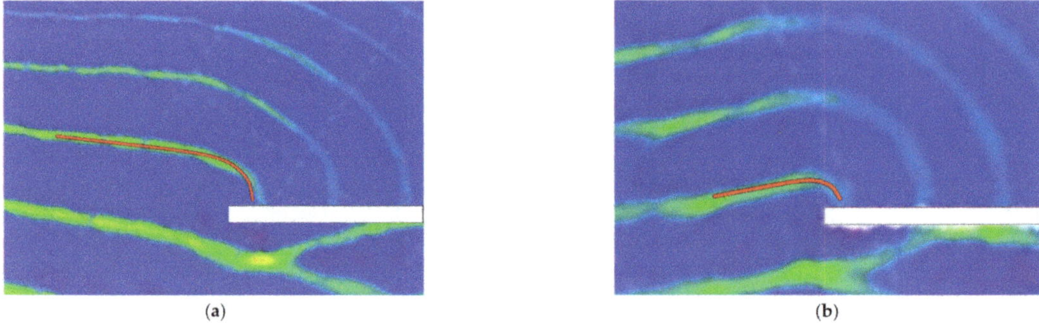

Figure 8. Hyperbolic tangent profile (red line) corresponds to relative wave front; both pictures have been obtained with the gentle slope configuration, (**a**) with wave direction −35° and wave period of 10 s and (**b**) with wave direction 25° and wave period of 10 s.

The behaviour outlined has been detected for each wave period and wave direction investigated, as shown in Figure 9. Moreover, this behaviour has been observed in both bottom configurations (gentle slope and flat bottom); consequently, it seems that the correlation between hyperbolic tangent profile and wave fronts is not influenced by the effect of refraction phenomenon. At the same time, although as a first approximation results indicate that wave front does not represent a static equilibrium profile, a more accurate inspection revealed that equilibrium beach profiles of single headland bays correspond to a simple translation of a wave front normal to the propagation line at the headland head: once shifted, it completely superimposes on the equilibrium profile (Figure 10). Specifically, this issue is elaborated and discussed in the next section.

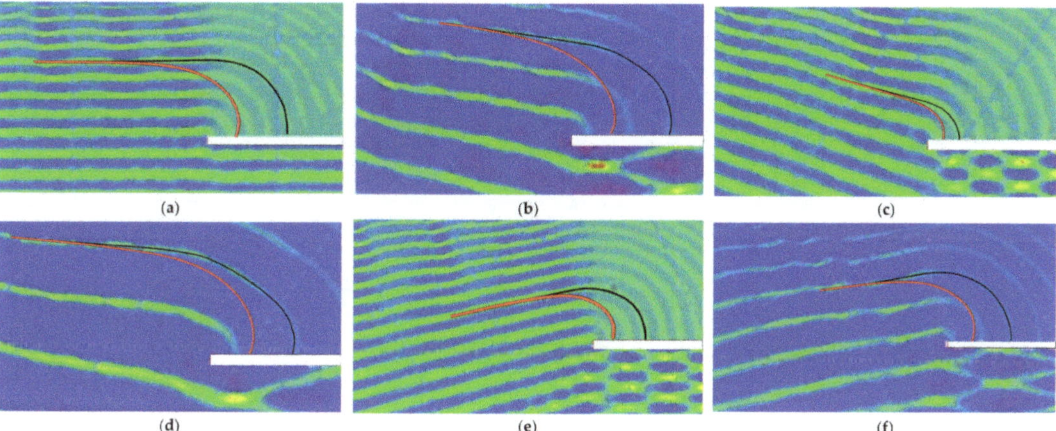

Figure 9. Comparisons between wave fronts (black line) and hyperbolic tangent profiles (red line): (**a**) wave direction 0°, wave period 6 s, gentle slope configuration; (**b**) wave direction −35°, wave period 10s, gentle slope configuration; (**c**) wave direction −35°, wave period 6s, gentle slope configuration; (**d**) wave direction −35°, wave period 15 s, gentle slope configuration; (**e**) wave direction 25°, wave period 6 s, gentle slope configuration; (**f**) wave direction 25°, wave period 10 s, gentle slope configuration.

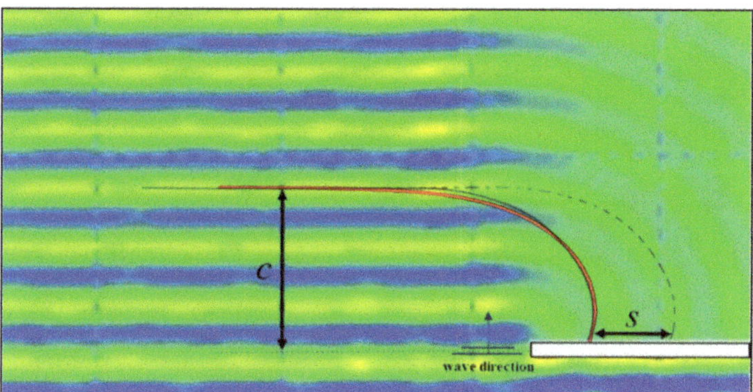

Figure 10. Definition sketch for parameters c (distance between the diffraction point and the asymptote of the profile) and s (distance needed to overlap the front on the profile).

Relationship between Wave Fronts and Equilibrium Profiles

As described in the above paragraph, numerical experiments demonstrated that, in contrast to what is usually supposed, wave fronts do not represent a static equilibrium profile. However, it is possible to establish a correlation between them. Regarding the fact that an equilibrium profile actually is a wave front translated perpendicularly to the wave direction at the headland head (Figure 10), and recognizing that wave period, wave direction and wave refraction do not influence this behaviour, it is possible to calculate a direct relationship between wave fronts and equilibrium profiles, called the *"wave-front-bay-shape equation"*.

Therefore, in order to derive this relationship, for each comparison between wave front and hyperbolic tangent profile, we measured the minimum distance between the diffraction point and the asymptote of the profile (distance c of Figure 10), and, also, the distance needed to overlap the front on the profile (distance s of Figure 10). These two distances have been standardised to the local wave length, L (the wave length at the diffraction point water depth), so obtaining c/L and s/L.

Moreover, it is necessary to take into account that wave fronts far from the diffraction point by less than one wave length coincide with the corresponding hyperbolic tangent profile. Therefore, it has been assumed that for a value of $c/L \leq 0.7$, it is not necessary to shift the wave front to overlap it on the profile; they are already superimposed (Figure 8).

Hence, for each configuration, results have been plotted on a graph where on the x-axis there is "$c/L - 0.7$" and on the y-axis there is "s/L". It can be noted that all results analysed tend to follow an increasing trend, which suggests that an equation to describe the correlation between diffractive wave fronts and static equilibrium profiles can be derived (Figure 11). The relationship is:

$$\frac{s}{L} = 0.0484\left(\frac{c}{L} - 0.7\right)^2 + 0.4694\left(\frac{c}{L} - 0.7\right) + 0.3567 \quad (15)$$

with a correlation coefficient R^2 of 0.9548.

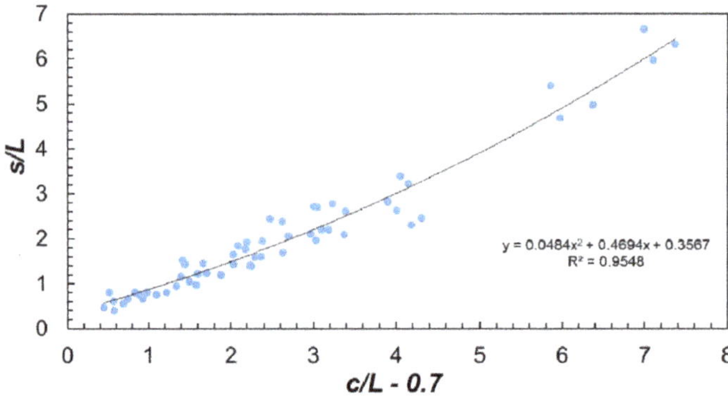

Figure 11. numerical simulations values of s/L in function of c/L, together with the best line.

5. Bagnoli Bay Case Study

To assess the results carried out with numerical experiments, a small beach has been taken into account. It is located on the Bagnoli coast, West of Naples, in southern Italy (Figure 12). The beach is delimitated on the right (looking offshore) by a revetment which protects the road behind, on the left by a little mound and behind by a seawall which marks out the road; it extends for about 190 m from North-West to South-East. This little bay appears to be fundamentally governed by a single headland located on the left side, next to the drain of Bagnoli (Figure 13). The orientation of the downcoast section of the bay is approximately 210 °N and its distance from the diffraction point is approximately 47 m (Figure 13).

As will be demonstrated in the next section, the case study can be considered a static equilibrium bay governed by a single headland. For this reason, the application to the Bagnoli bay is of particular interest, as it allows us to verify the behaviour observed with numerical experimentation and to check the validity of Equation (15).

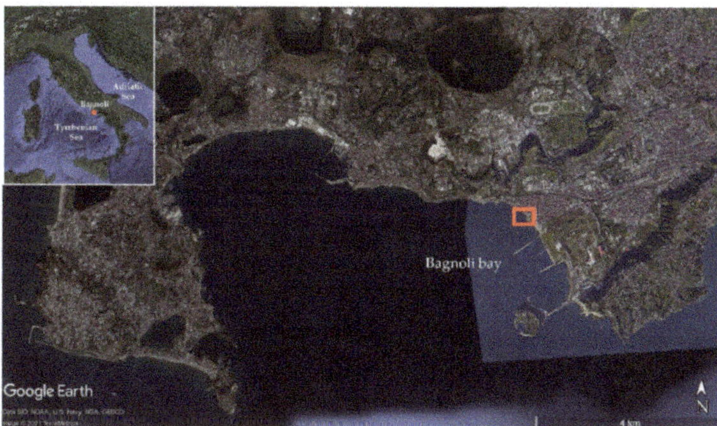

Figure 12. Location of the Bagnoli bay case of study.

Figure 13. Distance from diffraction point and shoreline orientation in downdrift section of the Bagnoli bay case study.

5.1. *Wave Climate*

The wave climate for the case study coast has been inferred from the MEDA-A buoy, an Acoustic Doppler Current Profile (ADCP), located approximately 800 m offshore from the bay (Figure 14a). The device is located at a depth of about 19 m below the low tide level, at a Latitude of 40°48.550′ N and a Longitude of 14°09.300′ E. Significant wave height, H_s, peak period, T_p, and azimuth of the mean wave direction, α, have been recorded in the period 2015–2018, with a large "white spot" interval detected, from 6 December 2016 to 22 March 2019. The histogram of wave direction for angular sectors of 22.5° N is shown in Figure 14b, where the offshore directed waves have been removed for sake of clearness. The graph exhibits how the bay is exposed to waves coming from a relatively narrow wave sector, included between 150 to 240° N, with a clear mode in the Southern quadrants (180°–210° N). It is important to highlight that, despite the short recorded time interval of the buoy (only three years), we employed its data, taking the advantage of being placed in the near-shore zone, bypassing all the procedures concerning the determination and propagation of the offshore wave climate (procedures which involve a certain degree of uncertainty, however).

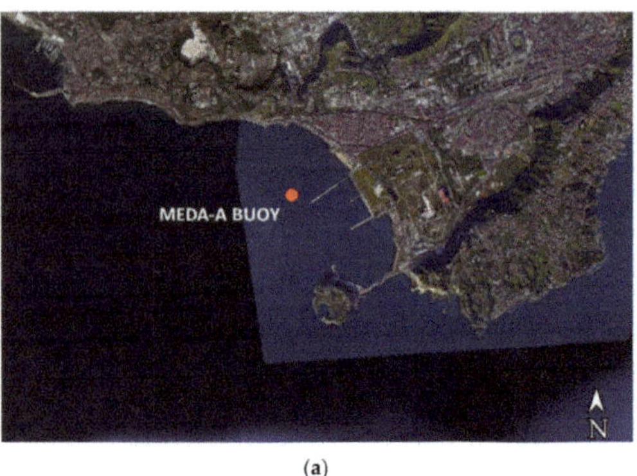

(a)

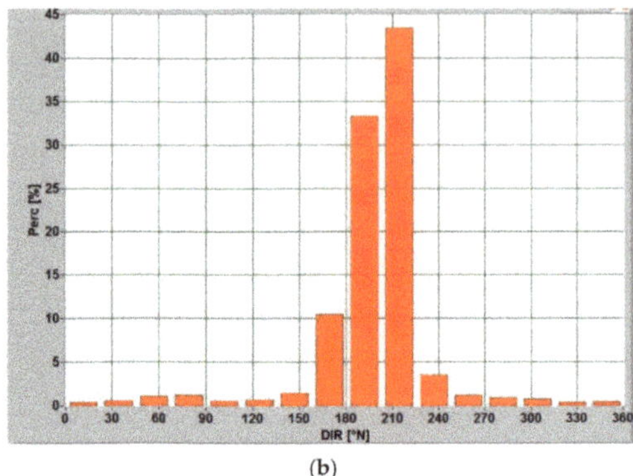

(b)

Figure 14. (a) Location of the MEDA-A buoy; (b) histogram of wave direction for angular sectors of 22.5° N.

5.2. LDR and Equivalent Wave

For the purposes of the present study, an understanding of longshore sediment transport is essential to sound headland control design practice, and for these scopes the concept of Littoral Drift Rose (LDR), [36,37] seems to represent a powerful resource. In fact, as stated previously, the long term sculpturing of headland bay beaches is related to the most persistent waves in their incidence and directions, since the straight section of the bay tends to lie normally to the wave rays. With minimizing the angle between dominant wave rays and the beach normal, the littoral drift progressively diminishes, so that the coast reaches its final equilibrium position. Therefore, such bays present their straight section normal to a dominant (or equivalent) sea state, embodied by the predominant swell on oceanic margins, and by the resultant energy vector in enclosed seas, where locally generated waves assume importance [5]. The LDR formulation is derived just from the energy vector concept [38], and, particularly, it holds the powerful property of estimating the climate equivalent (dominant) sea state, responsible for the sculpting of a stretch of coast; it is clear that it has a strong impact on the scopes of the article. Additionally, it is important to point out that, even though the equivalent wave concept is widely used in the field of practical coastal engineering, it lacks a firm theoretical basis. However, recent studies carried out by [39], demonstrated, through qualitative and quantitative analysis, how the equivalent wave concept may be reliable in explaining the long-term evolution of a stretch of coast. In light of this, the LDR for the case study area has been derived and the equivalent wave has been estimated and used as the wave attack that governs the plan-shape of the bay.

Now, as specified by [37], given a water climate represented by a series of N wave components, it is possible to determine the LDR, the compact polar representation of littoral transport potential for various shoreline orientations. For a segment of shoreline with outward normal azimuth β, it can be shown the net potential littoral drift rate, $Q(\beta)$, can be calculated as:

$$Q(\beta) = \sum_{\alpha_{0i}=\beta-\frac{\pi}{2}}^{\alpha_{0i}=\beta+\frac{\pi}{2}} p_i \cdot \frac{K \cdot (H_{s0,i})^{2.4} \cdot (T_{p0,i})^{0.2} \cdot g^{0.6}}{16 \cdot (s-1) \cdot (1-n) \cdot \pi^{0.2}} \, sin[2(\beta - \alpha_{0i})] \tag{16}$$

In the equation above, Q is intended as the in-place volumetric transport of sediment past a hypothetical plane perpendicular to the beach; additionally:

- K is a sediment transport coefficient;
- g is gravity;
- $s \approx 2.6$ is the ratio between the specific gravity of sediment and that of water;
- $n \approx 0.4$ is the in place porosity;
- $\gamma \approx 0.6$ is the breaker index (wave height to depth ratio);

As known, the LDR graph is a useful tool for interpreting littoral drift trends along a section of shoreline, and, more significantly, it is able to sum up the effects of the entire wave climate into a single equivalent wave component, of parameters $H_{s0,eq}$, $T_{p0,eq}$, $\alpha_{0,eq}$. This is to say:

$$Q(\beta) = \sum_{\alpha_{0i}=\beta-\frac{\pi}{2}}^{\alpha_{0i}=\beta+\frac{\pi}{2}} p_i \cdot \frac{K \cdot (H_{s0,i})^{2.4} \cdot (T_{p0,i})^{0.2} \cdot g^{0.6}}{16 \cdot (s-1) \cdot (1-n) \cdot \pi^{0.2}} \sin[2(\beta - \alpha_{0i})] \cong G_{eq} \cdot sin[2(\beta - \alpha_{0,eq})] \quad (17)$$

where:

$$G_{eq} = \frac{K \cdot (H_{s0,eq})^{2.4} \cdot (T_{p0,eq})^{0.2} \cdot g^{0.6}}{16 \cdot (s-1) \cdot (1-n) \cdot \pi^{0.2}} \quad (18)$$

The equivalent wave angle $\alpha_{0,eq}$, corresponds to the LDR node (null-point), as seen in Figure 15, and the magnitude (Equation (18)) can be easily inferred by using (for example) common harmonic regression techniques. The drift rose of Bagnoli bay has been derived using Equation (17) and the available wave data inferred through the MEDA-A Buoy. The range of shoreline orientations that exists at the site of interest has been also considered, and the net littoral drifts for each possible shoreline orientation have been calculated. LDR for the Bagnoli climate is shown in Figure 15: positive (transport to the right), and negative (transport to the left) lobes can be distinguished; moreover, the graph shows the null point; that is, the shoreline orientation at which no sediment transport is taking place.

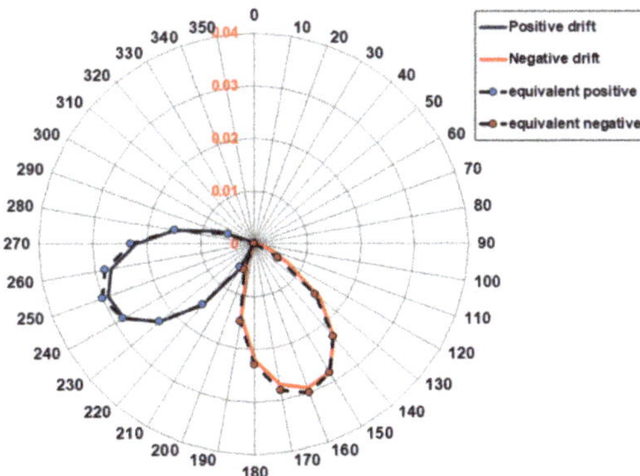

Figure 15. Comparison between Climate-LDR and Equivalent-LDR of the Bagnoli bay; blue solid line represents drift to right when looking offshore, while red solid line represents drift to left when looking offshore; dashed lobes represent the equivalent drifts. Littoral drift in cubic meters/s.

From the LDR, the equivalent wave component parameters have been estimated according to Equation (17). The equivalent direction corresponds to the null-point of the real LDR (205° N for the present case), while $H_{s,eq}$ and $T_{p,eq}$ are fitted to have the same littoral transport magnitude, obtaining $H_{s,eq}$ = 0.8 m and $T_{p,eq}$ =5.89 s (Equation (18)). Additionally, it is worth noticing that, as explained by [39], in [37] the authors do not define explicitly how to derive the equivalent wave height and period. While the equivalent wave direction is much easier to interpret, as it corresponds to the null point of the LDR, the same is not true for $H_{s,eq}$, and T_{eq}. However, the equivalent wave height can be easily inferred from Equation (18), considering that a relationship between wave height and period is established. As such, the equivalent period can be considered as the spectral peak period T_p; in fact, assuming that the energy distribution is that of a mean-JONSWAP spectrum (typical spectrum shape of Mediterranean Sea), wave height and period are related by:

$$T_p = 8.5\pi \sqrt{\frac{H_{m0}}{4\pi}} \quad (19)$$

On the other hand, the equivalent wave period can be considered as the "longer period" representative of swell-wave conditions. The authors of [40], in fact, argued the annual littoral transport to be driven by swells that follow the most intense storms and recognized that swells "arrive on a coast from persistent direction". This implicitly supports the idea that a dominant swell-wave attack for shoreline evolution may exist. However, [11] claimed that the swell-wave predominance is more common for ocean margins rather than for enclosed or semi-enclosed seas, where, on the contrary, the swell-effect is minimum. In light of this, only the peak period T_p has been taken into account in the present study. However, further investigation on the effects of swell wave could be carried out and verified in future research works. In Figure 15, the Climate-LDR and Equivalent-LDR are compared.

It is surprising to observe that the equivalent wave direction (205° N) is extremely close to the orientation of the downcoast section of the bay (illuminated zone, Figure 13), confirming the results obtained in [37] that the LDR equivalent wave is responsible of the sculpturing of the bay, in the long run, bringing it to its static equilibrium plan-form. Moreover, this is confirmed also by the wave climate mode shown in Figure 14b: waves with the largest percentage of occurrence are those comprised within the south west wave sector, in which the 205° equivalent direction can be detected. In fact, it can be said that the LDR equivalent wave approximately represents the average climate; in other words, it is the wave component that usually affects a given region from a certain direction. Therefore, when in narrow wave sector, as the present case of Bagnoli bay, we are used to observe mono-modal wave climate, so with a single high-frequency direction. In such conditions, the directional mode corresponds to the average one, which corresponds, in turn, to the LDR equivalent wave direction [39].

5.3. Static Equilibrium Condition

Before moving on the application of the methodology exposed in the previous paragraphs to the case study, it has been verified that the bay under study is in a static equilibrium condition. Therefore, we analysed the shoreline position over 10 years, specifically from 2008 to 2018. Data comes from the digitalization, in QGIS environment, of historical imageries of the area from Google Earth (Figure 16a).

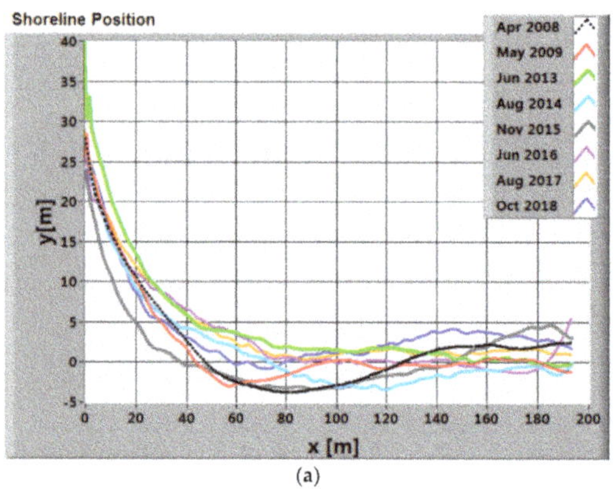

(a)

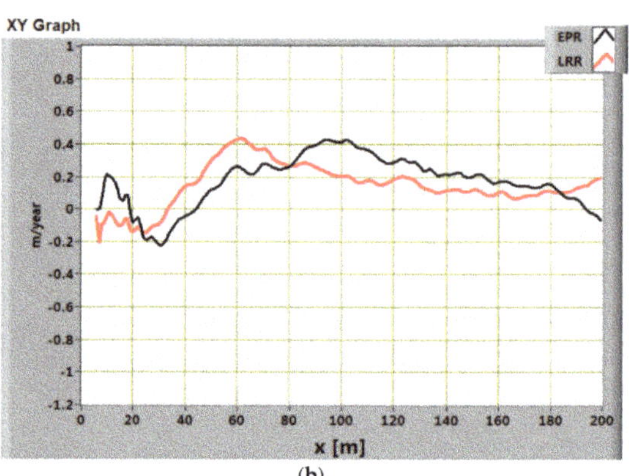

(b)

Figure 16. (**a**) y(x,t) coordinates of shorelines over time; (**b**) EPR (black line) and LRR (red line).

The *Linear Regression Rate (LRR)* and the *End Point Rate (EPR)* have been used as indicators of the rate of shoreline change. LRR corresponds to the slope of a least-square straight-line, fitted through the shoreline positions at the various available times; EPR, takes into account exclusively shoreline positions at the first and the last years concerned and represents shoreline movement during that time.

From the analysis of both LRR and EPR (Figure 16b), an erosive trend has been observed in the curved zone, meanwhile the linear stretch has been accreted. Therefore, this suggests that a long-shore transport occurs, which moves sediments from the shadow zone to the downcoast sector (illuminated zone). Nevertheless, the maximum rate of erosion and accretion are negligible, as they are approximatively 0.2 m/year and 0.4 m/year, respectively. Therefore, it can be asserted that the case study bay is in a static equilibrium condition.

6. Numerical Modelling of the Bay and Results

Numerical modelling of the bay has been performed through the wave driver MIKE 21 BW (DHI), which allows for obtaining the growth of the wave fronts leeward of the headland. In order to follow the method used in the numerical trial, the bay has been modelled in such a way that wave fronts could extend leeward the headland without any interference: the real bathymetry of the area has been employed until the diffraction point location, where the water depth is about 3 m, then a gentle slope of 1:100 has been adopted; the coastline has been shifted and cut landward and a sponge layer has been used, in order to avoid reflection phenomenon (Figure 17).

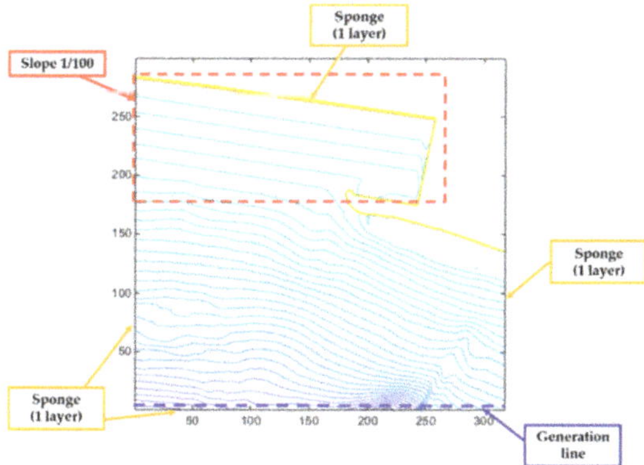

Figure 17. Model setup of case study bay.

Numerical simulations have been performed wave detecting the equivalent direction of 205° N, since it is representative of the effects of the entire wave climate on long-shore sediment transport [41]. As concerns wave period, the average measured omni-directional peak period (5.9 s) has been investigated.

Simulations have been carried out using regular waves, which wave height has been set at 0.8m; the breaking phenomenon has been neglected. Fine grid has been used (square cells with grid spacing 3 m), for which the orientation coincides with wave direction, with a time step of 0.1. A wave generation line has been used and wave absorbing sponge layers have been applied at the lateral boundaries. Grid geometric characteristics are summarized in Table 3.

Table 3. Grid geometric characteristics.

Grid Dimension [m^2]	Grid Spacing [m]	Grid Orientation [° N]
900 × 960	3	205

Comparison between equilibrium profile and wave fronts carried out through BW model, shown in Figure 18b, demonstrates that wave fronts' direction in the illuminated zone matches shoreline orientation, confirming the goodness of the LDR equivalent wave to represent the entire wave climate in the act of shape the beach planform. Specifically, simulations results confirmed the behaviour observed in the previous section: the shoreline planform of the headland bay is placed on more wave fronts, the downdrift section overlaps the wave front while the curved section is placed on the wave front closer to the headland (Figure 18a).

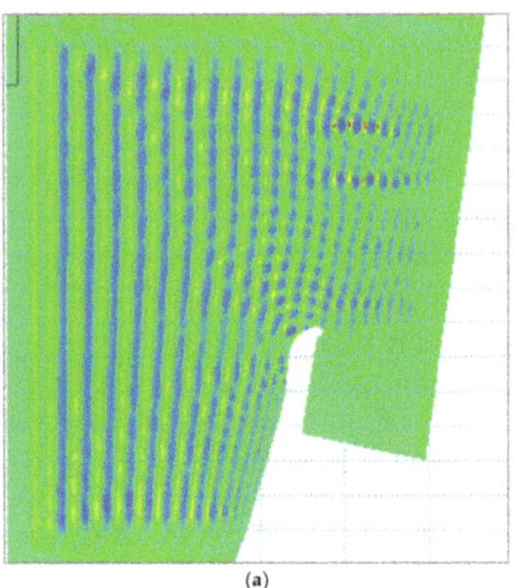

(a)

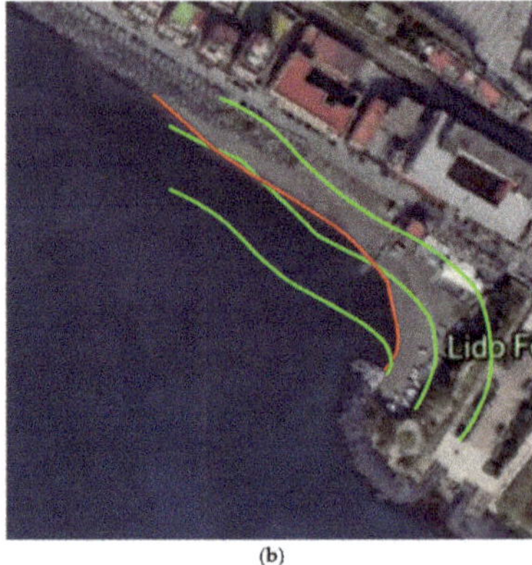

(b)

Figure 18. (**a**) Wave fronts generated by means BW; (**b**) comparisons between wave fronts (green lines) and shoreline (red line).

In order to attain an effective validation of the *wave-front-bay-shape equation*, Equation (15) has been applied. The minimum distance between the diffraction point and the asymptote has been measured (distance c in Figure 19a) and it has been standardised with respect to the local wave length, *c/L*. Applying Equation (15), we obtained the value of the shift, *s*, which represents the distance needed to superimpose the whole wave front on the shoreline. After shifting the wave front, as shown in Figure 19b, it is perfectly superimposed on the bay shoreline, thus verifying the correlation found between wave fronts and shoreline profile, reached by Equation (15).

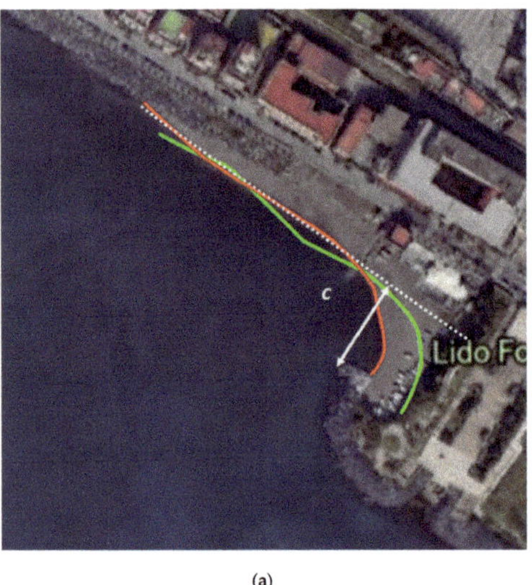

(a)

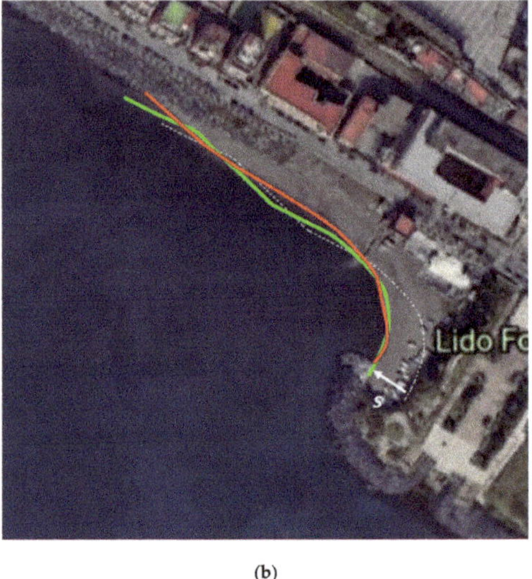

(b)

Figure 19. (**a**) Wave front (green line) that overlaps the shoreline (red line) in the downdrift section and the latter's distance, c, from the diffraction point; (**b**) wave front shifted by s, which was derived from Equation (15).

7. Conclusions

The main purpose of present paper is to establish a relationship between static equilibrium profiles of crenulated bays and wave front shapes and, more generally, understanding the correlation between bay shoreline response and wave forcings. The peculiar equilibrium plan-form (static or dynamic) of headland bays is assumed at a relatively long-term scale (e.g., annual to decadal) as a response to predominant wave direction; the downdrift segment of the bay, in fact, tends to align to the dominant direction of incoming waves, while, the curved up-coast segment is modelled by the diffracted wave fronts. In this regard, it is commonly believed by engineers that equilibrium beach profiles follow the wave front trend; however, this has been never fully proved so far. The most important works concerning headland bay beaches provided by the literature are empirical models, which simply describe the geometry of static equilibrium bays shape, neglecting the acting physical processes that sculpture the coast. In fact, they are merely of an empirical kind, lacking in a further insight on relationships between incident wave characteristics and beach shape. Therefore, in the lack of a model in which plan-shape is strongly correlated with wave characteristics, the project design of a new headland bay beach (necessary if the *headland control* practice has to be implemented) could, as a result, be extremely challenging.

To assess a possible correlation, numerical experiments have been carried out using the MIKE 21 Boussinesq Wave Module (BW), where wave fronts have been compared to the hyperbolic-tangent equilibrium profile, analysing the influence of wave direction, wave period and refraction phenomenon. Results proved that equilibrium profiles are located seaward compared to their relative wave fronts. Hence, designing a new beach following a wave front trends ensures a shoreline accretion in the sheltered zone (i.e., leeward to the headland). Additionally, a correspondence function, called the "*wave-front-bay-shape equation*" has been established, offering an easy application to engineering uses due to the simple geometric interpretation of its controlling parameters. The function seems to indicate that equilibrium beach profiles of a single headland bay correspond to a simple translation of wave front normal to the propagation line at the headland tip.

Moreover, the application of the "*wave-front-bay-shape equation*" to the case-study bay of the Bagnoli coast (south-west of Italy) has been performed. The numerical model has been set up, and the LDR equivalent wave concept has been used to embody the dominant wave attack that rules the long-term evolution of the little bay. Results confirm the behaviour observed from numerical experiments outcomes; the "*wave-front-bay-shape equation*" has been successfully verified, thus confirming that a correlation between equilibrium plan form and wave fronts can be found.

Nevertheless, the research still stands at a primary stage, and requires improvement and accurate verification in future research works in order to develop an enhanced guidance which could help in engineering and morphological practice.

Author Contributions: Conceptualization, M.B. and M.C.; Methodology, M.B. and S.T.; Software, S.T.; Supervision, M.C.; Writing—original draft, M.C.C. All authors have read and agreed to the published version of the manuscript.

Funding: This research received no external funding.

Institutional Review Board Statement: Not applicable.

Informed Consent Statement: Not applicable.

Data Availability Statement: Experimental data are available from the corresponding author upon request.

Conflicts of Interest: The authors declare no conflict of interest.

References

1. Inman, D.L.; Nordstrom, C.E. On the Tectonic and Morphologic Classification of Coasts. *J. Geol.* **1971**, *79*, 1–21. [CrossRef]
2. Short, A.D.; Masselink, G. Embayed and structurally controlled beaches. In *Handbook of Beach and Shoreface Morphodynamics*; Short, A.D., Ed.; John Willey & Sons: Hoboken, NJ, USA, 1999; pp. 230–249.
3. Hsu, J.R.C.; Yu, M.J.; Lee, F.C.; Benedet, L. Static bay beach concept for scientists and engineers: A review. *Coast. Eng.* **2010**, *57*, 76–91. [CrossRef]
4. Silvester, R. Stabilization of sedimentary coastlines. *Nature* **1960**, *188*, 467–469. [CrossRef]
5. van der Meer, J.W.; Briganti, R.; Zanuttigh, B.; Wang, B. Wave transmission and reflection at low-crested structures: Design formulae, oblique wave attack and spectral change. *Coast. Eng.* **2005**, *52*, 915–929. [CrossRef]
6. Calabrese, M.; Buccino, M.; Pasanisi, F. Wave breaking macrofeatures on a submerged rubble mound breakwater. *J. Hydro Environ. Res.* **2008**, *1*, 216–225. [CrossRef]
7. Srisuwan, C.; Rattanamanee, P. Modeling of Seadome as artificial reefs for coastal wave attenuation. *Ocean Eng.* **2015**, *103*, 198–210. [CrossRef]
8. Buccino, M.; del Vita, I.; Calabrese, M. Engineering modeling of wave transmission of reef balls. *J. Waterw.* **2014**, *140*. [CrossRef]
9. Silvester, R.; Ho, S.K. Use of crenulate shaped bays to stabilize coasts. In Proceedings of the 13th International Conference on Coastal Engineering, Vancouver, BC, Canada, 10–14 July 1972; ASCE: Reston, VA, USA, 1972. [CrossRef]
10. Silvester, R.; Hsu, J.R.C. *Coastal Stabilization: Innovative Concepts*; Prentice-Hall: Englewood Cliffs, NJ, USA, 1993; p. 578.
11. Silvester, R. *Coastal Engineering*, 2nd ed.; Elsevier: Amsterdam, The Netherlands, 1974.
12. Krumbein, W.C. Shore processes and beach characteristics. In *Technical Memorandum, vol. 3. Beach Erosion Board;* U.S. Army Corps of Engineers: Washington, DC, USA, 1944; p. 47.
13. Moreno, L.J.; Kraus, N.C. Equilibrium shape of headland-bay beaches for engineering design. In Proceedings of the Coastal Sediments, New York, NY, USA, 21–23 June 1999; ASCE: Reston, VA, USA, 1999.
14. Kemp, J.; Vandeputte, B.; Eccleshall, T.; Simons, R.; Troch, P. A modified hyperbolic tangent equation to determine equilibrium shape of headland bay beaches. In Proceedings of the 13th International Conference on Coastal Engineering, Baltimore, MD, USA, 30 December 2018. [CrossRef]
15. Hsu, J.R.C.; Evans, C. Parabolic Bay Shapes and Applications. *Proc. Inst. Civ. Eng. Part 2* **1989**, *87*, 557–570. [CrossRef]
16. Hsu, J.R.; Silvester, R.; Xia, Y.M. Generalities on static equilibrium bays. *Coast. Eng.* **1989**, *12*, 353–369. [CrossRef]
17. Hsu, J.R.; Silvester, R.; Xia, Y.M. Static equilibrium bays: New relationships. *J. Waterw. Port. Coast. Ocean Eng.* **1989**, *115*, 285–298. [CrossRef]
18. Buccino, M.; Daliri, M.; Dentale, F.; Di Leo, A.; Calabrese, M. CFD experiments on a low crested sloping top caisson breakwater. Part 1. nature of loadings and global stability. *Ocean Eng.* **2019**, *182*, 259–282. [CrossRef]
19. Buccino, M.; Daliri, M.; Dentale, F.; Calabrese, M. CFD experiments on a low crested sloping top caisson breakwater. Part 2. Analysis of plume impact. *Ocean Eng.* **2019**, *182*, 345–357. [CrossRef]
20. Buccino, M.; Daliri, M.; Calabrese, M.; Somma, R. A numerical study of arsenic contamination at the Bagnoli bay seabed by a semi-anthropogenic source. Analysis of current regime. *STOTEN* **2021**, *782*, 146811.
21. MIKE 21 Boussinesq Wave Module. Available online: http://manuals.mikepoweredbydhi.help/2017/Coast_and_Sea/MIKE21BW_Sci_Doc.pdf (accessed on 3 October 2016).
22. Dingenmans, M.W. *Water Wave Propagation over an Uneven Bottom*; World Scientific: Singapore, 1997.
23. Benedet, L.; Klein, A.H.F.; Hsu, J.R.C. Practical insights and applicability of empirical bay shape equations. In Proceedings of the 29th International Conference on Coastal Engineering, Lisbon, Portugal, 19–24 September 2004. [CrossRef]
24. Yasso, W.E. Plan geometry of headland bay beaches. *J. Geol.* **1965**, *73*, 702–714. [CrossRef]
25. Martino, E.; Moreno, L.J.; Kraus, N.C. Engineering guidance for the use of bayed-beach formulations. In Proceedings of the Coastal sediments, Sheraton Sand Key Resort, Clearwater Beach, FL, USA, 18–23 May 2003; ASCE: Reston, VA, USA, 2003.
26. Elshinnawy, A.I.; Medina, R.; Gonzalez, M. On the influence of wave directional spreading on the equilibrium planform of embayed beaches. *Coast. Eng.* **2018**, *133*, 59–75. [CrossRef]
27. Castelle, B.; Robinet, A.; Idier, D.; D'Anna, M. Modelling of embayed beach equilibrium planform and rotation signal. *Geomorphology* **2020**, *369*, 107367. [CrossRef]
28. Hurst, M.D.; Barkwith, A.; Ellis, M.A.; Thomas, C.W.; Murray, A.B. Exploring the sensitivities of crenulate bay shorelines to wave climates using a new vector-based one-line model. *J. Geophys. Res. Earth Surf.* **2015**, *120*, 2586–2608. [CrossRef]
29. Daly, C.J.; Bryan, K.R.; Winter, C. Wave energy distribution and morphological development in and around the shadow zone of an embayed beach. *Coast. Eng.* **2014**, *93*, 40–54. [CrossRef]
30. Madsen, P.A.; Murray, R.; Sørensen, O.R. A new form of the Boussinesq equations with improved linear dispersion characteristics. *Coast. Eng.* **1991**, *15*, 371–388. [CrossRef]
31. Madsen, P.A.; Sørensen, O.R. A new form of the Boussinesq equations with improved linear dispersion characteristics. Part 2: A slowly-varying Bathymetry. *Coast. Eng.* **1992**, *18*, 183–204. [CrossRef]
32. Madsen, P.A.; Sørensen, O.R.; Schäffer, H.A. Surf zone dynamics simulated by a Boussinesq type model. Part I: Model description and cross-shore motion of regular waves. *Coast. Eng.* **1997**, *32*, 255–287. [CrossRef]
33. Madsen, P.A.; Sørensen, O.R.; Schäffer, H.A. Surf zone dynamics simulated by a Boussinesq type model. Part II: Surf beat and swash zone oscillations for wave groups and irregular waves. *Coast. Eng.* **1997**, *32*, 289–319. [CrossRef]

34. Sørensen, O.R.; Schäffer, H.A.; Sørensen, L.S. Boussinesq type modelling using unstructured finite element technique. *Coast. Eng.* **2004**, *50*, 181–198. [CrossRef]
35. Svendsen, I.A. Mass flux and undertow in a surf zone. *Coast. Eng.* **1984**, *8*, 347–365. [CrossRef]
36. Walton, T.L.; Dean, R.G. Application of littoral drift roses to coastal engineering problems. In Proceedings of the Conference on Engineering Dynamics in the Surf Zone, Sydney, Australia, 1 January 1973; Institution of Engineers: Sydney, VIC, Australia, 1973; pp. 221–227.
37. Walton, T.L.; Dean, R.G. Longshore sediment transport via littoral drift rose. *Ocean Eng.* **2010**, *37*, 228–235. [CrossRef]
38. US Army Corps of Engineers. *Shore Protection Manual*; Coastal Engineering Research Centre, Government Printing Office: Washington, DC, USA, 1984.
39. Di Paola, G.; Ciccaglione, M.C.; Buccino, M.; Rosskopf, C.M. Influence of hard defence structures on shoreline erosion along Molise coast (southern Italy): A preliminary investigation. *Rend. Online Soci. Geol. Ital.* **2020**, 2–11. [CrossRef]
40. Silvester, R. Fluctuation in littoral drift. In Proceedings of the International Conference on Coastal Engineering, Houston, TX, USA, 3–7 September 1984; ASCE: Reston, VA, USA, 1984.
41. Buccino, M.; Di Paola, G.; Ciccaglione, M.C.; Del Giudice, G.; Rosskopf, C.M. A medium-term study of Molise coast evolution based on the one-line equation and "equivalent wave" concept. *Water* **2020**, *12*, 2831. [CrossRef]

 *geosciences*

Article

Vertical Land Motion as a Driver of Coastline Changes on a Deltaic System in the Colombian Caribbean

Juan Felipe Gómez [1,*], Eva Kwoll [1], Ian J. Walker [2] and Manoochehr Shirzaei [3]

1 Department of Geography, University of Victoria, Victoria, BC V8P 5C2, Canada; ekwoll@uvic.ca
2 Department of Geography, UC Santa Barbara, Santa Barbara, CA 93106, USA; ianjwalker@ucsb.edu
3 Department of Geosciences, Virginia Tech, Blacksburg, VA 24061, USA; shirzaei@vt.edu
* Correspondence: jfgomez@uvic.ca

Abstract: To face and properly mitigate coastal changes at a local level, it is necessary to recognize and characterize the specific processes affecting a coastline. Some of these processes are local (e.g., sediment starvation), while others are regional (e.g., relative sea-level change) or global (e.g., eustatic sea-level rise). Long tide gauge records help establish sea-level trends for a region that accounts for global (eustatic, steric) and regional (isostatic) sea-level changes. Local sea-level changes are also the product of vertical land motion (VLM), varying depending on tectonic, sedimentological, and anthropogenic factors. We investigate the role of coastal land subsidence in the present-day dynamics of an abandoned delta in the Colombian Caribbean. Satellite images and synthetic aperture radar acquisitions are used to assess decadal-scale coastline changes and subsidence rates for the period 2007–2021. We found that subsidence rates are highly variable alongshore. Local subsidence rates of up to −1.0 cm/yr correspond with an area of erosion rates of up to −15 m/yr, but coastal erosion also occurs in sectors where subsidence was not detected. The results highlight that local coastline changes are influenced by multiple, interacting drivers, including sand supply, coastline orientation and engineering structures, and that subsidence alone does not explain the high rates of coastal erosion along the study area. By the end of the century, ongoing coastal erosion rates of up to −25 m/yr, annual rates of subsidence of about −1 cm/yr, and current trends of global sea-level rise are expected to increase flooding levels and jeopardize the existence of the deltaic barrier island.

Keywords: sea-level rise; InSAR; Magdalena River; coastal changes; subsidence; mangroves; sediment compaction

Citation: Gómez, J.F.; Kwoll, E.; Walker, I.J.; Shirzaei, M. Vertical Land Motion as a Driver of Coastline Changes on a Deltaic System in the Colombian Caribbean. *Geosciences* **2021**, *11*, 300. https://doi.org/10.3390/geosciences11070300

Academic Editors: Germán Rodríguez, Carmen M. Rosskopf and Jesus Martinez-Frias

Received: 24 May 2021
Accepted: 15 July 2021
Published: 20 July 2021

Publisher's Note: MDPI stays neutral with regard to jurisdictional claims in published maps and institutional affiliations.

1. Introduction

The Intergovernmental Panel on Climate Change forecasts rates of sea-level rise (SLR) between 8 and 16 mm/yr by the end of the century, resulting in a global sea-level rise between 0.52 to 0.98 m for a high greenhouse gas emission scenario [1]. Adding to the predicted increases of global mean sea level (GMSL), the upward or downward movement of the land surface—also known as vertical land motion (VLM)—may exacerbate relative sea-level (RSL) changes at local scales. Changes in RSL result from the combined effect of the mean sea-level height (i.e., GMSL) and VLM [2], and it is measured relative to a local tide gauge benchmark [3]. The most common triggers of VLM are glacial isostatic adjustments (GIA), earthquakes and tectonics, aquifer compaction and sediment consolidation [3,4]. Where VLM takes place near the coast, it may modify the coastline by the emergence or submergence of the terrain [2], resulting in, for example, shoreline progradation or retreat, respectively. Within the various causes of VLM, this work focuses on subsidence, defined as the downward movement of the land surface with respect to a datum or point of reference [5], and how it relates to coastal morphodynamics. Subsidence is driven by factors such as natural sediment compaction [6], fault displacements [5], or human actions (e.g., groundwater extraction [7], withdrawal of petroleum and natural gas [8], soil desiccation [9]).

Deltas are common coastal landforms formed by the deposition and reworking of sediments from river systems where they enter larger bodies of water, such as oceans or lakes. Worldwide, most large deltas are subsiding at rates faster than the present global sea-level rise [10]. In the case that abundant sediment supply exists, a trade-off can be established between sediment delivery and subsidence in deltaic areas: the sediment deposition that triggers subsidence also helps compensate for or exceeds the subsidence. For instance, in some locations of the Mekong Delta in Vietnam, sediment accretion exceeds compaction rates, resulting in a net elevation gain of the delta surface [6]. However, when the delivery of fluvial sediment supply decreases, sediment compaction is no longer balanced by sediment deposition [6]. In this case, as the sediment matrix compresses and dehydrates, fluid pressure decreases and organic carbon-rich materials, if present, oxidize to carbon dioxide, resulting in a total mass and volume loss within the soil column [9] (Figure 1). Törnqvist et al. [11] noted that, adding to the sizeable natural component of subsidence due to isostatic flexure and the compaction of young sediments, subsidence caused by artificial drainage of wetlands and groundwater withdrawal can be up to an order of magnitude faster. For instance, using interferometric synthetic aperture radar (InSAR), VLM of up to −8 mm/yr have been measured in the coastal plain of the Selle River mouth (southern Italy) [12] and of up to −5 mm/yr in the Tiber Delta in Rome [13]. Additionally, Zhang et al. [14] reported average VLM rates of −5.1 mm/yr in the modern Yellow River Delta from 1992 to 2000.

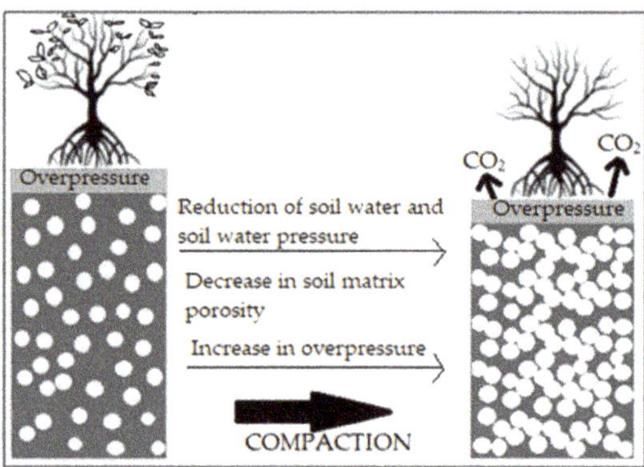

Figure 1. Properties of soil structure before (left) and after (right) compaction takes place. Emission of CO_2 occurs when dehydrated organic-rich soils are exposed to atmospheric oxygen (modified from Yuill et al. [8]).

Whilst the role of subsidence as a natural driver of delta growth and abandonment has been recognized and coined as the cyclic evolution of deltas [15], its influence in the evolution of coastal landscapes and the seaward progradation (i.e., accretion) or landward retreat (i.e., erosion) of coastlines has not been fully addressed. Despite subsidence being recognized as one of the key controls on coastline changes [13,16,17], the majority of research has focused on the added impact of current subsidence rates and rising sea levels for future scenarios of inundation and erosion (e.g., [12,18]). However, local-scale studies can help to better understand how VLM (subsidence or emergence) translates to coastline changes and related geomorphic responses. Accordingly, this study aims to support evidence-based policymaking by (i) testing the application of InSAR interferometry to estimate the rates and spatial distribution of VLM in a coastal setting, and (ii) linking VLM to recent coastline evolution derived from satellite imagery along a deltaic barrier.

Study Site

The study area is located in the Colombian Caribbean, between the municipalities of Ciénaga and Barranquilla (Figure 2). The area extends 70 km along the coastline eastward from the mouth of the Magdalena River. As the largest basin in Colombia, the Magdalena River and its tributaries cover 257,438 km^2 [19] and its discharge averages 10,287 and 4068 m^3/s during high and low flows, respectively [20]. Eight major shifts in delta location have been reported for the river since the Pliocene, the latest of which has occurred since the mid-Holocene, by way of a westward migration of a former river mouth located by the Ciénaga Grande de Santa Marta (hereafter CGSM) to its present location [21] (Figure 2). Adding to the natural displacement of the delta, human-made structures associated with a harbor built in the 1920s near the mouth of the river have confined its mouth to a single channel, thus further reducing the present influence of the Magdalena River in the study area [22]. Moreover, a highway built next to the coastline in the 1950s to connect the cities of Barranquilla and Santa Marta formed a causeway for the otherwise free interchange of water between ocean and lagoons, producing accelerated degradation and mortality of mangrove forest due to hypersalinization [23].

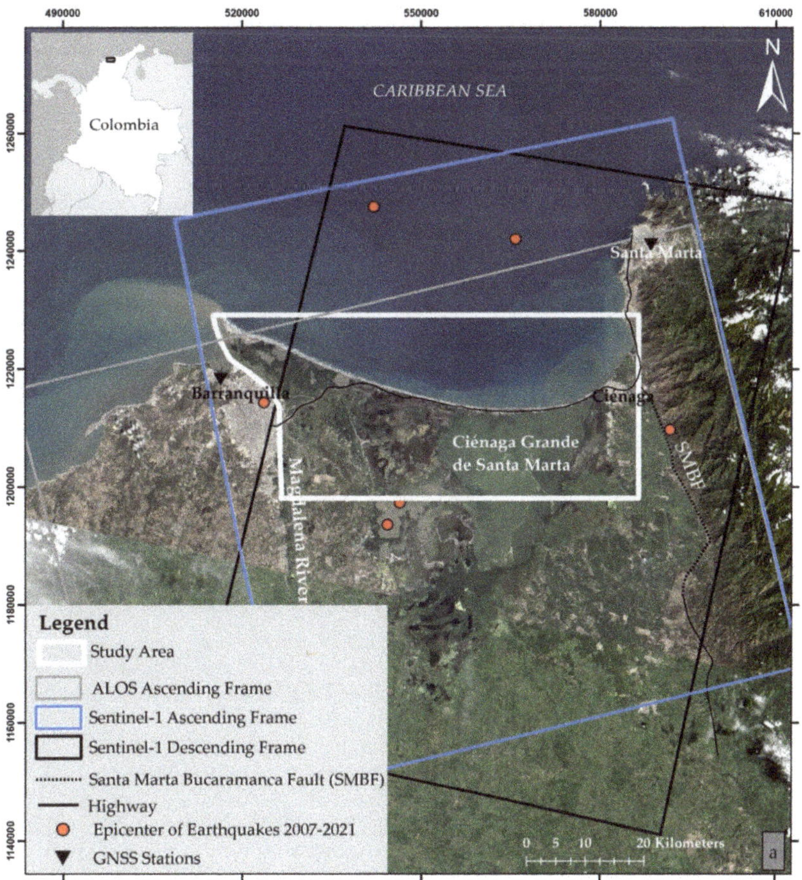

Figure 2. *Cont.*

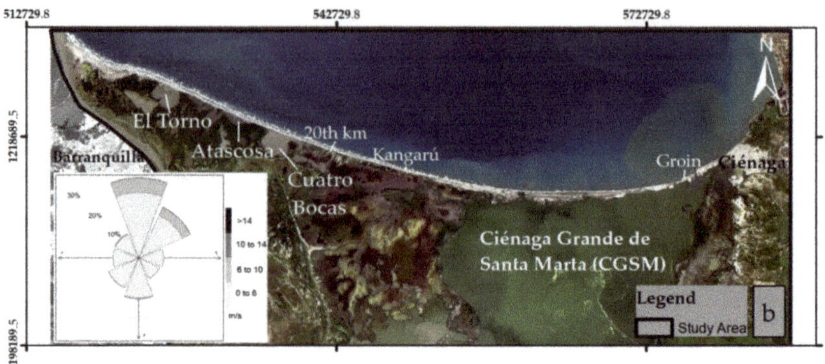

Figure 2. Study area and its surrounding wetlands and lagoons; (**a**) the epicenters of earthquakes and the coverage of ALOS and the Sentinel-1A/B ascending and descending radar data are included. The study site extends from the River Magdalena mouth (west) to the town of Ciénaga (east); (**b**) detail of the study area indicating the locations referred to in the manuscript. The wind rose displays the distribution of wind speed and direction for the study site.

The conditions described above have favored a transition of the landscape from fluvial to littoral-dominated process, portrayed by a palimpsest of fluvial and marine landforms such as channels, lagoons, oxbow lakes, salt plains, beaches, and dunes shaped by the predominant northeasterly trade winds [24] (Figure 2). Following Oertel [25], the landforms required to make up a barrier island system are present in the study area (i.e., mainland, backbarrier lagoon, barrier island, barrier platform, shoreface, inlets and inlet deltas). The sediments associated with these landforms, originated in marine and fluvio-lacustrine environments [26], range from poorly consolidated sands along the shorefront to laminated muds within lagoons and in the backshore. These Holocene sediments are bounded on the east side by the Santa Marta Bucaramanga Fault (SMBF), a regional-scale striking left-lateral structure that extends 374 km from the Caribbean coast to the eastern range of Colombia (Figure 2a) [27,28]. The fault marks a contrasting relief change between the Sierra Nevada de Santa Marta range east of the structure, and the flat topography of the Magdalena Delta floodplain to the west. According to the records of seismological activity in the area, there is no evidence of significant earthquakes associated with the SMBF with magnitudes (mb) larger than 4.9 since instrumental records are available (i.e., 1984). The few earthquakes registered for the observational period have magnitudes (mb) of less than 4.6 [29] (location of epicenters shown in Figure 2a).

Rapid coastline changes have been identified in the study area since the 1950s, affecting natural and human resources; specifically, coastal erosion rates increase alongshore in the same direction as the littoral drift (east–west) [24]. Chronic erosion is threatening the highway that connects the cities of Barranquilla and Santa Marta and, consequently, a rock rip-rap structure was built in 2014 to protect the highway from continuous erosive processes in a point known as the 20th km (the 0 km is located at the outskirts of Barranquilla) [30] (Figure 2). This hard structure has fixed the coastline in that area, but adjacent sectors keep rapidly retreating landward. The causes of coastal retreat have not been fully identified. A regional analysis of the sea-level trends for the Caribbean coast between 1950 and 2001 found a statistically significant sea-level rise trend of approximately 2 mm/yr [31]. Moreover, a sea-level rise of 5.3 mm/yr between 1950 and 2010 was measured at a tide gauge station located in Cartagena, approximately 100 km southeast of the study site [32], which indicates that, in addition to regional sea-level rise, there might be additional drivers contributing to the observed coastal retreat. Based on data from a global navigation satellite system (GNSS) station, subsidence velocities between −1.78 and −1.82 mm/yr have been quantified for Cartagena [33]. The influence of the rate of RSL change on barrier behavior is not completely understood, but some of the known effects of RSL rise include reworking of sand eroded from barrier fronts into transgressive dunes and shifts in sediment supply due to the modification of slope and course of rivers [34].

2. Materials and Methods

A two-fold approach was followed: (i) coastline changes were assessed using the USGS Digital Shoreline Analysis System (DSAS) tool [35] for 2010–2020, and (ii) VLM rates were quantified using Synthetic Aperture Radar (SAR) interferometric analysis of images acquired for the periods 2007–2011 and 2017–2021. The results of these approaches were illustrated and contrasted to establish the relationship between coastline changes and VLM.

2.1. Coastline Changes

Satellite images taken by GeoEye-1 and SkySat-1 platforms with 5 and 3 m ground pixel resolutions, respectively, were used to trace coastlines between 2010 and 2020. Using this imagery, past coastlines were visually delineated along the limit between wet and dry sediment. This well-established methodological approach [36–38] provides a proxy for the mean high water line (MHWL). Once a set of coastlines were established, changes were quantified using DSAS [35]. From the various statistics that DSAS produces, the average-of-rates (AOR) method, obtained by averaging the values resulting from the distance between each pair of available coastlines divided by the time between surveys, was selected to assess the rates of coastline changes over time. The AOR values were processed for transects spaced at 100 m intervals, perpendicular to a reference baseline traced parallel to the coastline. Accordingly, for two consecutive years, the accuracy of the coastline position depends on the image resolution (5 or 3 m), the georeferencing error (estimated to be 1 m), and a physical component of the error related to the magnitude of the tidal changes and slope of the beach (as intertidal range/tan (slope)) [17,39]. The latter was assessed by using an average beach slope of 9.0° (pers. obs. June/2019), and a maximum intertidal range of 0.4 m. Determining the quadratic sum of each of these components yielded an annual maximum uncertainty for coastline change estimates of $\pm$5.0 m/yr.

2.2. Vertical Land Motion Rates

SAR images from the Advanced Land Observing Satellite (ALOS) L-Band and Sentinel-1A/B C-Band satellites (search.asf.alaska.edu (accessed on 14 May 2021)) were gathered for the periods July 2007 to February 2011 and February 2017 to February 2021, respectively (see details in Table 1). The ALOS data were acquired along ascending orbit geometry, whereas ascending and descending orbits acquisitions were used for Sentinel-1. Interferograms for selected pairs of images were created and unwrapped using GAMMA, an InSAR processing software [40]. The analysis began with co-registering Single Look Complex (SLC) images to a reference image, which includes a standard matching algorithm using a digital elevation model (DEM), precise orbital parameters, and amplitude images [41]. For the Sentinel-1A/B datasets, the step above was followed by an enhanced spectral diversity (ESD) approach [42,43]. Using this dataset, a set of high-quality interferograms was generated, where only those interferograms with short perpendicular and temporal baselines were processed. A multi-looking operator of 20 and 4 pixels in range and azimuth was applied to obtain a ground resolution cell of approximately 46.0 m $\times$ 56.4 m (Table 1). The corresponding figures for ALOS are 6 and 3 pixels in range and azimuth, resulting in a ground resolution of 38.5 m $\times$ 23.6 m (Table 1).

Table 1. Parameters used to create interferograms during InSAR processing.

Condition	ALOS	Sentinel-1A/B
Acquisition dates	2007–2011	2017–2021
Path	143/144	77/142
Frame	200	32/555
Baseline distance (m)	1000	500
Temporal distance (days)	600	400
Azimuth looks	3	4
Range Looks	6	20
Images	13	50/53
Pairs selected	7	63
Incidence angle (°)	38.8	34.0
Heading angle (°)	348.8	347.9/192.0

To calculate and remove the effect of topographic phase and flat earth correction [44], a 1-arcsecond (~30 m) Shuttle Radar Topography Mission DEM [45] and precise satellite orbital information were used. To identify the elite (i.e., less noisy) pixels, only those pixels with average coherence larger than 0.65 were considered in the analysis, a value that has been used in coastal landscapes with conditions similar to the study area [12,46]. To retrieve the absolute (unwrapped) phase values, a minimum cost flow (MCF) algorithm adapted for sparsely distributed elite pixels was applied (see unwrapping in in Figure 3). Although the precise orbits are used, a few interferograms were still affected by a ramp-like signal, which was removed by fitting a second-order polynomial to their unwrapped phase [47]. Several wavelet-based filters were further applied to correct for effects of spatially uncorrelated topography error and topography correlated atmospheric delay [48] (see atmospheric and orbit correction in Figure 3). Subsequently, a re-weighted least square approach was iteratively applied to invert the corrected measurement of the unwrapped phase at each elite pixel and solve the surface deformation time series. The effect of residual atmospheric errors was further reduced by applying a high pass filter based on continuous wavelet transform to the time series of surface deformation at each elite pixel. Finally, an estimate of the long-term line-of-sight (LOS) deformation rate was obtained as the best-fitting linear slope (hereafter velocity) to the time series of surface deformation at each elite pixel. Together with the LOS velocities, an estimation of the residuals for each elite pixel is provided by plotting the standard deviation of the LOS velocities.

For the Sentinel-1A/B dataset, by combining the ascending and descending LOS velocities in Equation (1), the vertical velocities were retrieved in Cartesian coordinates over those areas where scatterers of both tracks overlap or are in close proximity [49].

$$dz \approx (DlosAsc + DlosDesc)/2\cos\theta \quad (1)$$

where DlosAsc and DlosDesc are the LOS velocities along the ascending and descending orbits, θ is the incidence angle, and dz is the projection of the displacement along the vertical Cartesian axis.

Only the Sentinel-1A/B dataset, therefore, yielded true estimates of VLM. In the case of ALOS data, a qualitative comparison was carried out between its LOS velocities and those obtained after processing the Sentinel-1A/B dataset. It should be noted that in using LOS velocities, the horizontal and vertical velocities are not separated for the observed period. The qualitative comparison between the velocities was made after performing an inverse distance weighted (IDW) interpolation model for transforming the discrete points representing VLM and LOS velocities of individual scatterers into a continuous surface. IDW has been used to interpolate InSAR scattered data points where a relationship or influence over neighboring data is not proven [49]. After running the IDW using a 120-m neighborhood search radius weighted with a cubic exponential power, a 100-m resolution, continuous velocity surface was obtained. A Moran's I test was used to support the choice of distances above for the IDW interpolation. The steps and parameters to obtain the VLM are summarized in Table 1 and Figure 3. Lastly, aggregations of both the VLM and coastline change transects at 1.5 and 5 km resolutions were performed to examine the effect of scale on our observations. The results are presented in the Supplementary Materials.

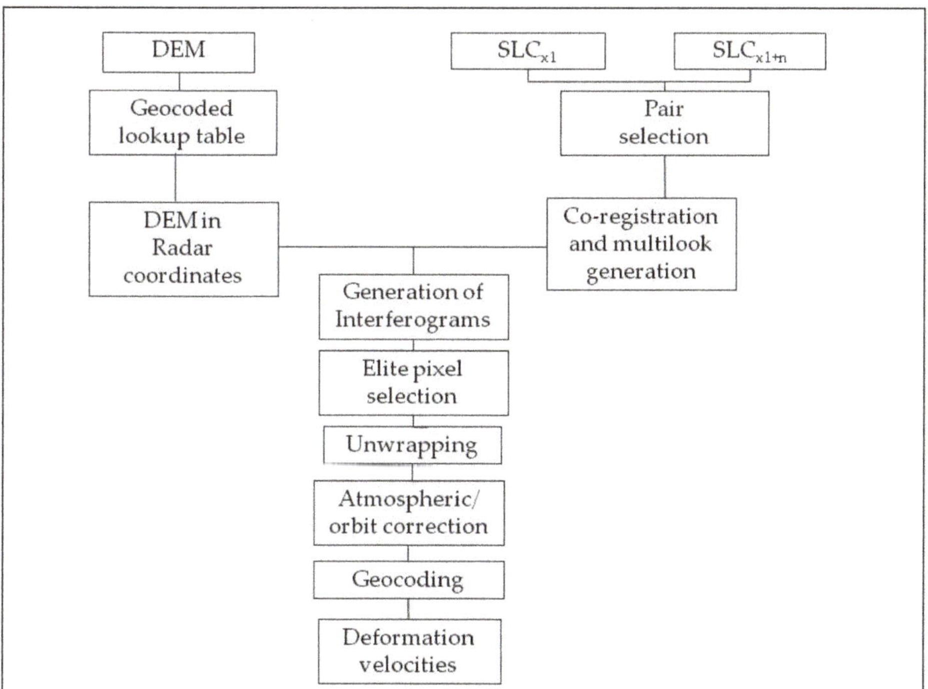

Figure 3. Flow diagram of interferometric processing of InSAR data (modified from Li et al. [50]).

Uncertainties in VLM rate estimates arise from the backscatter signal being affected by soil structure and moisture [51,52], resulting in apparent ground deformation values that may reach 0.5 radians [53]. Considering this value as the maximum uncertainty level, an annual uncertainty of ±0.2 cm/yr and ±0.7 cm/yr is expected for Sentinel-1A/B and ALOS, respectively. Adding to soil variability, the amount of radar wave penetration in areas with vegetation coverage depends on the structure and density of the canopy and the radar wavelengths; accordingly, wavelengths less than 10 cm (C-band, e.g., Sentinel-1A/B) mostly sense the upper portions of vegetation whilst wavelengths longer than 20 cm (L-band, e.g., ALOS) penetrate deeper into the canopy and can interact with the ground [54], yielding higher coherence values than C-band data [55]. Consequently, ALOS has proven to be more effective in reaching scatters below sparse vegetation [56,57].

Last, the deformation values obtained using SAR data were compared and validated with the velocity values from two permanent GNNS stations located in Barranquilla (11.0197°/−74.8496°) and Santa Marta (11.2253°/−74.1870°). Despite these stations being situated outside the study area, the satellite frames covered their locations, rendering them helpful to establish whether the trends found using these two approaches were alike (see Figure 2 for the location of GNSS stations).

3. Results

3.1. Coastline Changes

The average annual horizontal coastline change rates for 2010–2020 ranged from over −25 to +15 m/yr (Figure 4). Erosion rates increased alongshore in the same direction as the littoral drift (east–west). The only exceptions to the predominant erosive regime occurred on the westernmost segment of the study area where accretion rates larger than +10 m/yr were identified. Localized accretion was also observed on the east side of the mouth of the CGSM and next to two groins built west of the coastal town of Ciénaga in the early 2010s (see white and green colors in Figure 4).

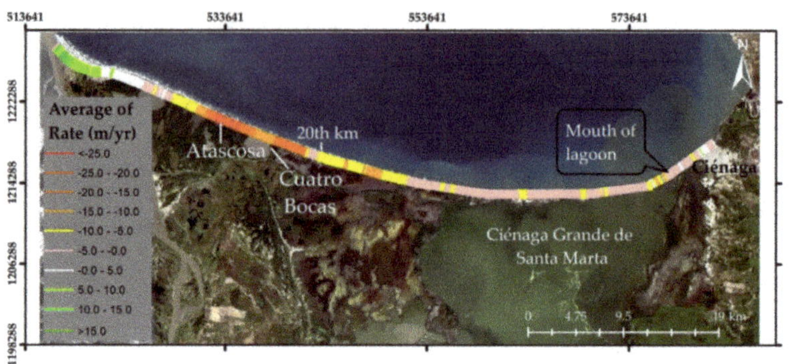

Figure 4. Coastline changes (m/yr) as derived from DSAS for the period 2010–2020.

From the mouth of the CGSM to the 20th km, erosion ranged between −1 and −10 m/yr (see pink and yellow transects in Figure 4). This erosion peaked along the stretch between the 20th km and La Atascosa lagoon (orange and red transects in Figure 4) and reached local average values up to −25 m/yr. Erosion rates progressively decreased westward from this lagoon and changed to accretion west of the El Torno lagoon (see green colors in Figure 4), where a levee, built at the mouth of the Magdalena River, hampers the westerly movement of sediment. The most significant erosion rate values for 2010–2020 occurred seaward of La Atascosa lagoon. Upon inspection of the imagery, it was observed that this largest average erosion rate was produced from the migration of an inlet after a breach of the lagoon in 2016.

3.2. Vertical Land Motion (VLM) and Line of Sight (LOS) Velocities

For both ALOS and Sentinel-1A/B data, vegetation coverage and the high dynamics of the coastline resulted in areas below the coherence threshold of 0.65 established in this work. Thus, there is a lack of information westward of Cuatro Bocas lagoon, where wetlands and vegetation are widespread (Figures 5 and 6). However, bare and sparsely vegetated terrain along the barrier island, particularly in the eastern and central areas where salt plains, stabilized dunes, and paleo-beaches are typical, provides a continuous strip of sufficient return signal, achieving the coherence threshold.

VLM velocities, calculated by combining the LOS velocity vectors of Sentinel-1A (i.e., ascending track) and Sentinel-1B (i.e., descending track) are shown in Figure 5. The Supplementary Materials show the same results of Figure 5 after aggregating the data in 1.5 and 5 km sections. In the coastal town of Ciénaga and northeast from this location, it is observed that VLM velocities are larger than −0.5 cm/yr, and decrease (i.e., become more negative) around the CGSM. West of Ciénaga and seaward from the CGSM, the signal is patchy. Subsidence rates are the largest westward from the location known as Kangarú (see Figure 2b for location), reaching values of up to −1.0 cm/yr along the stretch of coast between Kangarú and Cuatro Bocas lagoon, a sector characterized by erosion rates larger than 5 m/yr (see yellow and orange transects in Figure 5). In contrast, velocities larger than +0.5 cm/yr were observed west of the CGSM mouth (see dark blue in Figure 5).

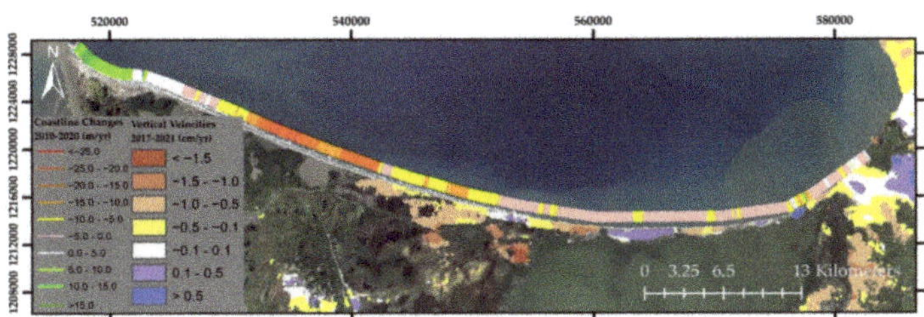

Figure 5. VLM rates for 2017–2021 as derived from Sentinel 1A/B data and coastline changes for 2010–2020 displayed with an arbitrary offset from the actual coastline for presentation purposes. Positive velocities (white and cold colors) represent stable areas and displacements toward the satellite, while negative velocities (warm colors) indicate displacement away from the satellite.

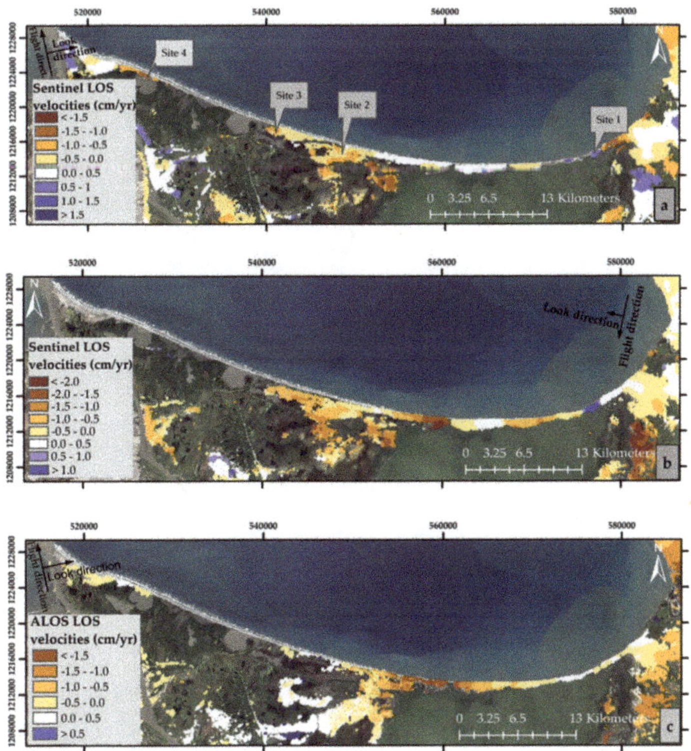

Figure 6. LOS velocities for (**a**) Sentinel-1A ascending track for 2017–2021; (**b**) Sentinel-1B descending track for 2017–2021, and (**c**) ALOS ascending track for 2007–2011. Positive velocities (white and cold colors) represent stable areas and displacements toward the satellite, while negative velocities (warm colors) indicate displacements away from the satellite. The location of the four time series described below is shown in (**a**).

LOS velocities from Sentinel-1A and B show the same trend as the overall VLM (Figure 5) along the deltaic barrier island, suggesting that the vertical component of land displacement dominates over the horizontal component at this incident angle (Figure 6a,b). LOS velocities related to scatterers in the coastal town of Ciénaga were around zero for all three tracks illustrated in Figure 6, and became smaller (i.e., subsidence increased), southward from the town, towards the eastern margin of the CGSM. Westward from the town Ciénaga, specifically in the stretch of coast between the mouth of the CGSM to Kangarú (see Figure 2b for locations), LOS velocities of Sentinel-1A showed a zone ranging from 0 to +0.5 cm/yr (white colors in Figure 6a), whereas both ALOS and Sentinel-1B indicated LOS velocities ranging between −0.5 and −1.5 cm/yr (see orange and brown colors in Figure 6b,c). Westward of Kangarú (close to site 2 in Figure 6a), the LOS velocities of ALOS, Sentinel-1A, and Sentinel-1B revealed negative values along the coastline and around the wetlands.

The standard deviation (S.D.) of the LOS velocities, illustrated in Figure 7, indicates that the smallest spread of the LOS velocities, given by the standard deviation values closer to zero, is found for the Sentinel-1B data (Figure 7b). For all tracks, the smallest standard deviation values occur in the town of Ciénaga (Figure 7), reflecting higher coherence values in urban settings. In the undeveloped areas, the largest S.D. values occur on the CGSM shores (see red colors in Figure 7), where the most negative LOS velocities within the study site occur (Figures 6 and 7). In general, it is observed that extreme LOS velocities, either positive or negative, coincide with high S.D. values (red colors in Figure 7).

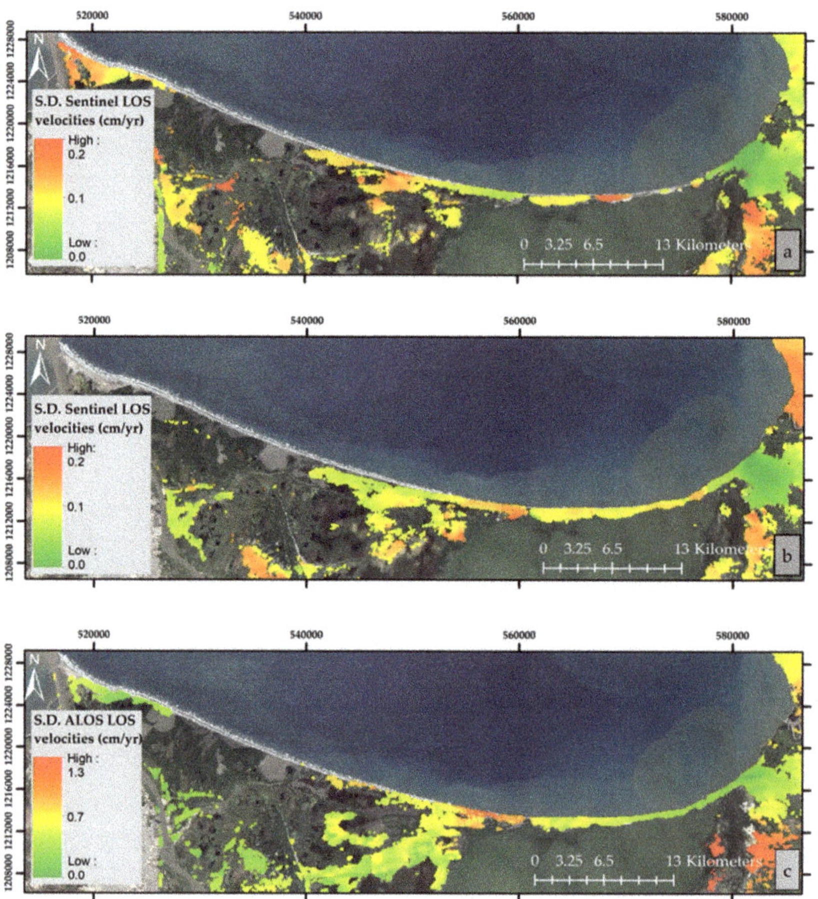

Figure 7. Standard deviation (S.D.) for each pixel of the estimated LOS velocities for (**a**) Sentinel-1A ascending track for 2017–2021; (**b**) Sentinel-1B descending track for 2017–2021, and (**c**) ALOS ascending track for 2007–2011. Red/green colors indicate high/low variability in the LOS velocities over time.

3.3. LOS Displacement Time Series

To inspect the trends of the land displacement over time, the LOS displacement time series of four sites are depicted in Figures 8–11, whose locations are shown in Figure 6a. Figure 8 shows the LOS displacements for site 1 and exemplifies the anthropogenic impact on VLM (Figures 6a and 8). In this sector, to face coastal erosion on the western side of the mouth of the CGSM, two groins were built in 2012. These coastal structures produced a gain of sediment on the updrift side of the groins, reflected by a slope value of 0.9 cm/yr for the best fitting line to the associated time series.

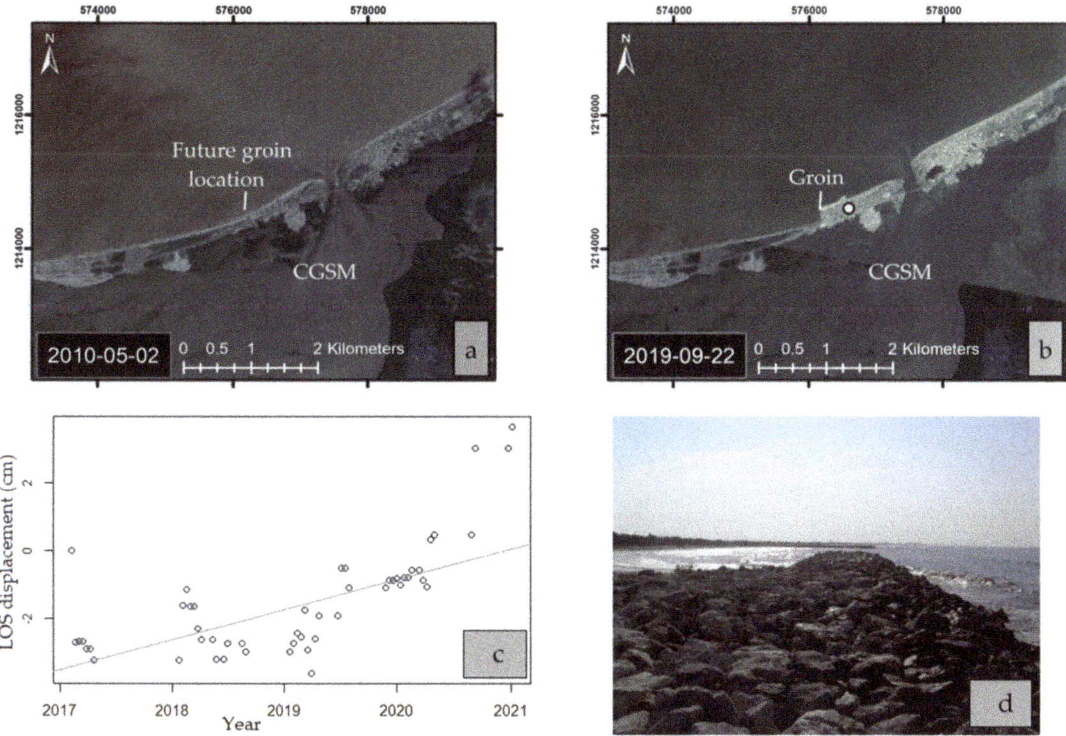

Figure 8. Satellite images before (**a**) and after (**b**) the groins were installed. Location of time series is indicated by white dot; (**c**) LOS displacement time series from Sentinel-1A for 2017–2021 (line represents the best fitting line with a slope of 0.9 cm/yr), and (**d**) photograph of the groins taken in 2014. Source of imagery: GeoEye images provided by Planet Labs Inc.

Figures 9 and 10 illustrate the time series of an area that was densely vegetated with mangroves until the 1980s (site 2 and 3, Figure 6a). This sector became the location of frequent overwashes that left behind sandy deposits, buried vegetation, and vegetated dunes scarped by erosive processes [24] (Figure 9c). Typical time series show gradual downward displacement with slopes of approximately −0.5 cm/yr (Figures 9b and 10b).

The westernmost time-series transect was located seaward from El Torno lagoon (site 4, Figure 6a). Although the shorefront in this area is erosive, no breachings have been observed in this lagoon during the last decade, and limited overwash deposits appear on satellite imagery. The area between the lagoon and the coastline consists of beach sands and embryo dunes sparsely covered with pioneer vegetation. LOS velocities for this site revealed a negative slope of −1.2 cm/yr for the linear fitting regression.

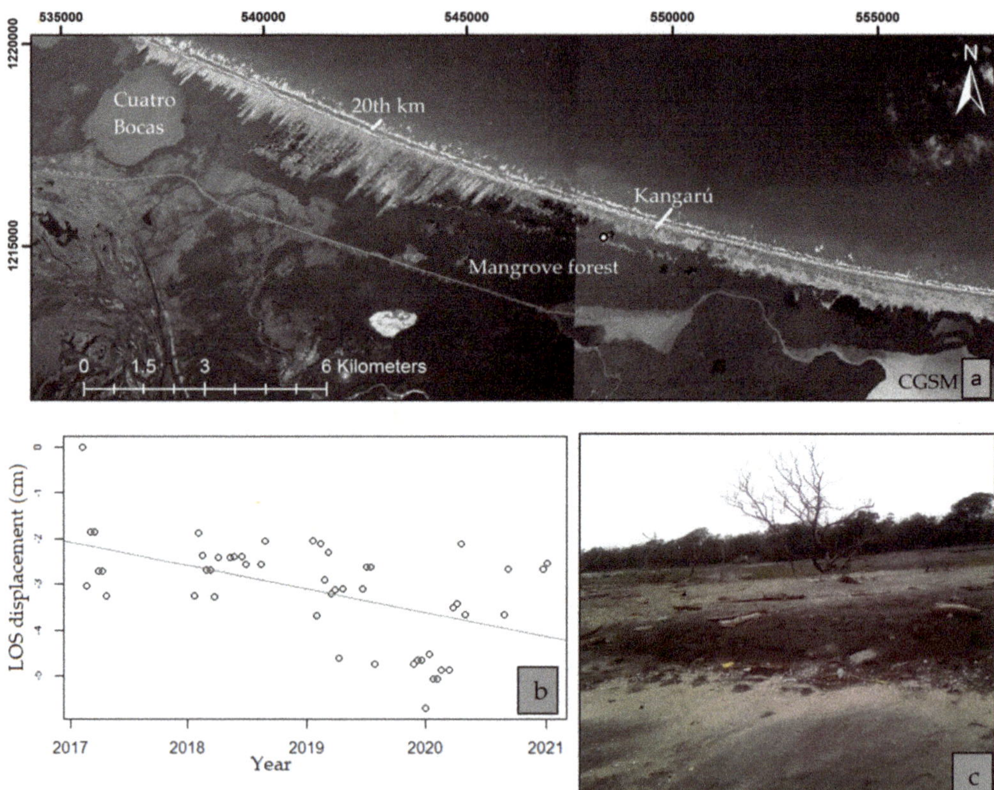

Figure 9. (**a**) Aerial photograph taken in 1953 showing the former mangrove forest before the highway was built. Parabolic dunes are seen in the backshore. Location of time series is indicated by white dot; (**b**) LOS displacement time series from Sentinel-1A for 2017–2021 (line represents the best fitting line with a slope of −0.52 cm/yr), and (**c**) shows standing dead trees on a former wetland.

Figure 10. (**a**) Area affected by coastal erosion seaward of the highway as observed in 2013. Location of time series is indicated by white dot; (**b**) LOS displacement time series from Sentinel-1A for 2017–2021 (line represents the best fitting line with a slope of −0.71cm/yr), and (**c**) photograph of the dunes landward from the 20th km. Source of imagery: Digital Globe provided GeoEye image.

Figure 11. (**a**) El Torno lagoon showing stabilized washover in 2013. Location of time series is indicated by white dot; (**b**) LOS displacement time series from Sentinel-1A for 2017–2021 (line represents the best fitting line with a slope of −1.2 cm/yr); (**c**) photograph of the beach seaward, and (**d**) next to the north shore of the lagoon. Source of imagery: Digital Globe provided GeoEye image.

3.4. Comparison of VLM and LOS Velocities and GNSS Data

The trends in vertical land movement measured at the GNSS stations located in Santa Marta and Barranquilla were contrasted with VLM and LOS velocity values from InSAR pixels located nearby these stations. Every four years, daily readings for each station were processed by the Deutsches Geodätisches Forschungsinstitut (DGFI) to provide long-term displacements for a network of permanent stations in South America. Roughly matching the time span for ALOS data, a solution for 2006–2011 compiled for the GNSS station in Santa Marta (Figure 2), indicated a vertical displacement of −0.52 cm/yr for that period [58]. For the same area, LOS velocities of −0.12 cm/yr were found for ALOS data. For the GNSS station in Barranquilla, vertical displacements of −0.14 cm/yr were reported for 2007–2009 [58], whereas a value of −0.31 cm/yr was obtained for the ALOS LOS velocities for 2007–2011. Sentinel data is not available for these years. For the period 2017–2021, VLM estimates using Sentinel-1A/B data near the Santa Marta station are −1.31 cm/yr. An updated processed solution for the GNSS stations to contrast the above values is not available yet.

4. Discussion

4.1. Subsidence in Deltas

Dating of modern deltas reveals that many major deltas worldwide were formed during the Holocene between 8,500 and 6,500 years before the present. The Fraser (Canada), Mississippi (United States), and Nile (Egypt) River deltas, to cite a few, were all formed within that period [59]. Accordingly, it has been suggested that following the initial rapid SLR during deglaciation, Holocene deltaic sequences began to accumulate as the rate of fluvial sediment input exceeded the declining rate of SLR in continental margins [59]. Compaction of the strata deposited during the Holocene has been described as a major driver of natural land subsidence in these dynamic environments [11,60]. In the case of the Magdalena River, we argue that downward trends in vertical land motion (VLM) reflect ongoing subsidence associated with modern fluvio-lacustrine sediments [26,61] that overlie relict sediments from a former delta that migrated westward after the mid-Holocene [21]. Low magnitude seismological activity associated with the SMBF indicates that VLM estimates reported in this work (Figure 5), at least during the period of observation (2007–2021), are not earthquake-driven, which suggests that one of the primary drivers of this process is sediment compaction of organic and mud deposits.

VLM in the study site is closely related to the Holocene history of the Magdalena delta and the supply of sediment from the drainage basin. Natural drainage displacements and levee constructions since the 1920s [22] have progressively diminished the amount of sediment delivered by the Magdalena River to the study area [22], disrupting the balance between sedimentation and natural compaction of sediments. More recently, the construction of the highway between Barranquilla and Santa Marta in the 1950s hampered the interchange of water between lagoons and ocean [23], indirectly causing subsidence by disturbing the mangrove forest.

Compared to previous assessments of RSL change for the region, which indicate current rising sea levels of approximately 0.5 cm/yr [32], local subsidence values of up to −1.5 cm/yr highlight the relevance of incorporating subsidence in the prediction of SLR and coastline positions. Earlier work aiming to correlate coastline erosion to sea-level rise, as measured from historical tide gauges, revealed that the average coastline change is two orders of magnitude greater than the rate of the sea-level rise [62]. Thus, by increasing the rate of local RSL, subsidence might increase future flooding risk associated with storms [55,63], resulting in the possible drowning of the deltaic barrier island.

4.2. Linking VLM and Coastal Erosion

The coastline changes and VLM findings are contrasting. Whereas erosion processes underlie most of the study area, the signal of VLM is highly variable along the study area. In other words, not all areas with annual shoreline erosion rates larger than the uncertainty threshold (i.e., ±5.0 m/yr) were associated with local subsidence. This indicates that subsidence is a factor that can exacerbate the impact of ongoing erosive processes. We explored the influence of scale by examining aggregations of the data at 1.5- and 5-km scales (see Supplementary Materials). At all scales considered, the sector with high coastline erosion rates of around −10 m/yr coincided spatially with subsidence rates of up to −1.0 cm/yr (i.e., stretch between Kangarú and Cuatro Bocas in Figure 5). Regions that exhibited coastal erosion within the range of uncertainty (± 5m/yr, e.g., seaward from the CGSM) showed a variable VLM signal at all scales. A similar interplay is described for a stretch of coast comprised of marshes and mud in the Mekong Delta in Vietnam, where secular coastal erosion values larger than −50 m/yr are paired with subsidence rates exceeding −1.5 cm/yr [17]. Similarly, high coastal erosion trends in an abandoned lobe of the Yellow River delta in China were associated with average subsidence rates of −0.7 cm/yr and up to −2.1 cm/yr for the period 1992–2000 [13]. These observations highlight the regional variability and complexity of drivers behind the coastal change that demand weighing the different drivers of change even within one deltaic system.

In addition to the natural compaction caused by sediment overloading, the site-specific subsidence rates observed along the stretch of coast between Kangarú and Cuatro Bocas might be triggered by the effect of localized dehydration and mortality of mangrove forest (Figure 1), following the highway construction in the early 1950s [23]. In this sector, as the mangrove decayed after the highway construction, soil salinity has hampered the establishment of vegetation [23], giving place to barren or sparsely vegetated swales (see Figure 9c). Yuill et al. [8] report that a large component of subsidence in organic-rich soils (e.g., peats associated with mangroves) is due to the exposure of soils to atmospheric oxygen, which reduces soil volumes by converting organic carbon into carbon dioxide gas (Figure 1). In coastal Louisiana, at least 60% of subsidence takes place within the uppermost 5–10 m, where woodpeat is widespread [11]. Subsidence in the former mangrove forest within the study area (Figures 9 and 10) may therefore be higher with this additional factor at play and is at its root caused by the anthropogenic alteration of the wetland. Other factors causing site-specific variability in subsidence rates are related to the uneven distribution of compressible sediments, hydraulic properties in aquifer systems, and groundwater extractions from aquifer systems [56,57]. In the Mississippi delta, the variability of the underlying Holocene lithology modulates compaction rates and their spatial variability [11]. Such factors may explain the variability in VLM and LOS velocity observations east of Kangarú (Figure 5). A lack of detailed knowledge of the Holocene stratigraphic sequence at this site hinders establishing any current relationship between subsidence rates and underlying substrates, but surficial geological cartography, developed by the Colombian Geological Survey [26], indicates that subsidence takes place in alluvial and fluvial-lacustrine sediments that make up a floodplain landform [24].

A rise in relative sea level as a product of either subsidence or an increase of GMSL creates more space for sediments to accumulate (i.e., accommodation space) [64], which is an underlying condition for transgression to occur. In other words, for any given interval of time, accommodation space is created by a rise in sea level and/or subsidence. Thus, where rates of accommodation space are larger than sediment deposition, there is a landward shift in sedimentary environments, an expansion of the subtidal zone [65], and a landward displacement of the barrier [34]. Although stratigraphic evidence of a transgressive succession was not gathered in this work, previous mapping of chronic erosion in conjunction with frequent overwashes between Kangarú and Cuatro Bocas [24] indicate that transgression might occur along this stretch of coast. Transgression was identified by two processes acting in tandem on the landscape: (i) storm overwash deposition resulting in temporal accretion of sediments landward of the contemporaneous shoreline, followed by (ii) a period of more gradual erosion interrupted by a new overwash cycle. Thus, we suggest that subsidence velocities of up to −1 cm/yr in the observed stretch of coast are a factor that underpins transgression of the coastline and the occurrence of overwashes, resulting in a landscape composed of scarped dunes and eroding beaches overlain by washover fans and surrounded by overwash channels. The spatial variability of VLM and LOS velocities east of Kangarú (see Figure 5) refrain us from extending the hypothesis of a transgressive shift of the coastline to the stretch of coast seaward from the CGSM.

4.3. Utility and Limitations of InSAR for Interpretations of Coastal Landscape Change

Within the study area, the analyses of four specific sites show the potential of InSAR as a tool to help explain local changes to the landscape. Although somewhat expected, coastline accretion and the upward movement of the terrain on the updrift side of a pair of groins built in 2012 (Figure 8 and site 1 in Figure 6a) stresses the reliability of the technique and its effectiveness in detecting VLM. It also illustrates that VLM estimates may include upward VLM related to crustal motion and the effect of sediment deposition and surface accretion. In a similar manner, in addition to subsidence, the finding of LOS velocities smaller than −1.0 cm/yr seaward of El Torno lagoon and near the coastline (Figure 11 and site 4 in Figure 6a), might be influenced by the erosive regime of the coastline. Indeed, Shirzaei et al. [55] indicate that, in dynamic landscapes, VLM estimates do include changes in surface elevation due to erosion or deposition and must be separated from other types of VLM. Thus, in addition to the locations mentioned above, areas next to sand beaches, such as the coastal town of Ciénaga and the westernmost extreme of the study site, the VLM velocities might be influenced by the removal or deposition of sand associated with erosion or accretion.

In contrast to the aforementioned locations, the locations of the time series for sites 2 and 3 are over 1 km landward from the coastline (see Figures 6a, 9 and 10), in sectors that were populated by abundant mangrove forest until the 1980s. Therefore, this area is prone to gas emission related to the oxidation of organic-rich soils. VLM along these locations ranges between −0.5 and −1 cm/yr (Figures 5, 9b and 10b). As the sites selected for the time-series profiles are not in close vicinity to the coastline, their velocity values are not influenced by the erosive regime of the coastline and can be considered to reflect subsidence only. Deltaic areas with similar subsidence rates to the values reported above have been described elsewhere [12–14].

Caveats of the InSAR methodology in this study also arise in vegetated areas when the surface backscattering properties due to growing vegetation change through time [66]. This factor is especially the case for wavelengths of less than 10 cm (e.g., Sentinel-1 A/B C-band) that interact with leaves and soft-stemmed vegetation. The constantly changing vegetation conditions result in temporal decorrelation and a reduction of the coherence level [66,67]. This issue was addressed by contrasting the outcomes of Sentinel-1A/B with ALOS, a satellite with a longer wavelength (L-band), which penetrates deeper in the canopy or ground surface [54]. When comparing LOS velocities for Sentinel-1A/B and ALOS ascending orbit, we found that, even though the latter provided absolute larger subsidence values, the trends for both platforms were aligned for most of the study area (Figure 6).

Given the caveats mentioned above, the absolute values of VLM should be interpreted with caution in the context of an application demanding high accuracy levels. However, despite the lack of temporal overlap between the Sentinel-1A/B and ALOS datasets, the outcomes of ALOS and Sentinel-1A/B were consistent in revealing areas prone to upward and downward movements with respect to a reference point in Ciénaga.

5. Conclusions

This study combined satellite images and InSAR data to examine the interplay of vertical land motion (VLM) (2007–2021) and coastline changes (2010–2020) along a deltaic barrier in the Colombian Caribbean. The findings are as follows:

- Annual coastal erosion rates increase in the same direction as the littoral drift (i.e., East-West direction), reaching a peak of −25 m/yr next to a lagoon known as Cuatro Bocas. In contrast, VLM patterns are highly variable along the study site, precluding linking coastal erosion rates to negative VLM values in general.

- VLM in the study area is linked to the Holocene History of the Magdalena River delta and the supply of sediment from the drainage basin. Anthropogenic alterations are known to have caused alteration in sediment supply, which is not sufficient to counterbalance sediment compaction.
- Although subsidence alone does not explain the high rates of coastal erosion along the study area, it is a factor that enhances erosive processes and is linked to the occurrence of overwashes and breaching of lagoons. The zones with the highest subsidence (up to −1.5 cm/yr) are areas surrounding the Ciénaga Grande de Santa Marta composed of marshes and mud. These local subsidence values are at least two times larger than gauge-measured rates of SLR.
- Local subsidence rates of up to −1 cm/yr were found in an area where mangrove forest was abundant before a highway was built in the 1950s. The likely primary drivers of subsidence in this sector are sediment compaction of the Holocene alluvium and gas emission followed by oxidation from organic-rich soils.
- Using InSAR close to the coastline is complex due to the ever-changing condition of the littoral, particularly in vegetated and highly dynamic coastlines. In those areas of the study site next (within ~100 m) to the coastline, the velocities obtained by combining Sentinel-1A/B interferograms might reflect the added effect of VLM of land surface change due to sediment accretion/erosion.

There is a lack of research that jointly discusses coastal dynamics and VLM at a local scale to date. Thus, this work provides a baseline to be used as a blueprint to track coastal dynamics and future ground motions once longer records of radar data become available. In addition to contributing to the explanation the chronic erosion of the study area, the outcomes of this project are relevant for engineers of a proposed expansion to the current highway.

Supplementary Materials: The following are available online at https://www.mdpi.com/article/10.3390/geosciences11070300/s1; Figure S1: VLM rates for 2017–2021 as derived from Sentinel 1A/B data and coastline changes for 2010–2020. Data was aggregated in (a) 1.5-km and (b) 5.0-km sections.

Author Contributions: J.F.G. performed the InSAR analysis with the support of E.K. and M.S.; J.F.G. performed coastal changes analysis with the assistance of I.J.W.; M.S. provided codes and expertise to perform InSAR analysis. All authors analyzed the results and contributed to the paper writing, editing and revisions. All authors have read and agreed to the published version of the manuscript.

Funding: This project was supported by funding provided to EK through an NSERC Discovery Grant. M.S. was supported by the National Aeronautics and Space Administration (NASA) grant 80NSSC170567.

Data Availability Statement: Processed data are available upon request to the first author.

Acknowledgments: The authors thank three anonymous reviewers for their constructive comments and feedback that improved the final manuscript. We are grateful to the following individuals and institutions for their input and support: Edwin Niessen, Alice Gentleman, Evaristo Rada, Ismael Fernández, Pro-Ciénaga Foundation, and to the staff of the Isla de Salamanca Park Way.

Conflicts of Interest: The authors declare no conflict of interest.

References

1. Church, J.A.; Clark, P.U.; Cazenave, A.; Gregory, J.M.; Jevrejeva, S.; Leverman, A.; Merrifield, M.A.; Milne, G.A.; Nerem, R.S.; Nunn, P.D.; et al. Sea Level Change. In *Climate Change 2013: Contribution of Working Group I to the Fifth Assessment Report of the Intergovernmental Panel on Climate Change*; Stocker, T.F., Oin, D., Plattner, G.K., Tignor, M., Allen, S.K., Boschung, J., Nauels, A., Xia, Y., Eds.; Cambridge University Press: Cambridge, UK, 2013.
2. Gregory, J.M.; Griffies, S.M.; Hughes, C.W.; Lowe, J.A.; Church, J.A.; Fukimori, I.; Gomez, N.; Kopp, R.E.; Landerer, F.; Cozannet, G.L.; et al. Concepts and Terminology for Sea Level: Mean, Variability and Change, Both Local and Global. *Surv. Geophys.* **2019**, *40*, 1251–1289. [CrossRef]
3. Pugh, D. *Changing Sea Levels: Effects of Tides, Weather and Climate*, 1st ed.; Cambridge Univesity Press: Cambridge, UK, 2004.
4. Pugh, D.; Woodworth, P. *Sea-Level Science: Understanding Tides, Surges, Tsunamis and Mean Sea-Level Changes*; Cambridge University Press: Cambridge, UK, 2014. [CrossRef]

5. Dokka, R.K. Modern-Day Tectonic Subsidence in Coastal Louisiana. *Geology* **2006**, *34*, 281–284. [CrossRef]
6. Zoccarato, C.; Minderhoud, P.S.J.; Teatini, P. The Role of Sedimentation and Natural Compaction in a Prograding Delta: Insights from the Mega Mekong Delta, Vietnam. *Sci. Rep.* **2018**, *8*, 11437. [CrossRef]
7. Minderhoud, P.S.J.; Middelkoop, H.; Erkens, G.; Stouthamer, E. Groundwater Extraction May Drown Mega-Delta: Projections of Extraction-Induced Subsidence and Elevation of the Mekong Delta for the 21st Century. *Environ. Res. Commun.* **2020**, *2*, 011005. [CrossRef]
8. Yuill, B.; Lavoie, D.; Reed, D.J. Understanding Subsidence Processes in Coastal Louisiana. *J. Coast. Res.* **2009**, *10054*, 23–36. [CrossRef]
9. Dixon, T.H.; Dokka, R.K. Earth Scientists and Public Policy: Have We Failed New Orleans? *Eos Trans. Am. Geophys. Union* **2008**, *89*, 96–97. [CrossRef]
10. Syvitski, J.P.M.; Kettner, A.J.; Overeem, I.; Hutton, E.W.H.; Hannon, M.T.; Brakenridge, G.R.; Day, J.; Vörösmarty, C.; Saito, Y.; Giosan, L.; et al. Sinking Deltas Due to Human Activities. *Nat. Geosci.* **2009**, *2*, 681–686. [CrossRef]
11. Törnqvist, T.E.; Wallace, D.J.; Storms, J.E.; Wallinga, J.; Van Dam, R.L.; Blaauw, M.; Derksen, M.; Klerks, C.J.; Meijneken, C.; Snijders, E.M. Mississippi Delta Subsidence Primarily Caused by Compaction of Holocene Strata. *Nat. Geosci.* **2008**, *1*, 173–176. [CrossRef]
12. Di Paola, G.; Alberico, I.; Aucelli, P.P.C.; Matano, F.; Rizzo, A.; Vilardo, G. Coastal Subsidence Detected by Synthetic Aperture Radar Interferometry and Its Effects Coupled with Future Sea-Level Rise: The Case of the Sele Plain (Southern Italy). *J. Flood Risk Manag.* **2018**, *11*, 191–206. [CrossRef]
13. Polcari, M.; Albano, M.; Montuori, A.; Bignami, C.; Tolomei, C.; Pezzo, G.; Falcone, S.; La Piana, C.; Doumaz, F.; Salvi, S.; et al. InSAR Monitoring of Italian Coastline Revealing Natural and Anthropogenic Ground Deformation Phenomena and Future Perspectives. *Sustainability* **2018**, *10*, 3152. [CrossRef]
14. Zhang, J.Z.; Huang, H.; Bi, H. Land Subsidence in the Modern Yellow River Delta Based on InSAR Time Series Analysis. *Nat. Hazards* **2015**, *75*, 2385–2397. [CrossRef]
15. Roberts, H.H. Dynamic Changes of the Holocene Mississippi River Delta Plain: The Delta Cycle. *J. Coast. Res.* **1997**, *13*, 605–627.
16. El-Fishawi, N.M. Coastal Erosion in Relation to Sea-Level Changes, Subsidence and River Discharge, Nile Delta Coast. *Acta Mineral. Petrogr.* **1989**, *30*, 161–171.
17. Anthony, E.J.; Brunier, G.; Besset, M.; Goichot, M.; Dussouillez, P.; Nguyen, V.L. Linking Rapid Erosion of the Mekong River Delta to Human Activities. *Sci. Rep.* **2015**, *5*, 14745. [CrossRef] [PubMed]
18. Shirzaei, M.; Bürgmann, R. Global Climate Change and Local Land Subsidence Exacerbate Inundation Risk to the San Francisco Bay Area. *Sci. Adv.* **2018**, *4*, eaap9234. [CrossRef] [PubMed]
19. Restrepo, J.D.; Kjerfve, B.; Hermelin, M.; Restrepo, J.C. Factors Controlling Sediment Yield in a Major South American Drainage Basin: The Magdalena River, Colombia. *J. Hydrol.* **2006**, *316*, 213–232. [CrossRef]
20. Torregroza-Espinosa, A.C.; Restrepo, J.C.; Correa-Metrio, A.; Hoyos, N.; Escobar, J.; Pierini, J.; Martínez, J.-M. Fluvial and Oceanographic Influences on Suspended Sediment Dispersal in the Magdalena River Estuary. *J. Mar. Syst.* **2020**, *204*, 103282. [CrossRef]
21. Romero-Otero, G.A.; Slatt, R.M.; Pirmez, C. Evolution of the Magdalena Deepwater Fan in a Tectonically Active Setting, Offshore Colombia. In *Memoir 108: Petroleum Geology and Potential of the Colombian Caribbean Margin*; Bartolini, C., Mann, P., Eds.; American Association of Petroleum Geologists AAPG/Datapages: Houston, TX, USA, 2015; pp. 675–707.
22. Von Erffa, A.F. Sedimentation, Transport Und Erosion an Der Nordkuste Kolumbiens Zwischen Barranquilla und Der Sierra Nevada de Santa Marta. *Bol. Investig. Cient. Colombo-Alem.* **1973**, *7*, 155–209.
23. Elster, C.; Perdomo, L.; Schnetter, M.-L. Impact of ecological factors on the regeneration of mangroves in the Ciénaga Grande de Santa Marta, Colombia. *Hydrobiology* **1999**, *413*, 35–46. [CrossRef]
24. Gómez, J.F.; Byrne, M.L.; Hamilton, J.; Islas, F. Historical Coastal Evolution and Dune Vegetation in Isla Salamanca National Park, Colombia. *J. Coast. Res.* **2016**, *33*, 632–641. [CrossRef]
25. Oertel, G.F. The barrier island system. *Mar. Geol.* **1985**, *63*, 1–18. [CrossRef]
26. Hernández, M.; Maldonado, I. *Geología de La Plancha 18-Ciénaga*; INGEOMINAS: Santafé de Bogotá, Colombia, 1999.
27. Idárraga-García, J.; Romero, J. Neotectonic study of the Santa Marta Fault System, Western foothills of the Sierra Nevada de Santa Marta, Colombia. *J. South Am. Earth Sci.* **2010**, *29*, 849–860. [CrossRef]
28. Paris, G.; Machete, M.N.; Dart, R.L.; Haller, K.M. *Maps and Database of Quaternary Folds and Faults in Colombia and Its Offshore Regions*; Open File 0284; United States Geological Survey: Denver, CO, USA, 2000; p. 61.
29. United States Geological Survey. Latest Earthquakes Report. Available online: https://earthquake.usgs.gov/earthquakes (accessed on 8 January 2021).
30. El Heraldo. Contratan Obras Del Kilómetro 19 Con Firma Edgardo Navarro Vives. Available online: https://www.elheraldo.co/local/contratan-obras-del-kilometro-19-con-firma-edgardo-navarro-vives-167730 (accessed on 10 May 2021).
31. Losada, I.; Reguero, B.; Mendez, F.; Castanedo, S.; Abascal, A.; Mínguez, R. Long-term changes in sea-level components in Latin America and the Caribbean. *Glob. Planet. Chang.* **2013**, *104*, 34–50. [CrossRef]
32. Andrade, C.A.; Thomas, Y.F.; Lerma, A.N.; Durand, P.; Anselme, B. Coastal Flooding Hazard Related to Swell Events in Cartagena de Indias, Colombia. *J. Coast. Res.* **2013**, *290*, 1126–1136. [CrossRef]

33. Andrade-Amaya, C.A.; Ferrero-Ronquillo, A.J.; León-Rincón, H.; Mora-Paez, H.; Carvajal-Perico, H. Sobre cambios en la línea de costa entre 1735 y 2011 y la subsidencia en la Bahía de Cartagena de Indias, Colombia. *Rev. Acad. Colomb. Cienc. Exactas Físicas Nat.* **2017**, *41*, 94–106. [CrossRef]
34. Van Heteren, S. Barrier Systems. In *Coastal Environments and Global Change*, 1st ed.; Masselin, K., Gehrels, K., Eds.; John Wiley & Sons: West Sussex, UK, 2014; pp. 194–226.
35. Thieler, E.R.; Martin, D.; Ergul, A. *The Digital Shoreline Analysis System: Shoreline Change Measurement Software Extension for ArcView*; Open File Report 03–076; Woods Hole Coastal and Marine Science Center: Woods Hole, MA, USA, 2003.
36. Boak, E.H.; Turner, I. Shoreline Definition and Detection: A Review. *J. Coast. Res.* **2005**, *214*, 688–703. [CrossRef]
37. Dolan, B.H.R. Storms and Shoreline Configuration. *J. Sediment. Res.* **1981**, *51*. [CrossRef]
38. Moore, L. Shoreline Mapping Techniques. *J. Coast. Res.* **2000**, *16*, 111–124.
39. Del Río, L.; Gracia, F.J. Error determination in the photogrammetric assessment of shoreline changes. *Nat. Hazards* **2012**, *65*, 2385–2397. [CrossRef]
40. Werner, C.; Wegmüller, U.; Strozzi, T.; Wiesmann, A. GAMMA SAR and interferometric processing software. In Proceedings of the ERS—ENVISAT Symposium, Gothenburg, Sweden, 16–20 October 2000.
41. Sansosti, E.; Berardino, P.; Manunta, M.; Serafino, F.; Fornaro, G. Geometrical SAR image registration. *IEEE Trans. Geosci. Remote Sens.* **2006**, *44*, 2861–2870. [CrossRef]
42. Yague-Martinez, N.; Prats-Iraola, P.; Gonzalez, F.R.; Brcic, R.; Shau, R.; Geudtner, D.; Eineder, M.; Bamler, R. Interferometric Processing of Sentinel-1 TOPS Data. *IEEE Trans. Geosci. Remote Sens.* **2016**, *54*, 2220–2234. [CrossRef]
43. Shirzaei, M.; Bürgmann, R.; Fielding, E.J. Applicability of Sentinel-1 Terrain Observation by Progressive Scans Multitemporal Interferometry for Monitoring Slow Ground Motions in the San Francisco Bay Area: Sentinel-1 Multitemporal Interferometry. *Geophys. Res. Lett.* **2017**, *44*, 2733–2742. [CrossRef]
44. Franceschetti, G.; Lanari, R. *Synthetic Aperture Radar Processing*, 1st ed.; CRC Press: Boca Raton, FL, USA, 1999. [CrossRef]
45. Farr, T.G.; Rosen, P.A.; Caro, E.; Crippen, R.; Duren, R.; Hensley, S.; Kobrick, M.; Paller, M.; Rodriguez, E.; Roth, L.; et al. The Shuttle Radar Topography Mission. *Rev. Geophys.* **2007**, *45*. [CrossRef]
46. Amato, V.; Aucelli, P.P.; Corrado, G.; Di Paola, G.; Matano, F.; Pappone, G.; Schiattarella, M. Comparing geological and Persistent Scatterer Interferometry data of the Sele River coastal plain, southern Italy: Implications for recent subsidence trends. *Geomorphology* **2020**, *351*, 106953. [CrossRef]
47. Shirzaei, M.; Walter, T. Estimating the Effect of Satellite Orbital Error Using Wavelet-Based Robust Regression Applied to InSAR Deformation Data. *IEEE Trans. Geosci. Remote Sens.* **2011**, *49*, 4600–4605. [CrossRef]
48. Shirzaei, M.; Burgmann, R. Topography correlated atmospheric delay correction in radar interferometry using wavelet transforms. *Geophys. Res. Lett.* **2012**, *39*, 01305. [CrossRef]
49. Vilardo, G.; Ventura, G.; Terranova, C.; Matano, F.; Nardò, S. Ground deformation due to tectonic, hydrothermal, gravity, hydrogeological, and anthropic processes in the Campania Region (Southern Italy) from Permanent Scatterers Synthetic Aperture Radar Interferometry. *Remote Sens. Environ.* **2009**, *113*, 197–212. [CrossRef]
50. Li, X.; Huang, G.; Kong, Q. Atmospheric Phase Delay Correction of D-InSAR Based on Sentinel-1A. *Int. Arch. Photogramm. Remote Sens. Spat. Inf. Sci.* **2018**, 955–960. [CrossRef]
51. Morrison, K.; Bennett, J.C.; Nolan, M.; Menon, R. Laboratory Measurement of the DInSAR Response to Spatiotemporal Variations in Soil Moisture. *IEEE Trans. Geosci. Remote Sens.* **2011**, *49*, 3815–3823. [CrossRef]
52. Morrison, K.; Bennett, J.C.; Nolan, M. Using DInSAR to Separate Surface and Subsurface Features. *IEEE Trans. Geosci. Remote Sens.* **2012**, *51*, 3424–3430. [CrossRef]
53. Molan, Y.E.; Lu, Z. Modeling InSAR Phase and SAR Intensity Changes Induced by Soil Moisture. *IEEE Trans. Geosci. Remote Sens.* **2020**, *58*, 4967–4975. [CrossRef]
54. Hensley, S.; Munjy, R.; Rosen, P. Interferometric Synthetic Aperture Radar (IFSAR). In *Digital Elevation Model Technologies and Applications: The DEM Users Manual*, 1st ed.; Maune, D.F., Ed.; American Society for Photogrammetry and Remote Sensing: Bethesda, MD, USA, 2001; pp. 143–206.
55. Shirzaei, M.; Freymueller, J.; Törnqvist, T.E.; Galloway, D.L.; Dura, T.; Minderhoud, P.S.J. Measuring, modelling and projecting coastal land subsidence. *Nat. Rev. Earth Environ.* **2021**, *2*, 40–58. [CrossRef]
56. Luo, Q.; Perissin, D.; Zhang, Y.; Jia, Y. L- and X-Band Multi-Temporal InSAR Analysis of Tianjin Subsidence. *Remote Sens.* **2014**, *6*, 7933–7951. [CrossRef]
57. Tosi, L.; Da Lio, C.; Strozzi, T.; Teatini, P. Combining L- and X-Band SAR Interferometry to Assess Ground Displacements in Heterogeneous Coastal Environments: The Po River Delta and Venice Lagoon, Italy. *Remote Sens.* **2016**, *8*, 308. [CrossRef]
58. Sánchez, L.; Seitz, M. Station positions and velocities of the SIR11P01 multi-year solution, epoch 2005.0. In Proceedings of the SIRGAS 2011 General Meeting, Heredia, Costa Rica, 8–11 August 2011. [CrossRef]
59. Stanley, D.J.; Warne, A.G. Worldwide Initiation of Holocene Marine Deltas by Deceleration of Sea-Level Rise. *Science* **1994**, *265*, 228–231. [CrossRef] [PubMed]
60. Teatini, P.; Tosi, L.; Strozzi, T. Quantitative evidence that compaction of Holocene sediments drives the present land subsidence of the Po Delta, Italy. *J. Geophys. Res. Space Phys.* **2011**, *116*, 08407. [CrossRef]
61. Colmenares, P.; Mesa, M.; Roncancio, J.; Pedraza, P.; Contreras, A.; Cardona, A.; Silva, C.; Romero, J.; Alvarado, S.; Romero, O.; et al. *Geología de La Plancha 18*; INGEOMINAS: Santafé de Bogotá, Colombia, 2007.

62. Leatherman, S.P.; Zhang, K.; Douglas, B.C. Sea level rise shown to drive coastal erosion. *Eos* **2000**, *81*, 55–57. [CrossRef]
63. Mazzotti, S.; Lambert, A.; Van Der Kooij, M.; Mainville, A. Impact of anthropogenic subsidence on relative sea-level rise in the Fraser River delta. *Geology* **2009**, *37*, 771–774. [CrossRef]
64. Catuneanu, O. *Principles of Sequence Stratigraphy*; Elsevier: Milan, Italy, 2006.
65. Mount, J.; Twiss, R. Subsidence, Sea Level Rise, and Seismicity in the Sacramento–San Joaquin Delta. *San Fr. Estuary Watershed Sci.* **2005**, *3*, 1–19. [CrossRef]
66. Pepe, A.; Calò, F. A Review of Interferometric Synthetic Aperture RADAR (InSAR) Multi-Track Approaches for the Retrieval of Earth's Surface Displacements. *Appl. Sci.* **2017**, *7*, 1264. [CrossRef]
67. Oliver-Cabrera, T.; Wdowinski, S. InSAR-Based Mapping of Tidal Inundation Extent and Amplitude in Louisiana Coastal Wetlands. *Remote Sens.* **2016**, *8*, 393. [CrossRef]

Article

Storm Driven Migration of the Napatree Barrier, Rhode Island, USA

Bryan A. Oakley

Department of Environmental Earth Science, Eastern Connecticut State University, Willimantic, CT 06226, USA; oakleyb@easternct.edu

Abstract: Napatree Point, an isolated barrier in southern Rhode Island, provides a case study of barrier spit migration via storm driven overwash and washover fan migration. Documented shoreline changes using historical surveys and vertical aerial photographs show that the barrier had little in the way of net change in position between 1883 and 1939, including the impact of the 1938 hurricane. The barrier retreated rapidly between 1945 and 1975, driven by both tropical and extra-tropical storms. The shoreline position has been largely static since 1975. The removal of the foredune during the 1938 hurricane facilitated landward shoreline migration in subsequent lower intensity storms. Dune recovery following the 1962 Ash Wednesday storm has been allowed due to limited overwash and barrier migration over the last several decades. Shoreline change rates during the period from 1945–1975 were more than double the rate of shoreline change between 1939 and 2014 and triple the rate between 1883 and 2014, exceeding the positional uncertainty of these shoreline pairs. The long-term shoreline change rates used to calculate coastal setbacks in Rhode Island likely underestimate the potential for rapid shoreline retreat over shorter time periods, particularly in a cluster of storm activity. While sea-level rise has increased since 1975, the barrier has not migrated, highlighting the importance of storms in barrier migration.

Keywords: barrier spit; overwash; barrier migration; shoreline change; hurricane; extra-tropical storm

Citation: Oakley, B.A. Storm Driven Migration of the Napatree Barrier, Rhode Island, USA. *Geosciences* **2021**, *11*, 330. https://doi.org/10.3390/geosciences11080330

Academic Editors: Gianluigi Di Paola, Germán Rodríguez, Carmen M. Rosskopf and Jesus Martinez-Frias

Received: 7 May 2021
Accepted: 31 July 2021
Published: 5 August 2021

1. Introduction

Coastal barrier spits and islands, which comprise approximately 10% of the world's coastline [1], are dynamic landforms that are impacted by storms and sea-level rise. Debate exists regarding the impacts of sea-level rise on shoreline change and barrier migration [2–6]. Over the millennia, sea-level rise and storms certainly drove the migration of shorelines across the continental shelf during the Holocene [5,7]. Barriers in the mid and northern Atlantic have largely been narrowing rather than migrating in the 20th century [8], and some authors argue the gradual recession of these areas is the result of sea-level rise, not storms [3,6]. Other authors have argued that roles of overwash and inlet dynamics in maintaining barrier systems drive shoreline changes at the decadal/century scale [5,9–11]. Difficulty remains in defining the direct cause and effect of sea-level rise and storms [12]. Gutierrez et al. [13] and Williams et al. [14] summarized likely responses of barriers to sea-level rise under various scenarios and concluded that it is virtually certain that barriers will experience morphological changes through erosion, overwash and washover fan deposition and the formation of inlets during storms. Subsequent research has focused on the various mechanisms related to barrier response to sea-level rise, including modelling studies focusing on the rate of sea-level rise, accommodation space and sediment availability [15,16] and the importance of tidal inlets and tidal deltas to barrier dynamics [17]. Understanding barrier dynamics in the face of potentially increased storminess (increased intensity) [18] and higher numbers of storms [19] is critical to understanding the response of barriers to forcings in the future.

This study documents the storm driven shoreline changes of the Napatree Point Conservation Area (NPCA), a 2.5 km headland bounded barrier spit located in southwest Rhode Island, extending across the mouth of the Little Narragansett Bay estuary (Figure 1). The barrier and adjacent headland are important for several key reasons: The barrier (1) Attenuates storm impacts (surge, waves) for the southern end of Little Narragansett Bay and adjacent shorelines and (2) provides a suitable habitat for several threatened or endangered species, including Piping Plovers (*Charadrius melodus*), American Oystercatchers (*Haematopus palliatus*) and Least Terns (*Sternula antillarum*). Napatree has been recognized by the Audubon Society as a Globally Important Bird Area [20]. (3) NPCA is managed as a conservation area and remains an important recreational destination with hundreds of visitors a day in the summer.

The objective of this study is to document historical changes in shoreline position and morphology of a barrier spit between 1883 and 2014 and examine the roles of storm frequency and magnitude on barrier migration. The erosion and subsequent recovery of foredunes is critical to understanding the style and rate of barrier retreat [21]. Increased storm frequency and sea-level rise are likely consequences of climate change [19], creating a feedback loop of increased frequency of overwash events and reduced periods of barrier recovery. This would lead to faster rates of shoreline migration as barriers potentially cross a geomorphic threshold [11,13,17]. Between 1938 and 1975 multiple storms (both tropical and extra-tropical) impacted Napatree Point, leading to overwash and 'roll-over' via washover fan deposition and migration of the barrier. The period before 1939 and after 1975 produced little to no net migration of the barrier, even with some moderate storm events. This suggests that the period of recovery was sufficient, such that the barrier was not overwashed in these events, even though similar storms overwashed the barrier in the past. The pattern of increased storm frequency with limited periods of recovery, even in the absence of rapid sea-level rise, provides at least a partial analog of future behavior of a barrier spit and furthers understanding of the importance of storm frequency in barrier migration.

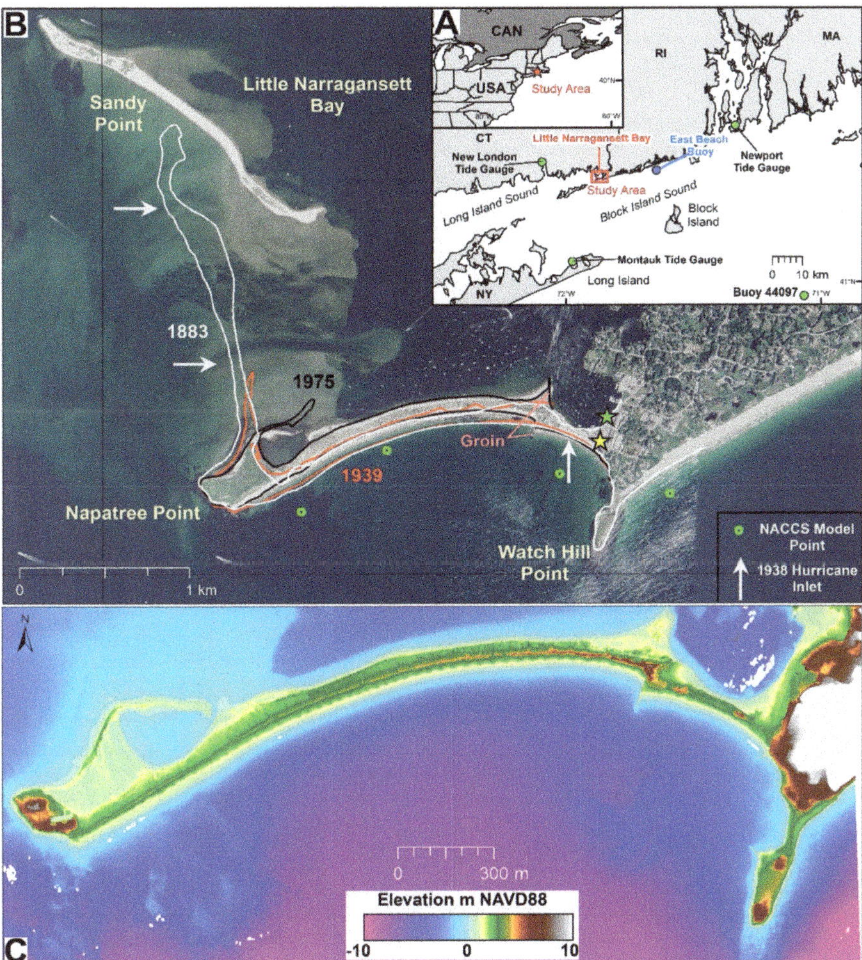

Figure 1. (**A**). Location map of Napatree Point. (**B**). A 2014 Digital orthophotograph [22] with past shoreline positions superimposed in 1883 (white), 1939 (red) and 1975 (black) [23–25]. Note that Sandy Point was connected to Napatree Point in 1883 but has migrated north since the 1938 hurricane. White arrows indicate the 1938 hurricane [26]. The yellow star shows the location that the photograph in Figure 5D was captured at. Green circles indicate the locations of the NACCS buoys discussed in the text. The green star shows the location of the U.S. Geological Survey water level gauge. (**C**). A 2018 LiDAR Topobathymetric model of Napatree point and the adjacent Watch Hill, elevation in meters above NAVD88 [27].

2. Materials and Methods

2.1. Study Area

The barrier spit portion of the NPCA is a welded barrier [28], bounded by Watch Hill Point to the east and Napatree Point to the west (Figure 1). The entirety of the point/barrier complex extends 2.5 km across Little Narragansett Bay. The barrier spit itself is 2 km long. It has an average width of 100 m and an average volume of 300 m^3 m^{-1}. Foredune heights range from 3.5 to 8 m MLLW (mean, 4.7 m MLLW), with a general increase in elevation from west to east. The headlands at each end of the barrier are composed of boulder diamict (glacial till) deposited as part of recessional end moraines during the Late Wisconsinan deglaciation [29,30]. Napatree Point headland ranges from 5 to 8 m in relief, and is unarmored, although the boulders eroded from the till bluffs at the far western end form a natural 'revetment' in the manner of FitzGerald et al. [31]. The eastern bluff (Watch Hill Point) is 5 to 6 m in relief along the shoreline; however, shoreline protection structures (sea walls and revetments) have been in place along much of this shoreline since at least 1939. Offshore of the headlands, cobble-gravel to boulder pavement extends approximately 500 m to a water depth here of 7 to 8 m. These boulder-gravel areas are interpreted to represent erosion of the glacial headlands as the shoreline retreated during transgression [32]. The shoreface offshore of the Napatree barrier is largely sandy, and sorted bedforms are visible on aerial photographs, Lidar and unpublished side-scan sonar imagery. One linear ridge of cobble-gravel to boulder pavement is visible in both the aerial imagery and lidar of Figure 1 along the east end of the barrier.

NPCA, like the rest of the Rhode Island South Shore (RISS), is a microtidal, wave-dominated shoreline in the classification of Hayes [33], and tidal range along the ocean shoreline (Block Island Sound) is 0.8 to 1.2 m. A U.S. Geological Survey gauge (Figure 1) installed in 2014 in the harbor behind the eastern end of the barrier does not report calculated tidal datums, but the range is similar to the open ocean conditions, suggesting tidal range is not significantly dampened within Little Narragansett Bay. Direct observations of wave conditions in Block Island Sound are limited, due to no long-term stations in the area. Approximately one year of observations between July 2010 and September 2011 reported that significant wave heights are typically <1 m with periods 5 to 11 s (see the location of the East Beach Buoy, Figure 1A) [34]. Fair weather waves approach predominantly from the southwest or south. Storm wave conditions during intense storms remain speculative; however, observations along the RISS in 10 m of water depth 18 km east of Napatree during Tropical Storm Irene (2011) measured significant wave heights of 4.1 m with a peak period of 10 s [34]. Modelled conditions during storms were calculated as part of the North Atlantic Coast Comprehensive Study (NACCS), and the station 200 m offshore of the Napatree barrier with approximately 6 m deep water (Figure 1) resulted in calculated the wave height for a storm with a 2-year return period of 4.7 m; the 10-year return period wave height was 5.4 m. The height was 5.7 m for a 100-year return period [35]. These values are lower for than the adjacent shoreline (east of Watch Hill Point, Figure 1), suggesting some reduction in wave energy relative to the RISS. Modelled wave heights (NACCS) are slightly lower at the east and west ends of the barrier; the calculated heights were 4.8 m for the east and west ends of the barrier during a 100-year return period storm. Two small groins were installed near the east end of the barrier, one on the southern (Block Island Sound) shoreline (100 m long, installed between 1948 and 1951) and one on the northern (Little Narragansett Bay) shoreline (160 m long, installed between 1945 and 1951) (Figure 1). Patterns of shoreline change suggest that longshore transport is towards the west on the southern shoreline and to the east on the northern shoreline. The longest sea-level record in southern New England is from the Newport, Rhode Island tide gauge (50 km northeast of Napatree (Figure 1)), with monthly mean sea-level values dating to 1930 (Figure 2). Linear regression through the data shows a reported rate of sea-level rise of 2.83 ± 0.16 mm yr^{-1} [36]. Two stations closer to Napatree with shorter records (New London, CT, 20 km northwest; 2.72 ± 0.21 mm yr^{-1} (1938–2020); Montauk, NY, 30 km southwest; 3.41 ± 0.25 mm yr^{-1} (1947–2020)) [37,38] show good correlations with the Newport gauge (see supplemental Figure S1) suggesting a regional rate of sea-level rise of ~3 mm yr^{-1} since 1930. Analysis of the record suggests the rate of sea-level rise at the Newport tide gauge has increased: between 1984 and 2014, it reached 4.1 mm yr^{-1} [39]. The 30 year trend suggested by Carey et al. [39] is consistent with the measured global rate of sea-level rise since 1993 (3.3 ± 0.4 mm yr^{-1}) [40] coupled with the vertical crustal motion of the region (-0.9 mm yr^{-1}) [41].

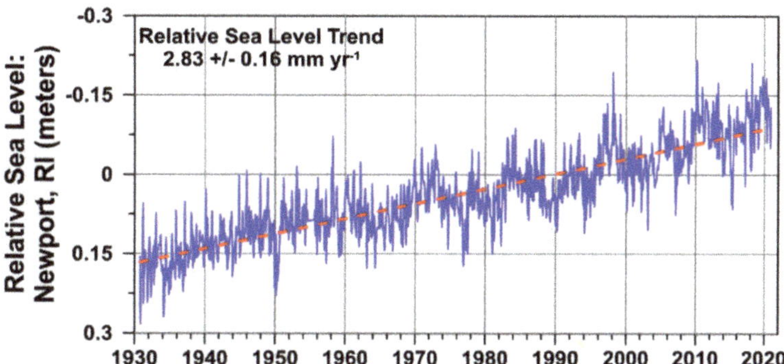

Figure 2. Monthly average sea-level at the Newport, RI tide gauge (blue line). The red dashed line is a linear fit through the data [36].

Previous studies have documented the historical shoreline change of the Rhode Island shoreline, including NPCA, the results of which are summarized in Table 1. Shoreline change values are reported here as negative when the shoreline migrated landward and positive when the shoreline migrated seaward. Boothroyd and Hehre [24] reported the overall shoreline change rates as the net movement between 1939 to 2004, with a mean of -1 m yr^{-1}. Hapke et al. [25] utilized T-sheet shorelines to extend the analysis to include the period between 1883 and 2004, and based on a linear regression of the shorelines, the average shoreline change was similar at -0.8 m yr^{-1}. Hapke et al. [25] also reported a short-term shoreline change between 1975 and 2000 of +0.4 m yr^{-1}. Boothroyd et al. [23] and Boothroyd and Hehre [24] reported the highest rates of shoreline change. Boothroyd et al. [23] calculated the rate using the most seaward shoreline in their study (1939) and the most landward shoreline (1975) based on the shoreline change envelope calculation in the Digital Shoreline Analysis System (DSAS), and the average rate for Napatree barrier was -1.4 m yr^{-1}. Boothroyd and Hehre [24] reported a similar rate between 1939 and 1985 (Table 1).

Table 1. Shoreline change rates for the Napatree barrier (transects 20 to 50) from previously published shoreline change maps (Boothroyd and Hehre et al. [24]; Boothroyd et al. [23]; Hapke et al. [25]) and this study.

Years	Method	Average Shoreline Change Rate (m yr^{-1})	Source
1939–1985	End Point Rate	−1.4	Boothroyd and Hehre, [24]
1985–2004	End Point Rate	0.0	Boothroyd and Hehre, [24]
1939–2004	End Point Rate	−1.0	Boothroyd and Hehre, [24]
1883–2004	Linear Regression	−0.8	Hapke et al. [25]
1975–2000	End Point Rate	0.4	Hapke et al. [25]
1939–2014	End Point Rate	−0.9	Boothroyd et al. [23]
1939–2014	Shoreline Change Envelope	−1.4	Boothroyd et al. [23]
1883–1939	End Point Rate	−0.2	This study
1939–1975	Linear Regression	−1.9	This study
1939–1975	End Point Rate	−1.9	This study
1975–2014	Linear Regression	0.2	This study
1975–2014	End Point Rate	0.1	This study
1883–2014	Linear Regression	−0.6	This study
1883–2014	End Point Rate	−0.6	This study

2.2. Storm History

Detailed summaries of the storm history of Southern New England and adjacent Long Island, NY [42–45] coupled with historical storm tracks (1860 and 2018) [46] and the water-level record at the Newport, RI tide gauge provided the basis for compiling the storm history for the NPCA (Figure 4). Ten hurricanes made landfall within 200 km of

Napatree during the study period (1883–2014) (Figure 3). Prior to the first shoreline map (1883), a hurricane made landfall 20 km east of Napatree in 1869; however, little damage was reported west of landfall (Ludlum, 1963). Category 1 hurricanes passed offshore or well to the east of Napatree in 1916, 1924 and 1936; one tropical storm made landfall in 1936 near the track of the 1938 hurricane [25,46]. The storm of record for the area is the hurricane of 1938, making landfall on 21 September 1938 approximately 200 km west of Napatree, placing the top right quadrant of the storm along the southern New England coastline. The 1938 hurricane caused widespread erosion of shorelines from Long Island, NY to Cape Cod, MA [26,47,48]. Storm surge measured at the Newport tide gauge was 2.9 m above MHHW. Impacts of the 1938 storm along the Rhode Island shoreline are summarized by Nichols and Marston [26] and Brown [48]. Napatree barrier was breached near the eastern end (Figures 1 and 5), forming an inlet which was subsequently filled by the town of Westerly [26]. The Sandy Point barrier, which extended north off the western end of Napatree Point, was also breached, and the formation of the inlet initiated the 1.5 km northern migration and rotation of Sandy Point barrier to the current location (Figure 1). The period from the 1938 hurricane to the Ash Wednesday storm of 1962 (inclusive) included two landfalling hurricanes (hurricane of 1944—surge elevation 1.2 m MHHW; and Hurricane Carol, 1954—surge elevation 2.1 m MHHW). Hurricane Carol (Figure 5) overwashed barriers along the Rhode Island south shore [49,50]. Other notable storms include the Ash Wednesday Hurricane of 1962 [51], the 'Blizzard of 1978' [52], Hurricane Gloria (1985) [53], Hurricane Bob (1991) [54], The Halloween Nor'Easter (1991) [55], a December 1992 extra-tropical storm [56], the Patriot's Day extra-tropical storm (April 2007) [57], Tropical storm Irene (2011) and Superstorm Sandy (2012) [58].

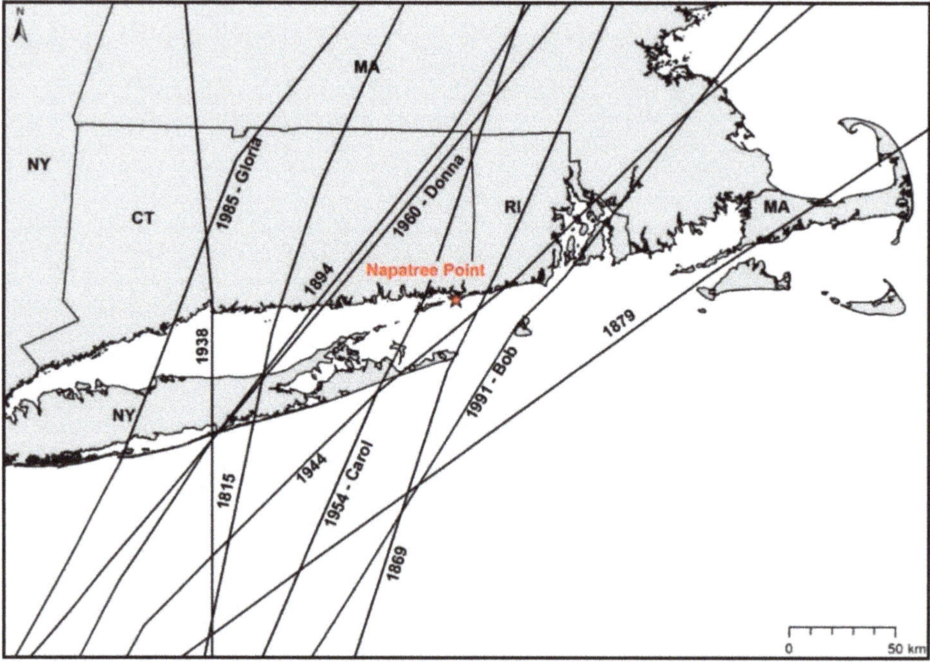

Figure 3. Storm tracks of hurricanes that made landfall within 200 km of Napatree Point (Storm tracks from NOAA Digital Coast).

The review of historical storms was supplemented with an analysis of water levels at the Newport tide gauge. Zhang et al. [6,59] identified storms where the water level exceeded two standard deviations, and this approach was modified slightly to use here. Analysis of the water levels at the Newport tide gauge showed two standard deviations is 0.8 m MHHW; however, this threshold value excluded several notable storms (Blizzard of 1978, December 1992), so the value was lowered to 0.7 m MHHW. The selected water level coincides with a storm with a return period of 4 to 5 years based on the annual exceedance probability curves for the Newport tide gauge [60]. Water levels at the Newport tide gauge exceeded 0.7 m MHHW 38 times between 1930 and 2018 (Figure 4). The storm water levels are summarized in Table S1 in the supplemental materials.

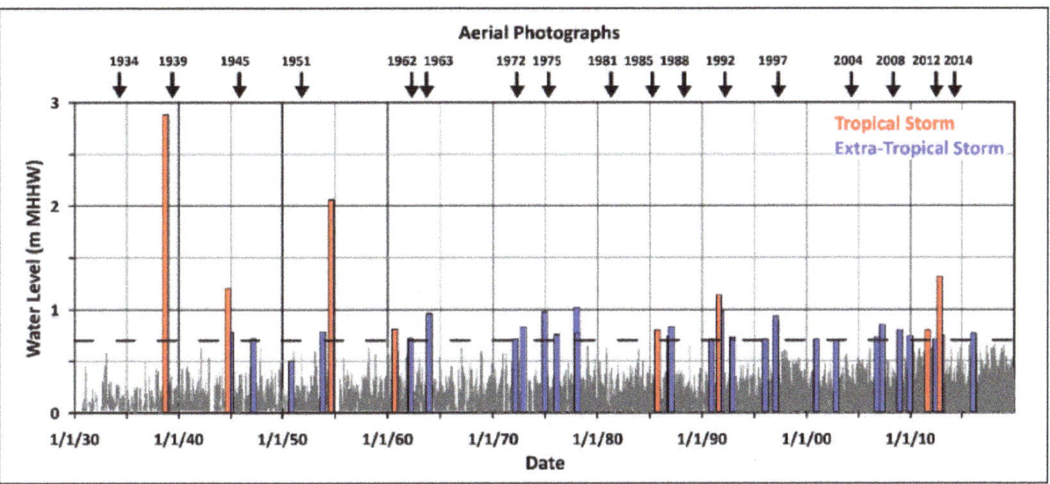

Figure 4. Water elevation levels for notable storms discussed in this study. The gray lines represent the hourly water levels at the Newport, RI tide gauge (Figure 1) [61]. The dashed black line is the threshold for storm detection used in this study (0.7 m MHHW). Red bars represent tropical storms; blue represent extra-tropical storms. Black arrows show the dates of aerial photographs (Figure S6; supplemental materials).

2.3. Shoreline Change Data

The primary data used to document changes in position of the Napatree barrier were historical high-water line shoreline positions [24,25,63] (Table 2). Shorelines were derived from T-Sheets (1883 and 1948), georeferenced vertical aerial photographs (1939, 1951, 1963, 1975, 1985 and 1992) and digital orthophotographs (2004, 2012 and 2014). Shoreline positions and annualized end-point rates using previously published shorelines [24,25,63] were reanalyzed at transects spaced 50 m alongshore relative to an onshore baseline. Reanalysis involved the inclusion of the 1883 and 1948 T-Sheets, which were not used in the original study by Boothroyd et al. [23], and examining individual shoreline pairs (e.g., 1883 and 1939; 1939 and 1948). Shoreline positional uncertainty was assessed in the original mapping by Hapke et al. (2010); Boothroyd and Hehre (2007); and Boothroyd et al. (2016). Total positional uncertainty (U_p) of a shoreline at each transect was calculated in these studies as the square root of the sum of five relevant uncertainty terms squared (Equation (1) for NOS T-Sheets; Equation (2) for historical aerial photographs)

$$U_p = \sqrt{(U_g^2 + U_d^2 + U_t^2 + U_{pd}^2)} \quad (1)$$

$$U_p = \sqrt{(U_g^2 + U_d^2 + U_a^2 + U_{pd}^2)} \quad (2)$$

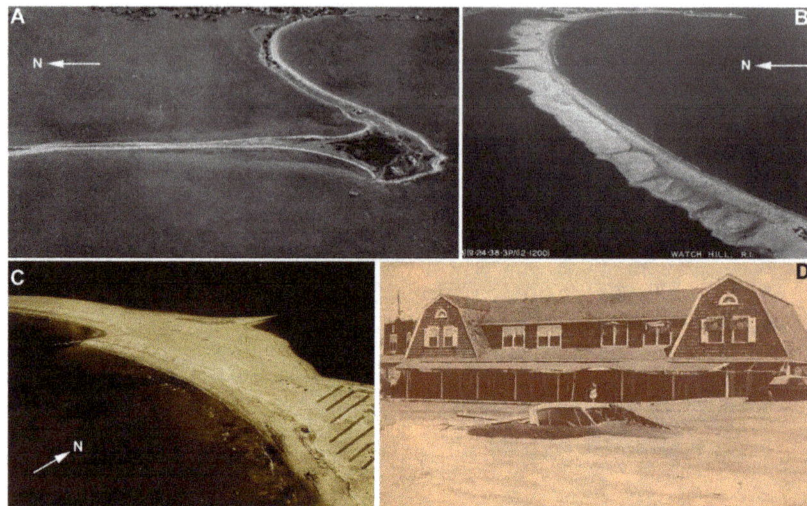

Figure 5. (**A**). Oblique aerial photograph of the Napatree Barrier and Sandy Point Barrier taken in 1936. Note the beach cottages and the presence of dune vegetation along the barrier. (**B**). Oblique aerial photograph of the Napatree Barrier taken on 24 September 1938, 3 days after the 1938 hurricane made landfall. North is to the left of both images, (**A**,**B**). Note the inlet at the east (far) end of the barrier and numerous surge channels across the barrier. (**C**). An oblique aerial photograph taken after Hurricane Carol (1954) of the eastern end of the NPCA (note the two groins installed between 1948 and 1951) (**D**). Car buried in a washover fan on the east end of the Napatree Barrier during Hurricane Carol (1954). This photograph is just out of frame to the east of Figure 5C. See Figure 1 for the location of the photograph [62]. View is towards the east.

Table 2. NOS T-Sheets (NOS-T), aerial photographs (AP) and digital orthophotos (DO) used in this study. Dates with an asterisk denote that a dataset was used to digitize a shoreline position. Time since storm indicates the number of days between the photograph/survey and last time the threshold water level (0.7 m MHHW) was exceeded at the Newport tide gauge.

Type	Date	Time Since Storm (Days)
NOS-T	1883 *	N/A
AP	4/15/1934	N/A
AP	5/15/1939 *	236
AP	10/15/1945	395
NOS-T	1948 *	>365
AP	11/15/1951 *	1718
AP	4/15/1962	39
AP	9/7/1963 *	549
AP	4/15/1972	56
AP	4/14/1975 *	133
AP	4/15/1981	1164
AP	3/22/1985 *	2601
AP	4/15/1988	448
AP	3/15/1992 *	136
AP	4/15/1997	95
AP	4/15/2004 *	526
AP	4/15/2008	365
AP	6/15/2012 *	292
AP	4/14/2014 *	473
AP	4/15/2018	796

These relevant terms include: georeferencing uncertainty (U_g), digitizing uncertainty (U_d), T-sheet uncertainty (U_t), aerial photograph uncertainty (U_a) and uncertainty in the position of the high-water line (U_{pd}) [25]. The total positional uncertainties for these shorelines are reported in Table 3. The uncertainty of the end point change between two shorelines (U_{pair}) was taken the square of the sum of positional uncertainty of the two shorelines (Equation (3)), and these were calculated as annualized rates (U_r) by dividing that results by the number of years between shoreline surveys (Equation (4)). Linear regression was applied to the shoreline positions for comparison with the long-term (1883–2004) shoreline change rates of Hapke et al. [23]. Barrier width was measured at each transect as the straight-line distance between the last high-tide swash shoreline of the bayside and ocean sides of the barrier.

$$U_{pair} = \sqrt{}(U_{p1}{}^2 + U_{p2}{}^2) \quad (3)$$

$$U_r = U_{pair}/(\text{year}_2 - \text{year}_1) \quad (4)$$

Table 3. Reported measurement and total positional uncertainties reported for the shorelines used in this analysis. Sources notes by the superscript next to the date. [1] Hapke et al. [25]. [2] Boothroyd and Hehre [24]. [3] Boothroyd et al. [23].

		Measurement Uncertainties				Meters	
Year	Type	Georeferencing (U_g)	Digitizing (U_d)	T-Sheet Survey (U_t)	Air Photo (U_a)	HWL (U_{pd})	Total Shoreline Position Uncertainty
1883 [1]	NOS T-Sheet	4	1	10		4.5	11.7
1939 [2]	VAP	3.6	1	-	3	4.5	6.6
1948 [1]	NOS T-Sheet	4	1	10		4.5	11.7
1951 [2]	VAP	3.7	1	-	3	4.5	6.6
1963 [2]	VAP	3.7	1	-	3	4.5	6.6
1975 [2]	VAP	2	1	-	3	4.5	5.9
1985 [2]	VAP	3.3	1	-	3	4.5	6.4
1992 [2]	VAP	3.1	1	-	3	4.5	6.3
2004 [2]	DO	-	1	-	3	4.5	5.5
2012 [3]	DO	-	1	-	3	4.5	5.5
2014 [3]	DO	-	1	-	1	4.5	4.7

2.4. Historical Aerial Photographs

Historical vertical and oblique aerial photographs, including photographs not used in the original shoreline mapping [23–25], were analyzed (Table 2). These photographs were not added to the shoreline database due to a combination of scale, photo quality (i.e., ability to distinguish a high-water line), poor georeferenced accuracy or high wave energy at the time of image collection; however, the images were still suitable for qualitative analysis of barrier morphology and vegetation cover. Vertical aerial images were georeferenced in ESRI ArcMap to the 2014 U.S. Geological Survey digital orthophotographs for qualitative comparison, and images were examined to document historical changes in geomorphology—notably, the presence of washover fans and/or the presence of dune vegetation. Additionally, context on pre- and post-1938 hurricane conditions was determined from oblique aerial photography (Figure 5).

3. Results

3.1. Shoreline Change

Analysis of subsequent shoreline pairs between 1883 and 2014 shows that the rate and magnitude of shoreline change on the Napatree varied over time, particularly following a major storm event that removed the foredune and overwashed the entire barrier. Shoreline change rate was lower and even slightly accretional during a subsequent period of foredune recovery. The results here show the changes between transects 20 and 50 located on the barrier and outside the influence of the groins. The average changes, maximum change, minimum change and annualized rate of change are summarized in Table 4. Transects 51 to 60, while located on the barrier, were not considered in the analysis, as the width has been altered by the installation of the groins and the construction of a concrete seawall on the bayside of the barrier adjacent to the parking lot.

Table 4. Average change in shoreline position, maximum change, minimum change, average annualized rate, positional uncertainty values (U_{pair}) and rate uncertainty values (U_r) for subsequent shoreline pairs on the Napatree barrier for transects 20–50.

Shoreline Pair	Average Change (m)	Max (m)	Min (m)	Annualized Rate (m yr^{-1})	U_{pair} (m)	U_r (m)
1883–1939	−9.4	−19.0	−1.0	−0.2	13.4	0.3
1939–1948	−24.1	−30.7	−13.3	−2.7	13.4	1.5
1948–1951	−13.4	−22.8	−1.5	−4.5	13.4	4.5
1951–1963	−21.9	−38.9	−13.2	−1.8	9.4	0.8
1963–1975	−10.1	−15.7	−0.7	−0.8	8.9	0.7
1975–1985	2.8	10.2	−1.7	0.3	8.7	0.9
1985–1992	7.6	12.7	3.7	1.1	9	1.3
1992–2004	−4.9	−11.7	−1	−0.4	8.4	0.7
2004–2014	3.1	7.8	−5.5	0.3	7.2	2.3

The shoreline change data (1883–2014) (Table 4) show that retreat of the Napatree barrier varied over time, with little change between 1883 and 1939, followed by landward shoreline migration between 1939 and 1975 and another period of little to no net change in position between 1975 and 2014. Figure 6 shows the shoreline positions for the entire Napatree Barrier over time. Figure 7 represents the average position of the barrier. Average landward migration of the shoreline here is reported as the change in distance from the baseline rather than the annualized rate. The error bars shown represent the standard deviation of the distance from the baseline for an individual shoreline between transects 20 and 50. These standard deviations are similar in magnitude to the shoreline positional uncertainty (see supplemental Table S2 for standard deviations).

The average retreat of the Napatree barrier between 1883 and 1939 was −9.4 m, which falls within the positional uncertainty of the shoreline pair (+/−13.4 m) (Table 4). Individual transects reported values ranging from essentially no change (+/−<3 m) to more than −20 m of change near the western end of the barrier (Figure 6). Examining the sequential shoreline pairs beginning in 1939 shows the highest overall amount of landward migration occurred between 1939 and 1948, with an average of −24.1 m of change along the barrier (−2.7 m yr^{-1}) and individual transects recording changes of more than −30 m. This was followed by average retreat of −13.4 m between 1948 and 1951 (the same value as the uncertainty of the shoreline pair), although some transects showed >22 m of retreat. A groin was installed between 1948 and 1951 on each side of the barrier at the east end (Figure 1) and appears to have impacted the shoreline position between transects 51 and 56 (Figure 6). The shoreline continued to retreat between 1951 and 1963. The average migration was 18 m (−1.5 m yr^{-1}). Western portions of the barrier retreated > −35 m. Retreat of the shoreline continued between 1963 and 1975. The average change was −10.1 m, and parts of the barrier retreated > 15 m. Overall, between 1939 to 1975 the shoreline retreated an average of −69.5 m, and some transects saw > −77 m of change. The average annualized shoreline change rate was −1.9 m yr^{-1}. These rates are substantially higher than other previously published rates of shoreline change (Table 1) [23–25]. The 1975 shoreline represents the most landward shoreline recorded. The shoreline position has fluctuated between 1975 and 2014 (largely within the uncertainty); however, the 2014 shoreline remains seaward of the 1975 shoreline along most of the barrier (Figures 6 and 7), suggesting the shoreline has been neutral to slightly progradational since 1975. These results match the reported short-term change between 1975 and 2000 (+0.28 m yr^{-1}; Table 2) derived from aerial photographs and airborne LiDAR as part of the national shoreline assessment [25].

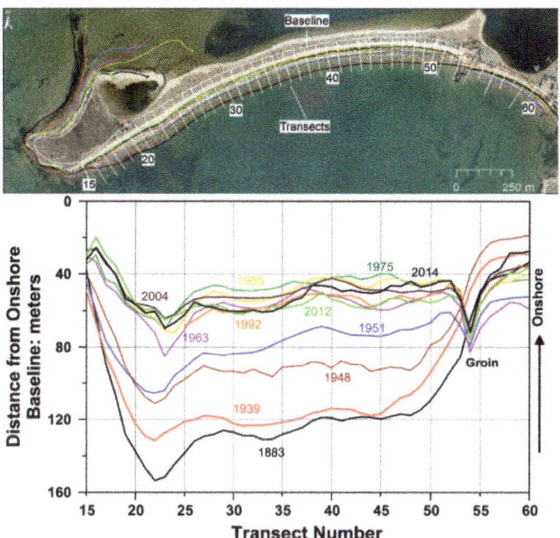

Figure 6. (**Top**) Digital orthophotograph of the Napatree Barrier showing the location of the transects used in this study. (**Bottom**) Distance from the onshore baseline for transects 15–60 for Napatree Point. The barrier begins at transect 20. Note the migration of the shoreline between 1939 and 1975, and the lack of change in position between 1975 and 2014. The apparent progradation of the shoreline at transect 54 represents the installation of the small groin between 1948 and 1951. The impact of the groin on shoreline position appears to be limited to transects 51 to 56.

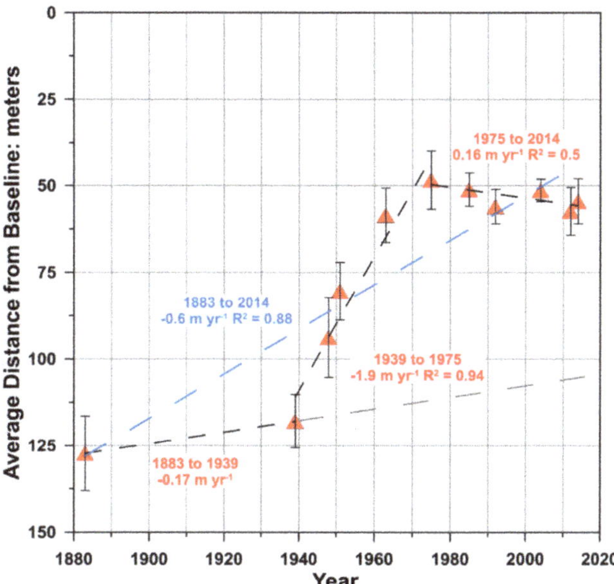

Figure 7. Average position of the southern shoreline of Napatree Barrier 1883–2014 expressed as distance from the onshore baseline. Triangles show the average position of the barrier (transects 20 to 50). Error bars are based on the standard deviation in distance from the baseline for that shoreline reported in Table 4. The dashed blue line represents a linear best fit through the average positions (1883–2014). The dashed gray line represents the continuation of the 1883–1939 trend.

3.2. Barrier Width

The variations in barrier width largely do not fall outside of the shoreline position uncertainties, so these results coupled with the aerial photographs should be viewed as qualitative. The average width of the barrier varied over time, ranging from <80 m to >105 m (Figure 8). The narrowest transects were measured in 1883 at transects 39–44 (average 56 m); the widest transects occurred near the groins near the eastern end of the barrier, >220 m at transect 53 in 2011 and 2014. Averaging the width of the barrier between transects 20 and 50 (outside groin influence) showed an increase in width between 1883 and 1963 and a slight decrease in width after 1963. In addition, there was an increase in width between 2004 and 2014 (red triangles, Figure 8). The width of the barrier at transects 26–30 is impacted by the migration and eventual welding of the small spit on the bayside of the barrier (visible in Figures 1, 6 and 9 and the supplemental materials); this is most apparent in the 2011–2014 shorelines. This resulted in an anomalously wide barrier in the 2011–2014 shorelines. The widening at transects 26–30 was not related to overwash and washover fan deposition, which is the inferred process that widened the barrier between 1939 and 1963. Excluding these transects (26–30) shows a decrease in barrier width between 1963 and 2014 (Figure 8).

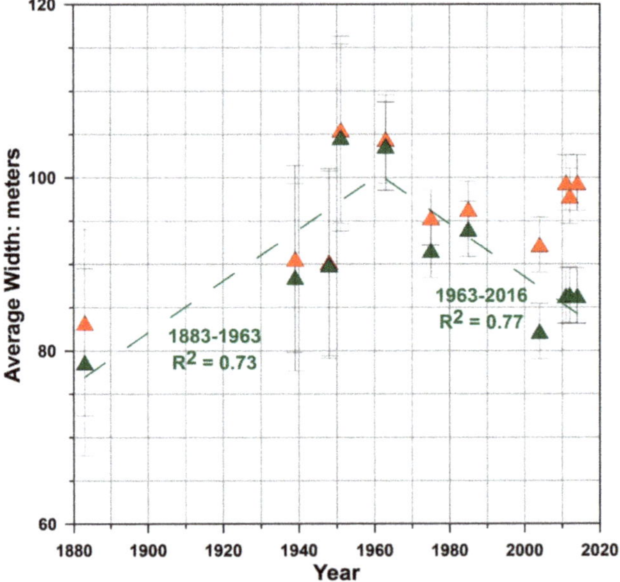

Figure 8. Average width of the barrier based on all transects (red triangles) and transects 20–25 and 31–50 (green triangles). Transects 26–30 were excluded because the dynamics of the small spit on the western bayside of the barrier (see Figure 1, Figure 6, Figure 9 and supplemental materials) alter the barrier width of the 2004–2018 shorelines. See the text for a more detailed discuss of why transects were excluded. The dashed green line represents linear best fit of the data between 1883–1963 and 1963–2016. Error bars are based on the shoreline position uncertainty reported in Table 3.

3.3. A Review of the Aerial Photography

Historical vertical aerial photographs (n = 19) from between 1934 and 2018 were examined for evidence of storm impacts. Figure 9 shows selected aerial photographs from 1934 (partial coverage of 830 m of the east end of the barrier), 1939, 1951, 1963, 1975, 1985, 1992 and 2014. All aerial photographs are available in the online supplemental materials. The road and houses that extended down the barrier are clearly visible in the 1934 (Figure 9) vertical aerial and 1936 oblique images (Figure 5A). Dune vegetation is visible between and behind houses, and while foredune height and continuity cannot be determined from these images, the 1936 oblique image (Figure 5A) shows that a single foredune ridge existed on the barrier, which was particularly well-formed along the western end and in front of some of the cottages. No evidence of recent overwash is visible in these photos. An oblique aerial photograph of the barrier was taken 3 days after the 1938 hurricane [64] and shows complete overwash of the barrier (Figure 5B) with numerous surge channels visible.

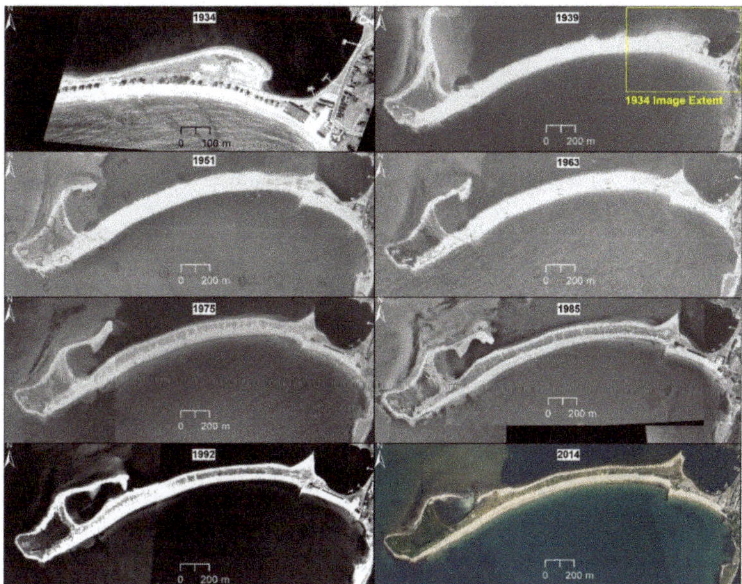

Figure 9. Napatree barrier 1934 through 2014. Only the eastern portion of the barrier was imaged in 1934. Note the houses and vegetation visible in 1934. Images shown from 1939, 1951 and 1963 all show evidence of recent washover fan deposition. Images from 1975 to 2014 show vegetation reestablished along the barrier. The 1992 images show some overwash on the western half of the barrier, but not full removal of the dunes. See the supplemental materials for all aerial images used in this study.

The first complete set of vertical aerial photographs of Napatree Point was collected in May 1939, 8 months after the hurricane of 1938 (Figure 9). The 1939 images show fresh (unvegetated) washover fans deposited across the barrier and into Little Narragansett Bay. Penetration distance across the barrier of the overwash is >100 m. The inlet formed in the eastern corner of the barrier (Figure 1) [26] had been filled in by this point; however, the inlet along the southern end of Sandy Point (Figure 1) remained open, and Sandy Point had begun to migrate to the north. Following the 1938 hurricane, the next photographs were collected in October 1945, one year after a hurricane in 1944. These photos also show fresh (unvegetated) washover fans along most of the barrier (See Supplemental Materials). Some patchy dune vegetation is visible on the western 100 m and eastern 500 m of the barrier, suggesting that some portions of the barrier may not have been overtopped completely. This pattern of some vegetation surviving along the western and eastern ends of the barrier was repeated in the November 1951 photographs (Figure 9), which were collected 1 year after a November 1950 extra-tropical storm. Vertical aerial images were not collected following Hurricane Carol (1954), so the direct assessment of impacts on Napatree Point Conservation Area could not be determined; however, significant overwash was photographed in the parking lot and road adjacent to the east end of the barrier (Figure 5C) [62].

April 1962 images collected 39 days after the Ash Wednesday storm show near complete overwash of the barrier, with again the exception being the eastern 500 m of the barrier. Images collected in September 1963 still show evidence of overwash (Figure 9); however, more dune vegetation is visible as compared to the 1962 images, suggesting that the barrier was beginning to recover from the Ash Wednesday storm. The next images available were collected in April 1972, and these show dune vegetation along the entire barrier, although some areas may show some minor washover fan deposition, possibly relating to an extra-tropical storm in February 1972 extra-tropical storm. April 1975 and 1976 (exact date unknown) imagery shows some unvegetated washover fans, although penetration distance across the barrier was a few 10s of meters, and largely focused near trails crossing the dune (Figure 9). These washover fans are likely related to extra-tropical storms in November 1972 and December 1974. The pattern continues in images collected in 1981, which show some slight overwash in the center of the barrier with penetration depth < 30 m. These images were collected three years after the Blizzard of 1978, which caused extensive erosion and overwash along much of the northeast [52]. Images taken in 1985 show no evidence of overtopping or overwash, and dune vegetation is visible along the entire barrier. April 1988 images show some slight overwash at trails crossing the dune, likely the result of Hurricane Gloria (September 1985), which produced washover fans elsewhere along the RISS [53].

March 1992 imagery shows moderate overwash of the western half of the barrier, with penetration of less than half the width of the barrier (30 to 40 m). These images were collected after Hurricane Bob (August 1991) and the Halloween Nor'easter (October 1991), so the impact of each individual storm cannot be parsed out; however, this shows the cumulative impact of these storms. Images collected after 1992 (1995, 1997, 2004) show no evidence of washover fan deposition. June 2012 images collected 292 days after Tropical Storm Irene (August 2011), show one small (25 m wide, penetration distance < 15 m) washover fan, although evidence of frontal erosion of the dune is seen as a retreat in the edge of vegetation. No vertical aerial photographs were collected following Hurricane Sandy (October 2012) for Napatree Point; however, oblique aerial images from March 2013 show the western half of the barrier was overwashed, and he eastern half largely showed evidence of frontal erosion of the dunes of up to 13 m (see 2014 images in the supplemental materials). Oblique aerial images collected by the author in August 2013 show little in the way of visible washover fans, and field observations at that time show rapid revegetation of the limited areas that were overwashed. Two sets of images collected in 2014 by the USGS (April) and NOAA (August) show little evidence of the Sandy overwash and most areas that saw washover fan deposition were revegetated, except for some isolated patches near the west end of the barrier (Figure 9).

4. Discussion

The combination of a historical aerial photograph review and shoreline change analysis provides a semi-quantitative summary of the impacts of storms on the Napatree Barrier. This combination of datasets allows for a review of historical storm events and the resulting impacts on the barrier. The results highlight the importance of storm timing and frequency relative to the timescale of foredune recovery. Based on storm surge elevation and regional coastal impacts, the most substantial storm to impact Napatree and the shoreline change analysis during the study period, was the 1938 hurricane. Washover fans are clearly visible in both oblique (Figures 5 and 9) and vertical aerial photographs following the 1938 hurricane. The post-storm photographs and 1939 vertical aerial images show complete removal of the foredune (and houses) along the barrier. While the historical photographs show clear impacts to the barrier, the net retreat of the shoreline over this period is less clear. Individual transects between 1883 and 1939 show changes ranging from essentially no change (+/−<3 m) to >20 m of retreat near the western end of the barrier (Figure 6). The average change along the barrier between 1883 and 1939 was −9.4 m. This falls within the positional uncertainty of the shoreline pair (+/−13.4 m; Table 4), so it cannot be said conclusively that the hurricane of 1938 produced net migration of the shoreline. The spring 1939 photographs were taken 236 days after the storm, so it is likely the active beach had largely recovered, and the high-water line position would not be considered storm impacted. The 1938 hurricane was likely the first time the barrier had been overwashed in at least several decades, so the dunes were likely well developed prior to the storm, and a vegetated foredune is visible in the 1934 and 1936 images (Figures 5 and 9).

Aerial photographs were not collected between 1939 and the 1944 hurricane, so the condition of the barrier and levels of dune recovery and revegetation cannot be assessed. Dune recovery when a barrier is completely overwashed can take up to a decade [65], so complete recovery following the 1938 hurricane was unlikely prior to the 1944 hurricane. Vertical aerial photographs from 1945 show that Napatree was overwashed presumably during the 1944 hurricane. The 1945 imagery was not used as part of the shoreline change analysis, as the photos were 'washed out,' and delineating a high-water line was problematic. Between 1939 and 1948, the shoreline retreat exceeded the positional uncertainty of the shorelines, with an average change of −24.1 m and a maximum change of −30.7 m (Table 4). The measured shoreline change, coupled with the washover fans visible in the 1945 imagery suggests that the 1944 hurricane resulted in a net migration of the barrier. A similar response was seen in 1951 imagery and the 1948 to 1951 shoreline pair. Fresh washover fans are visible in 1951 imagery, and the shoreline change between 1948 and 1951 averaged −13.4 m (the same value as the positional uncertainty); however, some transects exceeded −22 m of change (Table 4). Water levels at the Newport gauge did not exceed the storm threshold between 1948 and 1951. An extra-tropical storm in late November 1950 produced a 0.85 m storm surge; however, the storm peaked after high tide, so the maximum water level recorded was 0.5 m MHHW at the Newport tide gauge. Water levels were higher to the west of Napatree, peaking at 1.4 m MHHW at New London and 0.9 m MHHW at Montauk, suggesting that impacts of this storm were more substantial in western Rhode Island and eastern Connecticut. Elsewhere, the November 1950 extra-tropical storm produced washover fans and caused substantial damage along the New Jersey coastline [43]. Following the hurricane of 1944 coupled with an extra-tropical storm in 1947, foredune recovery was limited and the barrier remained low enough to be overwashed during this storm.

Aerial images were not collected between 1951 and 1962, so the impacts observed over that period are the cumulative results of Hurricanes Carol and Edna (1954), Hurricane Donna (1960), an extra-tropical storm in 1953 and the Ash Wednesday Storm (1962), among other smaller extra-tropical storms. As a result, the impacts of individual storms cannot be parsed out; however, the net result was migration of the barrier that exceeded the positional uncertainty. A shoreline was not derived from the 1962 shoreline, as it was considered too soon after the Ash Wednesday storm (39 days); however, between 1951 and 1963 the shoreline retreated an average of −21.9 m, with a maximum change of −38.9 m (Table 4). The timing of these storms following the 1950 extra-tropical storm continued to hinder foredune recovery, and each storm event that overwashed the barrier essentially reset the clock on dune recovery. The shoreline continued to retreat between 1963 and 1975 (average change −10.1 m; maximum change −15.7 m); however, the aerial photography shows that vegetation was largely reestablished at this point and migration via overwash had largely ceased, and it appears the barrier was narrowing. Barrier width increased between 1883 and 1963 and decreased between 1963 and 2016. The combination of shoreline change data and observations from the aerial imagery suggests that the increase in width was likely driven by the overwash of the barrier and deposition of washover fans on the back-barrier. Increased width via overwash and washover fan deposition was also noted in field surveys after the 1938 hurricane [26].

While the 1938 hurricane produced appeared to produce little net change in shoreline position compared to 1883, storms' impacts on the barrier were the impetus for the migration of the barrier that occurred between 1939 and 1975. The removal of the foredune increased the susceptibility of the barrier to overtopping in future storms, which allowed subsequent storms that likely would not have overwashed the barrier to overtop the partially recovered foredune, and deposit washover fans on the back barrier, 'rolling over' the barrier. The combination of field and LiDAR measurements provides insight into the elevation recovery from a moderate event (Hurricane Sandy). The majority of the Napatree barrier just exceeds the pre-Sandy volume and elevation 5.5 years after the storm, suggesting recovery from this event where the foredune was not completely removed took ~5 years [65]. Most of the vegetation was reestablished on the back barrier within 1 year following Sandy (as seen in 2014 aerial images and 2013 field surveys), and this pattern was likely repeated in smaller past storms, where washover was limited to a few 10s of meters across the barrier and dune vegetation can reestablish quickly.

No storm exceeded the threshold at the Newport tide gauge between November 1963 and December 1974. This 11-year period apparently allowed the dunes to recover enough elevation and volume to prevent wide-spread overwash of the barrier in subsequent moderate storm events. Based on the rates of recovery of the dune measured since Sandy [65], the foredune could have recovered ~1 m of elevation and ~30 m^3 m^{-1} of volume during that period. This is supported by work on other barriers, where recovery periods of <5 years to >10 years have been reported following storms that completely overwash the barrier [64,66,67]. It remains unclear if management practices (i.e., sand fences) were utilized in the study area in the past, and if so, how they impacted dune formation and recovery between 1963 and 1975. Sand fences have clearly been used along the eastern end of the barrier over the last few decades, with multiple levels to build the foredune up to 8 m (MLLW) just west of the groins. The overall increase in dune height from west to east along the barrier was likely driven by a combination of some sand fencing and sediment availability and the prevailing wind direction; the dominant wind directions are southwest (spring/summer) and northwest (fall/winter), both of which transport sand towards the eastern end of the spit. This likely explains the partial recovery and revegetation of the eastern 500 m of the barrier prior to 1975, which limited overwash in these areas. There is no record of beach replenishment at this site; the exception to this is filling of the inlet breach at the eastern end of the barrier following the 1938 hurricane.

Subsequent storms, including the Blizzard of 1978, which caused extensive erosion elsewhere in New England, did not appear to cause widespread overwash, and no localized overwash was apparent in 1981 aerial images (See supplemental materials). A similar response was observed for Hurricane Gloria (1985), which overwashed other areas of the RISS [53], and the combined impact of Hurricane Bob and the October 1991 extra-tropical storm, which had limited overwash of the dunes and localized washover fan penetration of 10 s of meters. Hurricane Sandy (October 2012) had a similar storm surge elevation as the 1944 hurricane, yet it in little change in position of the barrier. Overwash was limited to the western portion of the barrier, washover fans only extended across the barrier in a few locations and the eastern portion of the barrier remained largely in the collisional regime of Sallenger [68]. The 1944 hurricane had a greater impact because the barrier (specifically the foredune) had not recovered from the 1938 hurricane; the same storm event, if the foredune had fully recovered, would likely not have produced the same impacts on the barrier. Houser et al. [64,66] showed that a more erosive storm being followed by smaller storms can produce a larger cumulative impact and result in substantial shoreline retreat. The 1938 to 1975 period falls into the 'best-case' scenario of Houser et al. [64], where storms were clustered, followed by periods of recovery. Under these conditions, transgression is rapid, and dunes are small and discontinuous during the periods of storminess; however, the subsequent period of quiescence allows for the dunes to reestablish, and eventually return to pre-storm height [64]. Fenster and Dolan [7,69] reported a reversal in shoreline change trends, from a seaward migration to more landward migration between 1930 and 1970. This period of increased shoreline change was attributed to storm frequency—notably, an increase in extra-tropical storm frequency. The trend reversed (switched from erosional to accretional) around 1967–1968 [69], which falls between available aerial photographs for Napatree (1963 and 1972/1975). Donnelly et al. [43] reported a similar response to what occurred with the storms discussed here—overwash and washover fan deposition on back barrier salt marshes of Brigantine Island in New Jersey (250 km west of Napatree) in 1944,

1950 and 1962. This suggests that the paraglacial isolated and welded Napatree barrier behaved similarly to the barrier island chains along the Mid-Atlantic coast of the U.S.

Shoreline change rates calculated between 1939 and 1975 were higher than previously published for the Napatree Barrier (Table 1). The annualized rates calculated by Boothroyd et al. [23] between 1939 and 2014 and the USGS long term rate between 1883 and 2004 [25] are less than half the annualized rate observed at Napatree between 1939 and 1975 (Table 1). The USGS assessment [25] also included a short-term shoreline change rate for the period between 1975–2000, with reported shoreline change values reflecting progradation of the shoreline. The average (end point) rate was +0.4 m yr^{-1} for that period (Table 1). Taken together, this suggests that the long-term shoreline change rates underestimate actual rates of change, which are an order of magnitude higher following a large storm event. While long-term shoreline change rates may produce a better mean representation of the long-term trend [70], the observed shoreline migration between 1939 and 1975 are indicative of the change that can occur during a shorter interval with several impactful storms. This does not suggest that the storms are outliers; rather, if the shoreline migration occurs during a cluster of storms, the annualized rate of change is reduced by spreading the change out over a longer time with limited storm activity.

Coastal construction setbacks in Rhode Island are based on the shoreline change rates between 1939 and 2014. Specifically, the setback was calculated as 30× the long-term erosion rate. These rates do not include additional components to account for storm impacts [71]. The results of this work show that on this shoreline these rates are likely underestimated if a barrier experiences a series of significant storm events and can exceed the long-term trend here by a factor of three or more. Zhang et al. [6] argued that including storm-influenced shorelines in long-term shoreline change calculations leads to an overestimation of rates of shoreline change and could have consequences on coastal development based on setbacks. However, where storms drive the observed rates of change and dominate the shoreline change signal, the inverse would potentially be of more consequence, as setbacks calculated using low rates of change fail to calculate the actual risk to coastal properties over time. While the Napatree barrier is currently undeveloped, a hypothetical house constructed using the average long-term (1883–2014) shoreline change rate (-0.6 m yr^{-1}) would face substantially more risk if a similar storm sequence that was observed beginning in 1938 were to occur. Separating the shoreline response to storms from the long-term shoreline change record and understanding the risk to coastal development remain challenges for coastal managers. Numerous studies have outlined the likely response of barriers to sea-level rise [12–15,17,47]. These impacts include an increased rate of landward migration, more frequent overwash with the rising sea level and increased storminess, increased frequency of breaching and inlet formation/widening. The period between 1938 and 1975 provides at least a partial analog for the potential responses of other barrier spits within the glaciated northeast to a period of increased storminess. Headland separated barriers occur elsewhere within New England in both wave dominated and mixed energy regimes, and along Nova Scotia and Northern Germany [28]. Understanding the responses of these types of barrier systems to storms and a rising sea level fills an important gap in knowledge, as much of the existing literature focuses on barrier islands, particularly along the mid-Atlantic coast of the U.S.

A conclusive cause and effect summary linking all the observed shoreline changes to individual storms is not possible given the positional uncertainty and short-term changes observed. The lack of randomization, replication and direct measurements following storms also limit direct cause and effect attribution of the shoreline change to individual storms. However, the combination of shoreline change analysis and historical aerial photography here suggests that much of the change observed was the result of washover fan deposition during storms, which led to migration of the barrier. Extrapolating the shoreline trend between 1883 and 1939 (dashed grey line in Figure 7), which is admittedly limited to two data points, or the period between 1975 and 2014, suggests that the subsequent shoreline positions have not returned to the long-term trend. The interpretation here is that the storms produced net migration of the shoreline. If the change in the shoreline was driven by some other long-term factor (i.e., sea-level rise or changes in sediment supply), the shoreline should return to the long-term trend following storms [6]. Higher rates of sea-level rise over the last 40 years [39] do not coincide with any measurable shoreline retreat, and even correspond with a period of progradation (within the positional uncertainty) between 1975 and 2014. The responses of barriers under various sea-level rise scenarios have been the focus of recent modelling efforts, and the dynamics of the foredune have been shown in models to impact barrier retreat—namely, when the dunes are low, barriers are more susceptible to overwash, leading to migration of the barrier [21]. This leads to episodic retreat of the barrier, followed by periods of relative stability in barrier positions when the foredune recovers sufficiently to limit overwash. These results, both shown here at Napatree, reinforce the relationship between dune recovery and storm frequency reported in recent modelling studies [21].

Discussion of storm impacts here does not negate the impact of sea-level rise on barriers; however, the cumulative impact of storms between 1938 and 1975 likely far outweighs any response due to sea-level rise. The transgression observed over the span of a few decades could be indicative of future behavior of headland separated barriers if storm frequency increases in the future. Relative sea-level rise increases the vulnerability of the dunes to erosion, and this would exacerbate the impacts of a similar stormy period in the future. This combination of storms and sea-level rise could cause the barrier to cross a geomorphic threshold and lead to faster rates of transgression. Under moderate to high rates of sea-level rise and given the relatively consistent (low) back-barrier slope, the Napatree barrier could experience 'width drowning' in which the rapid overwash and migration of the barrier would outpace sediment transport from the shoreface, leading to barrier narrowing and possible barrier loss over the next few centuries [15]. Napatree, like the rest of the RISS, is sediment-starved, with little modern sediment on the shoreface [32]. Incision into the underlying glacial deposits is possible; however, this process has not yet been performed for this shoreline. Due to lacking an abundant sediment supply, Napatree is susceptible to increased storm frequency/overwash. An additional sediment added to the barrier is from either erosion of the small bluff at the east end and/or transport from the shoreface and back barrier during transgression. Transport of the sediment down the Pawcatuck River into Little Narragansett Bay is likely not a significant contributor of sediment to the Napatree Barrier [72].

What the morphology of the Napatree barrier will look like if continued transgression causes the spit to detach from the western headland remains unclear. A possible response may be for the barrier to rotate clockwise to a more northeast to southwest orientation, similarly to the migration of the Sandy Point barrier (Figure 1). This would somewhat follow the model of Orford et al. [73], where barriers between glacial till headlands are breached, and switch from swash-aligned ones to ones where the barrier is oblique to the headland and rapid migration is driven by longshore sediment transport. Future research should focus on coring and geophysical studies to examine the properties of the shoreface, and observations of physical processes coupled with models would be helpful to quantify sediment transport pathways on the shoreface. Sediment transport on the northern side of the barrier, including exchange between the barrier and deposits related to the former position of Sandy Point, and sediment lost from the system around the west end of Napatree Point, remains unknown. While Napatree lacks significant back barrier marsh and is not part of a restricted coastal lagoon where loss of marsh due to accelerated sea-level rise may significantly alter tidal characteristics [17], inlet formation and widening could affect sediment distribution, including the transport of sediment from the barrier and adjacent shoreface to a flood-tidal delta [11]. Transport of sediment into the tidal delta represents loss to the subaerial barrier system, as the sediment remains there until transgression of the barrier reaches the tidal delta [11]. This loss of sediment would be problematic for a starved system, particularly if an inlet remained open long enough to develop a flood-tidal delta. The overall response of the barrier here is illustrative of the challenges barriers face in periods of increased storminess, and the importance of storm frequency relative to dune recovery. It represents an important case study when considering storms' impacts on isolated, mainland attached barriers in future climate models.

Supplementary Materials: The following are available online at https://www.mdpi.com/article/10.3390/geosciences11080330/s1. Figure S1: Monthly sea-level elevations at the Newport, RI, New London, CT and Montauk, NY tide gauges. Figure S2: 1934 Partial vertical aerial photograph of the Napatree Barrier. Figure S3: 1939 Vertical aerial photograph of the Napatree Barrier. Figure S4: 1945 Vertical aerial photograph of the Napatree Barrier. Figure S5: 1951 Vertical aerial photograph of the Napatree Barrier. Figure S6: April 1962 Vertical aerial photograph of the Napatree Barrier. Figure S7: September 1963 Vertical aerial photograph of the Napatree Barrier. Original 9 × 9 Photograph scanned by Boothroyd and Hehre. Figure S8: April 1972 Vertical aerial photograph of the Napatree Barrier. Figure S9: April 1975 Vertical aerial photograph of the Napatree Barrier. Original 9 × 9 Photograph scanned by Boothroyd and Hehre. Figure S10: 1976 Vertical aerial photograph of the Napatree Barrier. Figure S11: April 1981 Vertical aerial photograph of the Napatree Barrier. Figure S12: March 1985 Vertical aerial photograph of the Napatree Barrier. Original 9 × 9 Scanned by Boothroyd and Hehre, 2007. Figure S13: April 1988 Vertical aerial photograph of the Napatree Barrier. Figure S14: March 1992 Vertical aerial photograph of the Napatree Barrier. Figure S15: Spring 1997 Digital Orthophotograph of the Napatree Barrier. Figure S16: April 2004 Digital Orthophotograph of the Napatree Barrier. Figure S17: Spring 2008 Digital aerial photograph of the Napatree Barrier. Figure S18: June 2012 Digital Orthophotograph of the Napatree Barrier. Figure S19: April 2014 Digital Orthophotograph of the Napatree Barrier. Figure S20: April 2018 Digital Aerial Photograph of the Napatree Barrier. Table S1: Water levels exceeding the 0.7 m MHHW threshold at the Newport, RI and New London tide gauges. Table S2: Average distance from baseline and standard deviation of shoreline position used in this analysis. See Table 3 for the sources of the shorelines.

Funding: This research received no external funding. Publication costs were provided via a Connecticut State University-American Association of University Professors Faculty Research Grant. The Watch Hill Conservancy provided in-kind support for field surveys.

Acknowledgments: This manuscript benefited from numerous discussions over the years regarding shoreline change mapping and processes, including Mark Borrelli, Rachel Henderson, Janet Freedman, Robert Hollis and Scott Rasmussen. Mark Borelli, Peter August, Janet Freedman and Nathan Vinhateiro, along with two anonymous reviewers provided helpful reviews of the manuscript.

Conflicts of Interest: The authors declare no conflict of interest. The funders had no role in the design of the study; in the collection, analyses, or interpretation of data; in the writing of the manuscript or in the decision to publish the results.

References

1. Stutz, M.L.; Pilkey, O.H. Open-ocean barrier islands: Global influence of climatic, oceanographic, and depositional settings. *J. Coast. Res.* **2011**, *27*, 207–222. [CrossRef]
2. Fenster, M.S.; Dolan, R.; Morton, R.A. Coastal storms and shoreline change: Signal or noise? *J. Coast. Res.* **2001**, *17*, 714–720.
3. Leatherman, S.; Zhang, K.; Douglas, B. Sea level rise shown to drive coastal erosion. *EOS Trans. Am. Geophys. Union* **2000**, *81*, 55–57. [CrossRef]
4. Pilkey, O.H.; Young, R.S.; Bush, D.M. Comment on "Sea level rise shown to drive coastal erosion". *EOS Trans. Am. Geophys. Union* **2000**, *81*, 436. [CrossRef]
5. Sallenger, A.H., Jr.; Morton, R.; Fletcher, C.; Thieler, E.R.; Howd, P. Comment on "Sea level rise shown to drive coastal erosion". *EOS Trans. Am. Geophys. Union* **2000**, *81*, 436. [CrossRef]
6. Zhang, K.; Douglas, B.; Leatherman, S. Do storms cause long-term beach erosion along the U.S. east barrier coast? *J. Geol.* **2002**, *110*, 493–502. [CrossRef]
7. Morton, R.A. Historical changes in the Mississippi-Alabama barrier-island chain and roles of extreme storms, sea level and human activities. *J. Coast. Res.* **2008**, *24*, 1587–1600. [CrossRef]
8. Leatherman, S.P. Migration of Assateague Island, Maryland, by inlet and overwash processes. *Geology* **1979**, *7*, 104–107. [CrossRef]
9. Morton, R.A. Factors controlling storm impacts on coastal barriers and beaches—A preliminary basis for near real-time forecasting. *J. Coast. Res.* **2002**, *18*, 486–501.
10. Morton, R.A.; Sallenger, A.H., Jr. Morphological impacts of extreme storms on sandy beaches and barriers. *J. Coast. Res.* **2003**, *19*, 560–573.

11. FitzGerald, D.M.; Hein, C.J.; Hughes, Z.; Kulp, M.; Georgiou, I.; Miner, M.D. Runaway Barrier Island Transgression Concept: Global Case Studies. In *Barrier Dynamics and Response to Changing Climate*; Moore, L.J., Murray, A.B., Eds.; Springer: Cham, Switzerland, 2018; pp. 3–56.
12. Williams, S.J. Sea-level rise implications for coastal regions. *J. Coast. Res.* **2013**, *63*, 184–196. [CrossRef]
13. Gutierrez, B.T.; Williams, S.J.; Thieler, E.R. *Potential for Shoreline Changes due to Sea-Level Rise along the U.S. Mid-Atlantic Region: U.S. Geological Survey Open-File Report 2007-1278*; U.S. Geological Survey: Reston, VA, USA, 2007.
14. Williams, S.J.; Gutierrez, B.T. Sea-level rise and coastal change: Causes and implications for the future of coasts and low-lying regions. *Shore Beach* **2009**, *77*, 13–21.
15. Ashton, A.D.; Lorenzo-Trueba, J. Morphodynamics of Barrier Response to Sea-Level Rise. In *Barrier Dynamics and Response to Changing Climate*; Moore, L.J., Murray, A.B., Eds.; Springer: Cham, Switzerland, 2018; pp. 277–304.
16. Lorenzo-Trueba, J.; Ashton, A.D. Rollover, drowning, and discontinuous retreat: Distinct modes of barrier response to sea-level rise arising from a simple morphodynamic model. *J. Geophys. Res. Earth Surf.* **2014**, *119*, 779–801. [CrossRef]
17. FitzGerald, D.M.; Fenster, M.S.; Argow, B.A.; Buynevich, I.V. Coastal Impacts Due to Sea-Level Rise. *Annu. Rev. Earth Planet. Sci.* **2008**, *36*, 601–647. [CrossRef]
18. Bender, M.A.; Knutson, T.R.; Tuleya, R.E.; Sirutis, J.J.; Vecchi, G.A.; Garner, S.T.; Held, I.M. Modeled Impact of Anthropogenic Warming on the Frequency of Intense Atlantic Hurricanes. *Science* **2010**, *327*, 454–458. [CrossRef]
19. Emanuel, K.A. Downscaling CMIP5 climate models shows increased tropical cyclone activity over the 21st century. *Proc. Natl. Acad. Sci. USA* **2013**, *110*, 12219–12224. [CrossRef] [PubMed]
20. National Audubon Society. National Audubon Society Important Bird Areas in the U.S.: Napatree Point/Sandy Point. Available online: https://www.audubon.org/important-bird-areas/napatree-pointsandy-point (accessed on 21 September 2020).
21. Reeves, I.R.B.; Moore, L.J.; Murray, A.B.; Anarde, K.A.; Goldstein, E.B. Dune Dynamics Drive Discontinuous Barrier Retreat. *Geophys. Res. Lett.* **2021**, *48*, e2021GL092958. [CrossRef]
22. RIGIS. April 2014 Rhode Island Statewide High Resolution Orthoimages. In *Rhode Island Geographic Information System (RIGIS) Data Distribution System. Environmental Data Center*; University of Rhode Island: Kingston, Jamaica, 2015.
23. Boothroyd, J.C.; Hollis, R.J.; Oakley, B.A.; Henderson, R. *Shoreline Change Maps for Washington County Rhode Island Depicting Shoreline Change from 1939–2014*; Rhode Island Geological Survey: Kingston, RI, USA, 2016.
24. Boothroyd., J.C.; Hehre, R.E. *Shoreline Change Maps for the South Shore of Rhode Island*, Map Folio 2007-2 ed.; Rhode Island Geological Survey: Kingston, RI, USA, 2007.
25. Hapke, C.J.; Himmelstoss, E.A.; Kratzmann, M.G.; List, J.H.; Thieler, E.R. *National Assessment of Shoreline Change: Historical Shoreline Change along the New England and Mid-Atlantic Coasts*; US Geological Survey: Reston, VA, USA, 2010.
26. Nichols, R.L.; Marston, A.F. Shoreline changes in Rhode Island produced by hurricane of 21 September 1938. *Bull. Geol. Soc. Am.* **1939**, *50*, 1357–1370. [CrossRef]
27. USACE. 2018 USACE NCMP Topobathy Lidar: East Coast. 2020. Available online: https://www.fisheries.noaa.gov/inport/item/55881 (accessed on 3 August 2021).
28. FitzGerald, D.M.; Van Heteren, S. Classification of paraglacial barrier systems: Coastal New England, USA. *Sedimentology* **1999**, *46*, 1083–1108. [CrossRef]
29. Schafer, J.P. Surficial Geologic Map of the Watch Hill quadrangle, Rhode Island-Connecticut. In *U.S. Geological Survey Geological Quadrangle Map GQ-410*; US Geological Survey: Reston, VA, USA, 1965.
30. Stone, J.R.; Shafer, J.P.; London, E.H.; DiGiacomo-Cohen, M.; Lewis, R.S.; Thompson, W.B. *Quaternary Geologic Map of Connecticut and Long Island Sound Basin.: U.S. Geological Survey Geologic Investigations Series Map I-2784, Scale 1:125,000, 2 Sheets and Pamphlet*; US Geological Survey: Reston, VA, USA, 2005; pp. 1–72.
31. FitzGerald, D.M.; Baldwin, C.T.; Ibrahim, N.A.; Sands, D.R. Development of the northwestern Buzzards Bay Shoreline, Massachusetts. In *Glaciated Coasts*; FitzGerald, D.M., Rosen, P.S., Eds.; Academic Press: San Diego, CA, USA, 1987.
32. Oakley, B.A.; Murphy, C.; Varney, M.; Hollis, R.J. Spatial Extent and Volume of the Shoreface Depositional Platform on the Upper Shoreface of the Glaciated Rhode Island South Shore. *Estuaries Coasts* **2019**, 1–20. [CrossRef]
33. Hayes, M.O. Barrier island morphology as a function of tidal and wave regime. In *Barrier Islands from the Gulf of St. Lawrence to the Gulf of Mexico*; Leatherman, S.P., Ed.; Academic Press: New York, NY, USA, 1979; pp. 1–27.
34. WHG. *Wave, Tide and Current Data Collection, Washington County, Rhode Island: Report to the U.S. Army Corps of Engineers by the Woods Hole Group*; US Army Corps of Engineers: New England District, MA, USA, 2012.
35. USACE. *North Atlantic Coast Comprehensive Study: Resilient Adaptation to Increasing Risk*; Main Report; US Army Corps of Engineers: New England District, MA, USA, 2015.
36. NOS. Mean Sea Level Trend Station 8452660 Newport, Rhode Island. Available online: https://tidesandcurrents.noaa.gov/sltrends/sltrends_station.shtml?stnid=8452660 (accessed on 20 May 2021).
37. NOS. Mean Sea Level Trend Station 8461490 New London, CT. Available online: https://tidesandcurrents.noaa.gov/sltrends/sltrends_station.shtml?id=8461490 (accessed on 20 May 2021).
38. NOS. Mean Sea Level Trend Station 8510560 Montauk, New York. Available online: https://tidesandcurrents.noaa.gov/sltrends/sltrends_station.shtml?id=8510560 (accessed on 20 May 2021).
39. Carey, J.C.; Moran, S.B.; Kelly, R.P.; Kolker, A.S.; Fulweiler, R.W. The declining role of organic matter in New England salt marshes. *Estuaries Coasts* **2017**, *40*, 626–639. [CrossRef]

40. Beckley, B.; Ray, R.; Zelensky, N.; Lemoine, F.; Yang, X.; Brown, S.; Desai, S.; Mitchum, G. *Global Mean Sea Level Trend from Integrated Multi-Mission Ocean Altimeters TOPEX/Poseidon Jason-1 and OSTM/Jason-2 Version 5*; NASA Physical Oceanography DAAC: Pasadena, CA, USA, 2020. [CrossRef]
41. Karegar, M.A.; Dixon, T.H.; Engelhart, S.E. Subsidence along the Atlantic Coast of North America: Insights from GPS and late Holocene relative sea level data. *Geophys. Res. Lett.* **2016**, *43*, 3126–3133. [CrossRef]
42. Boose, E.R.; Chamberlin, K.E.; Foster, D.R. Landscape and regional impacts of historical hurricanes in New England. *Ecol. Monogr.* **2001**, *71*, 27–48. [CrossRef]
43. Donnelly, J.; Bryant, S.S.; Butler, J.; Dowling, J.; Fan, L.; Hausmann, N.; Newby, P.; Shuman, B.; Stern, J.; Westover, K.; et al. A backbarrier overwash record of intense storms from Brigantine, New Jersey. *Mar. Geol.* **2004**, *210*, 107–121. [CrossRef]
44. Ludlum, D.M. *Early American Hurricanes*; American Meteorological Society: Boston, MA, USA, 1963; p. 198.
45. Pore, N.A.; Barrientos, C.S. *Storm Surge*; New York SeaGrant Institute: Stony Brook, NY, USA, 1976.
46. NOAA. *Historical Hurricane Tracks*; NOAA Office for Coastal Management: North Charleston, SC, USA, 2020; Volume 2020.
47. Ashton, A.D.; Donnelly, J.P.; Evans, R.L. A discussion of the potential impacts of climate change on the shorelines of the Northeastern USA. *Mitig. Adapt. Strateg. Glob. Chang.* **2008**, *13*, 719–743. [CrossRef]
48. Brown, C.W. Hurricanes and shore-line changes in Rhode Island. *Geogr. Rev.* **1939**, *29*, 416–430. [CrossRef]
49. Donnelly, J.P.; Bryant, S.S.; Butler, J.; Dowling, J.; Fan, L.; Hausmann, N.; Newby, P.; Shuman, B.; Stern, J.; Westover, K. 700 yr sedimentary record of intense hurricane landfalls in southern New England. *Geol. Soc. Am. Bull.* **2001**, *113*, 714–727. [CrossRef]
50. Shaw, A.; Hashemi, M.R.; Spaulding, M.; Oakley, B.; Baxter, C. Effect of Coastal Erosion on Storm Surge: A Case Study in the Southern Coast of Rhode Island. *J. Mar. Sci. Eng.* **2016**, *4*, 85. [CrossRef]
51. Dolan, R. The Ash Wednesday Storm of 1962: 25 Years Later. *J. Coast. Res.* **1987**, *3*, 2–5.
52. Leatherman, S.; Zaremba, R.E. Overwash and aeolian processes on a U.S. north.east coast barrier. *Sediment. Geol.* **1987**, *52*, 183–206. [CrossRef]
53. Boothroyd, J.C.; Gibeaut, J.C.; Dacey, M.F.; Grant, J.A.; Blais, A.G.; Pickart, D.S.; Gricus, C.; Szak, C. The geological impact of Hurricane Gloria: South shore of Rhode Island. In *Proceedings of the Geological Society of America*; Geological Society of America: Boulder, CO, USA, 1986; p. 5.
54. Cheung, K.F.; Tang, L.; Donnelly, J.P.; Scileppi, E.M.; Liu, K.B.; Mao, X.Z.; Houston, S.H.; Murnane, R.J. Numerical modeling and field evidence of coastal overwash in southern New England from Hurricane Bob and implications for paleotempestology. *J. Geophys. Res. Earth Surf.* **2007**, *112*, F3. [CrossRef]
55. FitzGerald, D.M.; Van Heteren, S.; Montello, T.M. Shoreline Processes and Damage Resulting from the Halloween Eve Storm of 1991 along theNorth and South Shores of Massachusetts Bay, U.S.A. *J. Coast. Res.* **1994**, *10*, 113–132.
56. Butman, B.; Sherwood, C.R.; Dalyander, P.S. Northeast storms ranked by wind stress and wave-generated bottom stress observed in Massachusetts Bay 1990–2006. *Cont. Shelf Res.* **2008**, *8*, 1231–1245. [CrossRef]
57. Marrone, J.F. Evaluation of impacts of the Patriots' Day storm (15–18 April 2007) on the New England coastline. In *Solutions to Coastal Disasters 2008*; American Society of Civil Engineers: Reston, VA, USA, 2008; pp. 507–517.
58. Blake, E.S.; Kimberlain, T.B.; Berg, R.J.; Cangialosi, J.P.; Beven, J.L., II. *Tropical Cyclone Report Hurricane Sandy (AL182012); 22–29 October 2012*; National Hurricane Center: Miami, FL, USA, 2013.
59. Zhang, K.; Douglas, B.C.; Leatherman, S.P. Twentieth-Century Storm Activity along the U.S. East Coast. *J. Clim.* **2000**, *13*, 1748–1761. [CrossRef]
60. NOS. *Annual Exceedance Probability Curves for Station 8452660 Newport, RI (Newport Tide Gauge)*; National Ocean Service: Silver Spring, MD, USA, 2021.
61. NOAA. Verified Water-Level Data at the Newport, RI Tide Gauge (Station 8452660). 2020. Available online: https://tidesandcurrents.noaa.gov/stationhome.html?id=8452660 (accessed on 20 May 2021).
62. Hale, S.O. *Hurricane Carol Lashes Rhode Island*; Providence Journal Co.: Providence, RI, USA, 1954; p. 80.
63. USAAC. 1938 Hurricane Damage Photos: 23–24 September 1938, 118th Photographic Section of the U.S. *Army Air Corps and the 43rd Division of the Connecticut National Guard. Connecticut State Library Aerial Photograph Collection*. 1938. Available online: https://libguides.ctstatelibrary.org/hg/aerialphotos (accessed on 7 March 2020).
64. Houser, C.; Barrineau, P.; Hammond, B.; Saari, B.; Rentschler, E.; Trimble, S.; Wernette, P.; Weymer, B.; Young, S. Role of the Foredune in Controlling Barrier Island Response to Sea Level Rise. Barrier Dynamics and Response to Changing Climate. In *Barrier*; Moore, L.J., Murray, A.B., Eds.; Springer: Berlin/Heidelberg, Germany, 2018; pp. 175–207.
65. Oakley, B.A. Response and recovery of welded barrier: Napatree Point Conservation Area following 'Superstorm' Sandy measured using LiDAR and field surveys. In Proceedings of the American Shore and Beach Preservation Association, Long Beach, CA, USA, 13–16 October 2020.
66. Houser, C.; Hamilton, S. Sensitivity of post-hurricane beach and dune recovery to event frequency. *Earth Surf. Process. Landf.* **2009**, *34*, 613–628. [CrossRef]
67. Morton, R.A. Texas Barriers. In *Geology of Holocene Barrier Island Systems*; Davis, R.A., Jr., Ed.; Springer: New York, NY, USA, 1994; pp. 74–114.
68. Sallenger, A.H., Jr. Storm Impact Scale for Barrier Islands. *J. Coast. Res.* **2000**, *16*, 890–895.
69. Fenster, M.S.; Dolan, R. Large-scale reversals in shoreline trends along the U.S. mid-Atlantic coast. *Geology* **1994**, *22*, 543–546. [CrossRef]

70. Dolan, R.; Fenster, M.S.; Holme, S.J. Temporal Analysis of Shoreline Recession and Accretion. *J. Coast. Res.* **1991**, *7*, 723–744.
71. RICRMC. *Rhode Island Coastal Resources Management Program: Rhode Island Coastal Resources Management Council;* Rhode Island Coastal Resources Managament Council: Wakefield, RI, USA, 1995.
72. Boothroyd, J.C.; Friedrich, N.E.; McGinn, S.R. Geology of microtidal coastal lagoons: Rhode Island. *Mar. Geol.* **1985**, *63*, 35–76. [CrossRef]
73. Orford, J.D.; Carter, R.W.G.; Jennings, S.C. Coarse clastic barrier environments: Evolution and implications for quaternary sea level interpretation. *Quat. Int.* **1991**, *9*, 87–104. [CrossRef]

 geosciences

Article

Geomorphologic Recovery of North Captiva Island from the Landfall of Hurricane Charley in 2004

Emma Wilson Kelly and Felix Jose *

Department of Marine & Earth Sciences, The Water School, Florida Gulf Coast University, Fort Myers, FL 33965, USA; ewkelly8615@eagle.fgcu.edu
* Correspondence: fjose@fgcu.edu

Abstract: Hurricane Charley made landfall on the Gulf Coast of Florida on 13 August 2004 as a category 4 hurricane, devastating North Captiva Island. The hurricane caused a breach to occur to the southern end of the island, which naturally healed itself over the course of three years. By 2008, the cut was completely repaired geomorphologically. LiDAR data analysis shows the northern half of the island has been subjected to persistent erosion from 1998–2018, while the southern half experienced accretion since 2004, including the complete closure of the "Charley cut". The maximum volume of sediment erosion in the northern sector of the island (R71–R73) from 2004–2018 was −85,710.1 m^3, which was the source of southern accretion. The breached area of the island (R78b–R79a) obtained 500,163.9 m^3 of sediments from 2004–2018 to heal the cut made by Hurricane Charley. Along with LiDAR data analysis, Google Earth Pro historical imageries and SANDS volumetric analysis confirmed the longshore transport of sediments from the northern to the southern end of the island. Winter storms are mainly responsible for this southerly longshore transport and are hypothesized to be the main factor driving the coastal dynamics that restored the breach and helps in widening the southern end of North Captiva Island.

Keywords: North Captiva Island; barrier island; Hurricane Charley; sediment transport

Citation: Kelly, E.W.; Jose, F. Geomorphologic Recovery of North Captiva Island from the Landfall of Hurricane Charley in 2004. *Geosciences* **2021**, *11*, 358. https://doi.org/10.3390/geosciences11090358

Academic Editors: Gianluigi Di Paola, Germán Rodríguez, Carmen M. Rosskopf and Jesus Martinez-Frias

Received: 13 July 2021
Accepted: 20 August 2021
Published: 25 August 2021

1. Introduction

North Captiva is a low-lying barrier island located along the Gulf coast of Florida and is highly susceptible to morphological changes. During the hurricane season and winter storms, natural erosion of frontal beaches and transport of sediments is increased due to higher wave energy, a phenomenon that is also often observed and documented throughout the Caribbean and in tropical and subtropical regions around the world [1,2]. When Hurricane Charley made landfall near the island as a Category 4 hurricane on 13 August 2004, it caused a significant breach of approximately 0.5 km towards the southern end [3]. Interestingly, this breach naturally healed itself over the course of only about three years, without the help of any artificial nourishment. Since 1998, the National Oceanic and Atmospheric Administration (NOAA) along with the US Army Corps of Engineers (USACE) have conducted a series of LiDAR mapping surveys for the study area, along with the rest of the coastal United States [4]. These data sets are available in the public domain and have been proven to be an effective tool in monitoring long-term evolution of barrier islands and frontal beaches. This study sought to further examine the shoreline evolution, beach erosion characteristics and winter-storm induced longshore sediment transport that could have led to North Captiva Island's rapid and natural recovery, including the healing of the "Charley cut"; the breach that occurred from the landfall of Hurricane Charley. In order to do this, an Arc GIS-based modeling approach was used to quantify the morphological changes of the island from the landfall of Hurricane Charley and subsequent years until 2018, when the latest LiDAR survey was conducted for the region.

Hurricane Charley

On 9 August 2004, a tropical depression developed south-southeast of Barbados, and approaching Jamaica two days later, became Hurricane Charley, reaching a Category 4 status on 13 August 2004, when it made landfall on the southwest coast of Florida [5]. Landing near Cayo Costa just north of Captiva Island, Charley had maximum sustained winds nearing 150 mph [6]. This is where the hurricane reached peak intensity, and as it traversed across the state of Florida, it left destruction in its wake. Moving into the Atlantic shortly after its initial landfall, Charley re-strengthened and then weakened to a lesser hurricane when it hit South Carolina, lessening still to a tropical storm by the time it reached southeastern North Carolina [5]. Across the state of Florida, maximum rainfall was measured to be just above 5 inches from gauges, although radar-estimated precipitation was as high as 8 inches [7]. In total, Hurricane Charley was responsible for ten deaths in the United States along with twenty-five indirect deaths, and an additional five in Cuba and Jamaica. The total damage across Florida and the Carolinas is estimated to be 6.8 billion dollars in insured losses [5].

The 2004 hurricane season was extraordinary for the state of Florida. Of the 9 hurricanes named in the season, 5 of them made landfall, and 4 of them (Charley, Frances, Jeanne, and Ivan) battered the state [7]. The higher hurricane activity in the time period of 1995–2004 has been attributed to warmer sea surface temperatures in the Atlantic along with reduced wind shear over the deep tropics [7].

2. Materials and Methods

2.1. LiDAR Data and GIS

Beginning in 1998, the National Oceanic and Atmospheric Administration (NOAA) in partnership with the US Army Corps of Engineers (USACE) have conducted light detection and ranging, also known as LiDAR mapping surveys of beaches and nearshore areas on a national scale [8,9]. The data from these surveys have been shown to be effective tools for monitoring the long-term evolution of barrier islands and coastal environments. These datasets are available in the public domain at https://coast.noaa.gov/dataviewer/#/ (accessed on 25 January 2021).

Using NOAA Digital Coast archives, classified LiDAR data were extracted for the years of 1998, 2007, 2010, 2015 and 2018, as well as two additional LiDAR sets for 2004; post-Hurricane Charley and post-Hurricane Ivan [10]. As the data from 1998 were not classified and having low spatial resolution, it has been removed from further processing and analysis. The LiDAR data from each consecutive year were filtered for ground elevation and bathymetry, representing reasonable elevation points down the coastal environment. Using LiDAR data processing tools in ArcGIS, terrain models for each year were generated with a spatial resolution of 2 m and 5 levels of pyramid structure for optimal zooming [11]. From these terrain models, raster DEMs were generated using the 3D Analyst tool in ArcGIS.

To create the DEMs (see Figure 1), LAS datasets for each of the subsequent years were needed [9]. LAS files were extracted from the LiDAR datasets, and the statistics of the LAS datasets were calculated. Next, the dataset was added to ArcMap and the classification codes were edited and filtered so that only the ground and bathymetric elevation data would be incorporated. This was done to exclude data points that may have shown tree canopies, vegetation, and building elevations that would interfere and skew the results.

The next step in this mapping process was to create a terrain model from the LiDAR data. This terrain model is a multiresolution TIN-based surface stored as features, created by importing the ground and bathymetry LiDAR data points to a multipoint feature class. "LAS to Multipoint", a 3D Analyst tool, was used to create this feature. Next, the multipoint feature class was combined with a 2D shapefile of the study area [11].

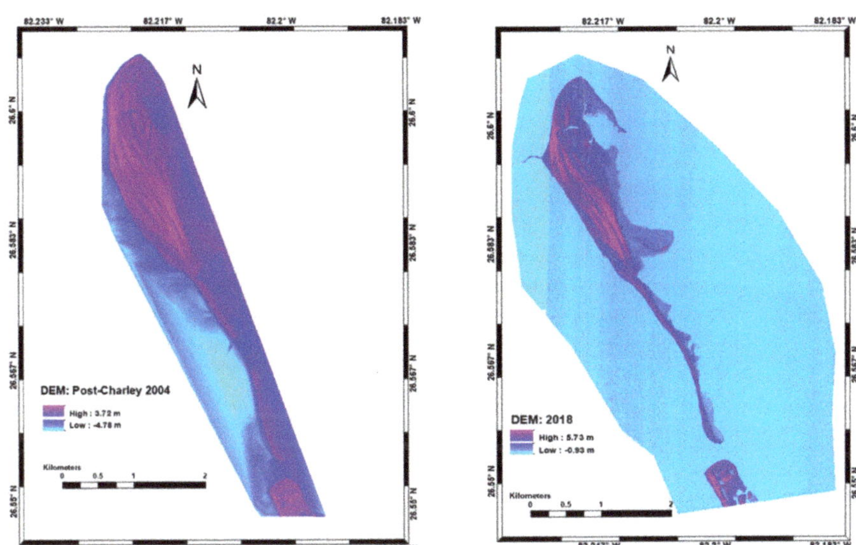

Figure 1. Digital Elevation Models (DEMs) for the North Captiva island using LiDAR data from 2004 and 2018.

Florida Department of Environmental Protection (FDEP) has been collecting shoreline data for the entire state since the early 1970s and they established R-monuments along the coast at an interval of 304.8 m (1000 ft) for periodically measuring the beach topography to assess the long-term erosional/accretional trends.

2.2. DEM Skill Assessment

To verify the accuracy of the DEMs generated from the terrain dataset, the DEMs from the study area were compared to those already generated by NOAA and provided in the public domain. Specifically, the datasets from 2015 were compared by extracting beach profiles from each R-monument and plotting them into Excel, where profile graphs could be generated and overlaid with one another to compare the accuracy of the data to that of DEM's generated by NOAA. The two datasets showed very good agreement across all of the R-monuments down the study area, confirming the reliability of the approach and the accuracy of the DEM development [11].

2.3. Beach Elevation Profiles

From the elevation model rasters of the study area generated for each year, beach profiles were extracted. By overlaying a shapefile of the R-monuments located on the North Captiva coastline, markers for intervals along the coast were able to be displayed and served as guides for profile extraction for each year studied. Lines extending from east to west were interpolated, and from this, a point profile and point graph were displayed showing the data points along the path. These data points were exported into Excel, where scatterplots were produced based on the data from each R-monument and year in the study. Compiling data from each year in the study and along each R-monument provided information on the beach elevation changes across the study area.

2.4. Google Earth Pro Historical Imagery

An additional aspect of this study was representing changes to the coastline of North Captiva Island and observing the different erosional and accretion patterns using Google Earth Pro. Beginning in 1994 and using historical satellite imagery base maps from Google Earth Pro archives, it was possible to outline the coastline of North Captiva Island up until 2019. Using the path and polygon features, and designating each year with a different colored outline, it is possible to see the changes to the geomorphology of the study area, specifically the areas of sediment erosion and accretion throughout the years. As the bay side of the barrier islands are fringed with mangroves and/or marsh vegetation, shoreline change analysis using historical imageries has been limited to the Gulf side of the island only.

2.5. Sediment Volume

To analyze the shoreline transformational behavior, SANDS Assessment Management software can be used to quantify the volume of sediments eroded and accreted from shorelines [12]. In the volumetric analysis of the North Captiva coastline, SANDS was used to quantify these sediment changes. This analysis was carried out from 2004 Post Hurricane Charley through 2018. The beach volume change, measured in cubic meters, was estimated for the island segment delimited by transects referenced by neighboring R-monuments between the years in the study area [11]. For the breached location, towards the southern end, additional transects were established for an increased resolution in volumetric analysis.

3. Results

3.1. North Captiva Historic Imagery

Considering the Google Earth Pro historical coastline mapping, erosion on the northern portion of the island and a comparable sediment accretion on the southern end of the island can be observed consistently from 1994–2019. In 1994, North Captiva Island had a very narrow northern coastline (Figure 2) that widened moving south towards the mid-section of the island, narrowing again towards the very southern tip of the island. Since then, the northern portion of the island has experienced steadily increased erosion, with the southern portion after the landfall of Hurricane Charley has experienced steady accretion. While Hurricane Charley has had the greatest physical impact on the shoreline of North Captiva, other factors also have contributed to the changes in its morphology. Additional storms that hit the southwest Florida region such as Ivan in 2004, and more frequent winter storms also played a significant role in the shoreline evolution of this barrier island [13,14].

Figure 2. *Cont.*

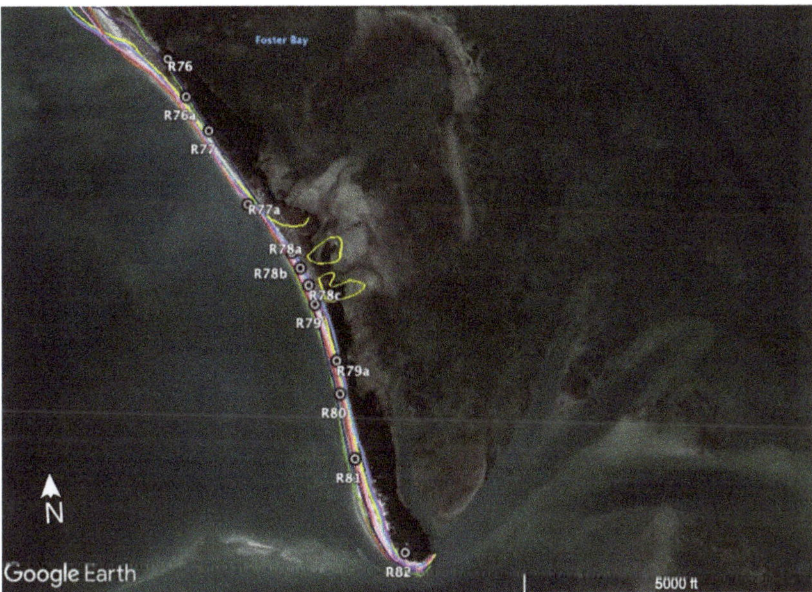

Figure 2. Google Earth image of North Captiva Island, depicting coastline from 1994–2019, denoted by different colored paths. R-monuments are overlaid on map, R67–R82.

Sediment transport since the landfall and resulting damage of Hurricane Charley in 2004 is evident in the evolving coastline of North Captiva island, specifically the healing of the southern cut and overall widening of the southern tip. By 2008, the southern portion of the coastline was consistently widening, showing the greatest width post-Charley by 2019. The modern coastline of North Captiva Island fits into this sediment transport pattern as well, as the island continues to erode in the north and widen in the south from sediment transport via longshore currents where the breach once was [7].

3.2. Beach Elevation Profiles

Elevation profiles of North Captiva's coastline from 2004–2018 are grouped based on their corresponding R-monuments, as well as approximate location on the island based on the section of island it is found in. Profiles are displayed from the northern-most region, mid-region, the breached area, and the southern tip. Figures 3–6 show the changes in elevation between 2004 post-Charley and 2018.

Figure 3 shows profiles extracted from the northernmost sector of the study area, including R-monuments 67 and 69. These profiles show a relatively stable beach with little geomorphological changes occurring from 2004–2018, aside from slight sediment accretion following Hurricane Charley and greater erosion in 2018. The orientation of this shoreline from north northeast to south southwest helps to maintain the coastal stability for this region.

Figure 4 shows profiles from the mid region of the study area including R72 and 74, which show significant erosional behavior for the region. After a slight sediment accretion post-Charley, since 2010 the shoreline has been subjected to significant erosion. In some areas along R72, specifically 75 m and 150 m onshore, the height of the beach has been eroded by about 1 m. Along R74, similar patterns are observed at 300 m and 440 m onshore where the height of the coast decreased by about 1 m since 2007 and 2010. High erosion from this area could be explained by greater exposure of this section of the island to northwesterly waves during winter storms.

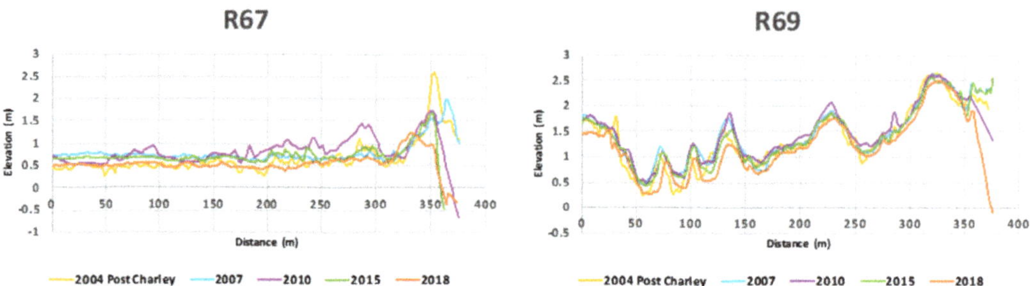

Figure 3. R67 and R79 elevation profiles with respect to NAVD88 of North Captiva Island for 2004 post-Charley through 2018.

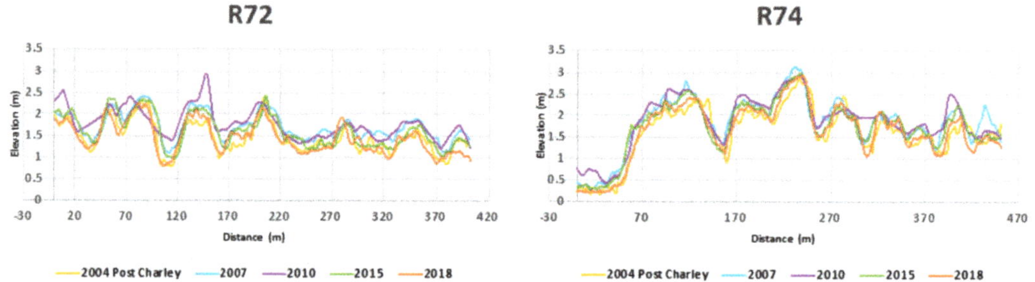

Figure 4. R72 and R74 elevation profiles with respect to NAVD88 of North Captiva Island for 2004 post-Charley through 2018.

Figure 5 depicts profiles R78a and 78c, representing the area on the island that was breached during Hurricane Charley made landfall. Data from 2004 shows evidence of the cut, as can be seen where the elevation of the island in 2004 falls below sea level on both the profiles from R-78a and for R-78c. Evidence of island recovery can be seen as the elevation increases from about −1 m to almost 2 m since 2004 post-Charley. Both of these peaks can be seen from 200–250 m. Evidence of sediment accretion from the elevation profiles began as early as 2007, showing the natural healing of the island that occurred. Sediment eroded from the beach section just north has been transported southward and helped in rebuilding this section.

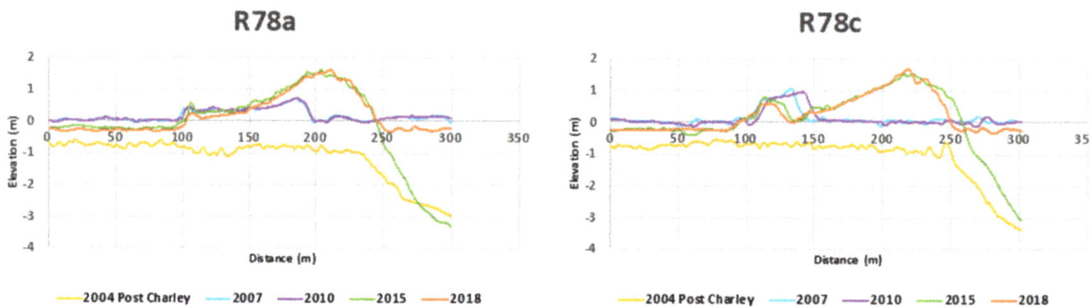

Figure 5. R78a and R78c elevation profiles with respect to NAVD88 of North Captiva Island for 2004 post-Charley.

Figure 6 shows the southern end of the study area denoted by R-monuments 79a and 80. The southern end of the island is the recipient of materials eroded from the north and transported southward. Sediment accretion can be seen in the increase in elevation of the study area since 2004. This elevation increase spikes in 2015 and 2018.

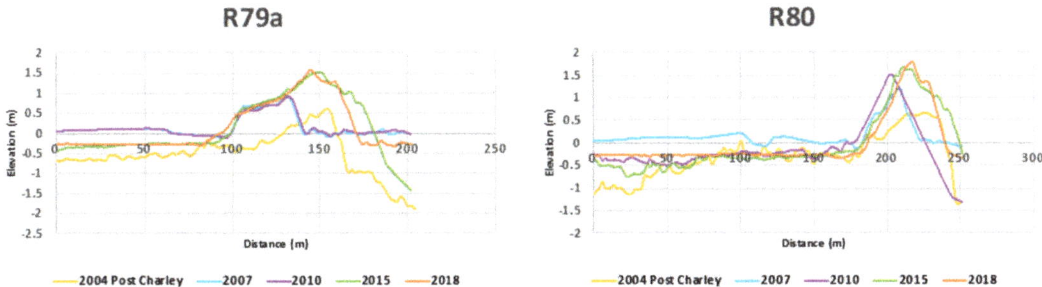

Figure 6. R79a and R80 elevation profiles with respect to NAVD88 of North Captiva Island for 2004 post-Charley through 2018.

Hurricane Charley had the most notable impact on the region denoted by R-monuments 78–79. This is the area that was breached when Hurricane Charley made landfall. Additionally, it is also the region that was most impacted by sediment transport since 2004, recovering in some areas approximately 2.5 m of elevation without the influence of artificial beach recovery.

3.3. Volumetric Changes

The calculated sediment volume changes for the northern and the mid sections of the North Captiva Island, between R67–R76 (north), from 2004–2018 are provided in Table 1. The extremely low and negative values of percentage change show the magnitude of erosion from the northern stations. The northernmost section of the barrier island has remained relatively stable since 2004, whereas from R-monuments 71–74, persistent erosion has been observed, as can be seen by the negative percentage change in sediment volume.

Table 1. Sediment volume calculations between R-monuments in the northern portion of North Captiva Island for 2004 to 2018. Beach volume change is estimated using a master profile (MP) kept at 2 m below NAVD88 datum.

Change between Locations		2004PC-2018—Volume Changes above MP 30 August 2004 to 15 August 2018	
Location 1	Location 2	Vol Diff (m^3)	% Change
R-67	R-68	28,221.73	6.27
R-68	R-69	16,785.05	2.44
R-69	R-70	67,938.88	10.98
R-70	R-71	61,228.89	7.4
R-71	R-72	−34,835.72	−3.31
R-72	R-73	−50,874.36	−5.44
R-73	R-74	−17,488.78	−2.19
R-74	R-75	−2676.5	−0.31
R-75	R-76	76,413.59	8.07
		144,712.78	Av = 2.66%
			Min = −5.44%
			Max = 10.98%

Percentage change shown in Table 2 shows the large-scale sediment accretion in the southern end of North Captiva island since 2004 (R76–R81). The magnitude of this accretion increases more towards the south, excluding the southernmost station R81.

Cumulatively, the breached area received 1,151,899.11 m^3 of sediments since the damage it sustained after Hurricane Charley; during 2004–2018. This massive transport of sediments from the north both helped to heal the cut created by Hurricane Charley, along with allowing for the overall expansion of the southern section of the barrier island.

Table 2. Sediment volume calculations between R-monuments in the southern portion of North Captiva Island for 2004 to 2018. See Table 1 caption for the master profile (MP) description.

Changes between Locations		2004PC-2018—Volume Changes above MP	
		30 August 2004 to 15 August 2018	
Location 1	Location 2	Vol Diff (m^3)	% Change
R-76	R-77	166,533.26	26.46
R-77	R-78	485,428.17	68.43
R-78	R-78b	166,307.04	105.69
R-78b	R-78c	106,824.82	113.81
R-78c	R-79	94,069.20	133.45
R-79	R-79a	299,269.88	129.89
R-79a	R-80	152,641.70	110.83
R-80	R-81		
		1,471,074.07	Av = 86.07%
			Min = 26.46%
			Max = 133.45%

4. Discussion

The geomorphology of beaches and barrier islands along southwest Florida can be significantly altered by hurricanes and winter storms by means of frontal beach erosion, overwash deposits, migration of foredunes, and the transport of sediments. Persistent transport of beach sediments can cause either erosion or accretion. The landfall of Hurricane Charley on North Captiva Island in 2004 created a cut to the southern end of the island, and since then, both localized sediment erosion and accretion have been observed on the barrier island when surveying the island until 2018 using LiDAR. Sediments were removed from the northern portion of the island after the hurricane in 2004 (R67–R76) and transported to the south to heal the island breach and build back up the width of the southern portion of the island (R76–R82). Most of this sediment redistribution can be attributed to strong southerly long-shore transport during the winter storms and the passage of tropical storms and hurricanes. Another study on barrier island erosional patterns, conducted on the Chandeleur Islands, Louisiana, supports the analysis that shoreline erosion and sediment transport are greatly influenced by changing wind and wave energy as a result of hurricanes and winter storms [15]. Additionally, a study conducted on a Dutch barrier island further backs this analysis that sediment transport on barrier islands is highly impacted by storm frequency [16].

As a low-lying barrier island, North Captiva is a coastal environment that is very susceptible to geomorphological changes. Alterations to the physical outline of the coast, elevation changes of the beach and fordunes, and percentage changes of sediments are all pieces of evidence that support the argument that the northern portion of North Captiva Island has experienced greater erosion of sediments since 2004, sediments which were naturally transported down the longshore current and used to heal the cut inflicted on the southern part of the island when Hurricane Charley made landfall.

Recovery of North Captiva is most evident in the southern portion of the island where Hurricane Charley inflicted the most damage in 2004. LiDAR mapping immediately after the hurricane shows that at the southern breach, a section of the island is underwater.

Only three years after Hurricane Charley, at the area of the southern breach, the island sat above sea level once again, and by 2018, it had gained over 2 m of height. The sediment loads that likely supplied renourishment to the southern end of the island likely came from the northern part of North Captiva Island, the coastline of which has been eroding consistently for almost two decades. Winter storms likely played a significant role in the redistribution of these eroded sediments along the study area. Overall, the damage inflicted to the island in Hurricane Charley was significant, but the subsequent erosion of the northern beaches and longshore transport of sediments to the damaged area was able to naturally heal the cut.

Similar patterns in erosion and accretion as a result of tropical cyclones and winter storms have been frequently observed and documented across the Caribbean Sea also [17,18]. In several small islands across the Eastern Caribbean, significant erosion of accretionary features such as spits and tombolas has been reported over the past several decades, especially on islands impacted by recent hurricanes [5]. A study conducted on Colombian Caribbean beaches found that storms with cold front characteristics were equally, and in most cases, more damaging than hurricanes. The erosional effect of these winter storms was found to have greater magnitude and additionally, remain for longer periods of time. As a result, coastal environments were often unable to fully recover from sediment loss before the next storm season [19]. Another study was conducted to compare the damage inflicted by Hurricane Wilma in 2005 to an exposed beach in Cancun and a beach fronted by a fringing reef in Puerto Morelos. Widespread erosion was observed at Cancun after the hurricane, whereas Puerto Morelos experienced substantial accretion of about 30 m on this beach [17,20]. Similar to the accretion of sand in the southern portion of North Captiva Island, accretion at Puerto Morelos is thought to be the contribution of sand from the northern beaches transported during storms [17].

Chronic erosion occurred to majority of the coastal environments mentioned can also be attributed to other anthropogenic factors such as coastal development, sand mining, coastal construction, and land clearing. Beach erosion in those cases were further exacerbated by increased wave energy from storms [5,16,18,21]. However, practices such as those mentioned above could prove to be detrimental to coastal environments that are most impacted by winter storms and hurricanes because of the way that they can potentially degrade natural barriers. Additional studies bring to light the importance of preserving natural protective barriers in coastal regions, such as coral reefs, mangroves, sand dunes and spits. All of these natural structures provide coastal environments with the valuable ecosystem service of coastal buffering and protection. In many cases, these structures may disperse wave energy and prevent sediment erosion. In Puerto Morelos, the presence of fringing coral reefs not only protected the beach from Hurricane Wilma, but also contributed to inducing coastal growth [20].

5. Conclusions

Over the decades, North Captiva Island has evolved significantly as a result of tropical storms, hurricanes, and frequently occurring winter storms, as well as seasonal changes to the wave energy experienced in the Gulf of Mexico. Specifically, after the impact of Hurricane Charley, North Captiva Island experienced significant morphological changes and has sustained remarkable shoreline evolution. North Captiva was split apart in the southern end, but the cut was healed naturally by sediment nourishment. The breached section received 1,151,899.11 m^3 of sediments delivered from the northern portion of the island from 2004–2018. Additionally, the breached area of the island saw a notable increase in elevation of approximately 2.5 m since 2004. Since Hurricane Charley, a summary of the observed and measured changes to the island includes increased volume of sediments in the southern end of the island, heightened elevation, and stabilization of the island where the breach occurred, and overall widening of the southern shoreline.

Future studies of sediment transport on North Captiva Island should further explore where alternative sources of sediment may come from, and the role that sediments in the Gulf of Mexico, or on neighboring islands, may play in island re-nourishment. The effects of rising sea level should also be monitored closely in relation to island reshaping and the ability to recover from storms [13]. Additionally, future studies detailing sediment transport should consider aspects of land usage and the alteration of natural land barriers where applicable.

Author Contributions: Data curation, E.W.K.; Formal analysis, F.J. All authors have read and agreed to the published version of the manuscript.

Funding: This research received no external funding.

Data Availability Statement: Not applicable.

Conflicts of Interest: Authors declare no conflict of interest.

References

1. Claudino-Sales, V.; Wang, P.; Horowitz, M.H. Effect of Hurricane Ivan on Coastal Dunes of Santa Rosa Barrier Island, Florida: Characterized on the Basis of Pre- and Poststorm LiDAR surveys. *J. Coast. Res.* **2010**, *26*, 470–484. [CrossRef]
2. Cambers, G. Beach Changes in the Eastern Caribbean Islands: Hurricane Impacts and Implications for Climate Change. *J. Coast. Res.* **1997**, *24*, 29–47.
3. Weisberg, R.H.; Zheng, L. Circulation of Tampa Bay driven by buoyancy, tides, and winds, as simulated using a finite volume coastal ocean model. *Fla. Sci.* **2006**, *69*, 152–165. [CrossRef]
4. Jose, F.; Savarese, M.; Gross, A.M. Impact of Hurricane Irma on Keewaydin Island, Coastal Sediments, Southwest Florida Coast. *Coast. Sediments* **2019**, 1142–1151. [CrossRef]
5. Pasch, R.J.; Brown, D.P.; Blake, E.S. *Tropical Cyclone Report;* National Hurricane Center: Miami, FL, USA, 2004.
6. Wang, R.; Manausa, M.; Cheng, J. *Hurricane Charley Characteristics and Storm Tide Evaluation;* Florida Department of Environmental Protection, Bureau of Beaches and Coastal Systems: Tallahassee, FL, USA, 2005.
7. Sallenger, A.H.; Stockdon, H.F.; Fauver, L.; Hansen, M.; Thompson, D.; Wright, W.; Lillycrop, J. Hurricanes 2004: An Overview of Their Characteristics and Coastal Change. *Estuaries Coasts* **2006**, *29*, 880–888. [CrossRef]
8. Houser, C.; Wernette, P.; Rentschlar, E.; Jones, H.; Hammond, B.; Trimble, S. Post-storm beach and dune recovery: Implications for barrier island resilience. *Geomorphology* **2014**, *234*, 54–63. [CrossRef]
9. Priestas, A.M.; Fagherazzi, S. Morphological barrier island changes and recovery of dunes after Hurricane Dennis St. George Island, Florida. *Geomorphology* **2009**, *114*, 614–626. [CrossRef]
10. National Oceanic and Atmospheric Administration (NOAA). Available online: https://coast.noaa.gov/dataviewer/#/ (accessed on 25 January 2021).
11. Harvey, N.; Gross, A.; Jose, F.; Savarese, M.; Missimer, T.M. Geomorphological impact of Hurricane Irma on Marco Island, Southwest Florida. *Springer Nat.* **2020**. [CrossRef]
12. Pye, K.; Blott, S.J. Assessment of beach and dune erosion and accretion using LiDAR: Impact of the stormy 20132914 winter and longer term trends on the Sefton coast, UK. *Geomorphology* **2016**, *266*, 146–167. [CrossRef]
13. Wang, P.; Roberts Briggs, T.M. Chapter 10- Storm-Induced Morphology Changes along Barrier Islands and Poststorm Recovery. *Coast. Mar. Hazards Risks Disasters* **2015**, 271–306. [CrossRef]
14. Houser, C.; Hamilton, S. Sensitivity of post-hurricane beach and dune recovery to event frequency. *Earth Surf. Process. Landf.* **2009**, *34*, 613–628. [CrossRef]
15. Georgios, I.Y.; Schindler, J.K. Wave forecasting and longshore sediment transport gradients along a transgressive barrier island: Chandeleur Islands, Louisiana. *Geo-Mar. Lett.* **2009**, *29*, 467–476. [CrossRef]
16. Wesselman, D.; de Winter, R.; Engelstad, A.; McCall, R.; van Dongeren, A.; Hoekstra, P.; Oost, A.; van der Vegt, M. The effect of tides and storms on the sediment transport across a Dutch barrier island. *Earth Surf. Process. Landf.* **2017**, *43*, 579–592. [CrossRef]
17. Silva-Casarin, R.; Marino-Tapia, I.; Enrique-Ortiz, C.; Mendoza-Baldwin, E.; Escalante-Mancera, E.; Ruiz-Renteria, F. Monitoring Shoreline Changes at Cancun Beach, Mexico: Effects of Hurricane Wilma. *Coast. Eng.* **2006**, 3491–3503. [CrossRef]
18. Acuna-Piedra, J.F.; Quesada-Roman, A. Multidecadal biogeomorphic dynamics of a deltaic mangrove forest in Costa Rica. *Ocean Coast. Manag.* **2021**, *211*, 105770. [CrossRef]
19. Cueto, J.; Otero, L. Morphodynamic response to extreme wave events of microtidal dissipative and reflective beaches. *Appl. Ocean Res.* **2020**, *101*, 102283. [CrossRef]
20. Marino-Tapia, I.; Enrique-Ortiz, C.; Silva-Casarin, R.; Mendoza-Baldwin, E.; Escalante-Mancera, E.; Ruiz-Renteria, F. Comparative Morphodynamics Between Exposed and Reef Protected Beaches Under Hurricane Conditions. *Coast. Eng. Proc.* **2014**. [CrossRef]
21. Anfuso, G.; Rangel-Buitrago, N.; Correa Arango, I.D. Evolution of Sandspits Along the Caribbean Coast of Colombia: Natural and Human Influences. *Sand Gravel Spits* **2015**, *12*, 1–19. [CrossRef]

Article

Spatiotemporal Variability of Extreme Wave Storms in a Beach Tourism Destination Area

Daniel Guerra-Medina [1] and Germán Rodríguez [1,2,*]

[1] Departamento de Física, Universidad de Las Palmas de Gran Canaria, 35017 Las Palmas de Gran Canaria, Spain; dguerramedina@gmail.com

[2] Instituto Universitario de Estudios Ambientales y Recursos Naturales, Universidad de Las Palmas de Gran Canaria, 35017 Las Palmas de Gran Canaria, Spain

* Correspondence: german.rodriguez@ulpgc.es

Abstract: This study explores the spatiotemporal variability of extreme wave storms around the Canary archipelago, with special focus on the southern coastal flank of Tenerife island, a strategic beach tourism destination of large socioeconomic importance. To this end, experimental and simulated data of winds and waves are used to study the severity, seasonality, and directionality of wave storms with considerable potential to cause significant impact on beaches. Furthermore, tidal experimental records are employed to test the joint occurrence of wave storms and significantly high sea levels. Long-term statistical analysis of extreme wave storms at different locations reveals a complex spatial pattern of wave storminess around the islands and in the southern flank of Tenerife, due to the intricacy of the coastline geometry, the presence of deep channels between islands, the high altitude and complex topography of the islands, and the sheltering effects exerted by each island over the others, depending on the directionality of the incident wave fields. In particular, south of Tenerife, the energy content and directionality of wave storms show substantial spatial variability, while the timing of extreme wave storms throughout the year exhibits a marked seasonal character. A specific extreme storm is examined in detail, as an illustrative case study of severe beach erosion and infrastructure damage.

Keywords: extreme wave storms; tidal levels; tourist beaches; beach erosion; infrastructure damage; socioeconomic impacts; Tenerife; Canary Islands

Citation: Guerra-Medina, D.; Rodríguez, G. Spatiotemporal Variability of Extreme Wave Storms in a Beach Tourism Destination Area. *Geosciences* **2021**, *11*, 237. https://doi.org/10.3390/geosciences11060237

Academic Editors: Jesus Martinez-Frias and Fabrizio Antonioli

Received: 27 April 2021
Accepted: 27 May 2021
Published: 31 May 2021

1. Introduction

Coastal areas are generally densely populated and have high strategic value from a socioeconomic point of view. However, the coast in general, and beaches in particular, constitute systems with highly nonlinear, complex dynamics and are strongly vulnerable to the individual or joint action of different types of natural hazards, which can lead to erosion or flooding processes with significant negative socioeconomic repercussions (e.g., [1]). This is especially true in the case of areas heavily dependent on beach tourism, such as the Canary Islands and, in particular, the southern flank of the Tenerife island [2]. The islands are a tourist destination of world importance, notably in Europe. The archipelago received more than 15 million tourists in 2019 and, according to the Canary Islands Institute of Statistics [3], tourism visiting the islands generates 32% of the Canary Islands' GDP and 30% of jobs. These percentages are even higher on the southern coast of Tenerife, where a large fraction of tourism received in the islands, some six million tourists per year, is concentrated in a few municipalities.

The great majority of tourists visiting the island are looking to enjoy the climate and the beaches (e.g., [4]). In this sense, it is important to bear in mind that beaches are dynamic complex systems that evolve and change their characteristics depending on the hydrodynamic conditions to which they are subjected, particularly during severe wave conditions. Extreme wave storms represent risky events for the natural environment and human activities on the coast (e.g., [5]). In particular, they have the potential to produce significant beach erosion episodes in relatively short periods, resulting in loss of sand, beach retreat, and the consequent undesirable reduction in beach dimensions. This kind of impact can be temporal or even semi-permanent, depending on the nature of the beach, but, in any case, extends over variable but considerable periods of time, because post-storm beach recovery by long-period swell onshore sediment transport is generally a slow process. Accordingly, it is evident that the occurrence of severe wave storms may have strong negative impacts on tourism activities.

The effects of wave storms on a beach depend on a substantial number of factors, such as their severity, duration, directionality, and seasonality, as well as on the likelihood of occurrence during periods of high (tidal and non-tidal) sea water level elevation, among others. All of the above highlights the importance of having a good understanding of the space–time variability of the extreme wave storms affecting a given coastal zone, as an essential support tool for the development of management strategies that reduce the socio-economic impacts caused by these abnormally severe events.

The study of wave conditions in the Canary archipelago has focused mainly on the northern and northwestern edges of the islands to explore the potential use of waves as a renewable energy resource (e.g., [6,7]). However, the characterization of long-term wave conditions, including high severity episodes, at the southern flanks has received much less attention, despite its potential negative impact on beaches and surrounding areas and the corresponding socioeconomic implications.

In view of the above, the primary focus of this study is to explore the wave climate on the southern coast of Tenerife, with emphasis on stormy conditions, because of the socioeconomic importance of this stretch of coast. Specifically, we focus our interest on those storms with a remarkable capacity to cause damage, mainly in terms of coastal erosion and/or flooding. As a first step towards achieving this goal, the space–time behavior of the wave climate along the outer edges of the Canary Islands is examined to understand the general wave conditions reaching the coasts of the archipelago.

The paper is structured as follows. After justifying the need to study extreme wave storm characteristics, in both space and time domains, in the southern coastal strip of Tenerife island, a brief description of the geographical and meteoceanic characteristics of the study area, as well as the main characteristics of experimental data used in the study, is provided in Section 2. The methodological approaches used to examine the spatial and temporal variability of extreme wave storms are introduced in Section 3. Section 4 presents the results derived from the study and their discussion, including the detailed study of a selected storm, as an illustrative case of severe beach erosion and infrastructure damage, highlighting the evolution of wind, wave, and tidal conditions, and showing evidence of its socioeconomic impacts. Finally, Section 5 summarizes the main findings of the study.

2. Study Area and Datasets

2.1. Study Area

The Canary archipelago consists of seven major islands and several islets and constitutes a Spanish autonomous community located on the Northwest African continental shelf, in the Eastern Central Atlantic, off the Saharan coast at a minimum distance close to 100 km, measured from Fuerteventura, the easternmost island. Gran Canaria (GC) and Tenerife (T) are the two most populated islands, together constituting more than 80% of the total population (over 1,750,000 inhabitants). The rest of the residents are mostly concentrated in Fuerteventura (F), Lanzarote (L), and La Palma (P), while the minor islands of El Hierro (H) and La Gomera (G) are barely populated. The archipelago is approximately centered at the coordinates (28° N, 15° E) and extends around 450 km from east to west between 27° and 30° of northern latitude, as shown in Figure 1. The enlarged image shows the relative position of each island within the archipelago and with respect to the African continent, as well as the complex geometry of the islands' coast. In addition, it shows an enlarged illustration of the southern side of Tenerife. An important aspect, not observed in the figure, is the altitude of the islands. Lanzarote and Fuerteventura, the islands closest to Africa, are significantly flat, but the altitude is over 1500 m in the further west islands, reaching the maximum height (3715 m) at Pico del Teide, Tenerife, the largest (2034 km^2) and highest of the islands.

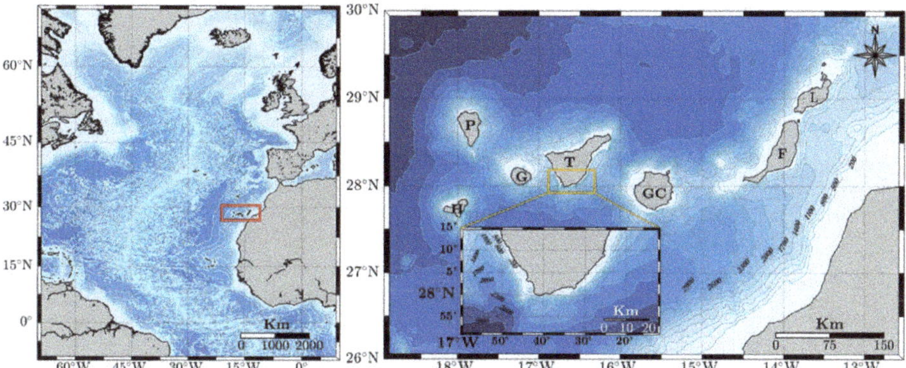

Figure 1. Geographical location of the Canary Islands (**left**) and zoom of the archipelago, with the African coast as a reference in the (**right**) lower corner, as well as an expansion of the southern edge of Tenerife island.

Due to its geographical location, in the southern edge of the Azores High, the Canaries are within the fairly regular Trade Winds belt. The trade winds regime exhibits a clear seasonal pattern throughout the year, governed by the relative intensity and location of the Icelandic Low and Azores High pressure systems. During summer, the dominant trade winds blow with moderate or weak intensity from the N-NNE directional sector, with frequencies between 90% and 95%, while, during winter, NE trade winds blow with lower intensity and frequency, over 50% (e.g., [8]).

Naturally, the wave climate in the archipelago is strongly related to the above atmospheric conditions and, consequently, wave conditions are rather mild, Thus, the northern edge of the island is the most exposed to wave action, while there are significant spatial variations around the islands' coastline [9,10]. Furthermore, wave conditions undergo a clear seasonal pattern in the most energetic areas, the northern and western sides of the archipelago, with mild wave conditions from April to October and more severe situations from November to March. Regarding wave storm conditions on the north side of the archipelago, it has been observed that extreme wave events also exhibit a statistically significant seasonal behavior [11]. The tidal regime in the islands is mesotidal, with a semidiurnal tide pattern and a tidal range oscillating approximately between 0.5 m and 3 m, and a mean value close to 1.5 m [12]. Furthermore, meteorological residuals are almost negligible, ranging within ±20 cm, approximately, but with a modal value that is almost null [13].

2.2. Datasets

The investigation is based on four datasets of different nature. On the one hand, datasets including oceanographic and meteorological information have been provided by the Spanish Port authority and include wind and wave data obtained from the coupling of wind and wave numerical models (hindcasting), wind and wave observations recorded in-situ by means of meteoceanic buoys, and mean water level information registered with a tide gauge. On the other hand, in the absence of more rigorous sources of information, evidence on the impact of wave storms in the coastal zone of interest has been obtained by using the digital information database JABLE [14], created by the University of Las Palmas de Gran Canaria, which includes an enormous volume of historical and current press produced in the Canary Islands from 1808 to the present. This digital platform allows searching on a specific topic with keywords by island, locality, period, etc., among more than 7 million pages from more than 700 newspapers, newsletters, bulletins, gazettes, journals, magazines, and other serial publications.

The database containing wind and wave information obtained by using wind and wave numerical models is referred to as SIMAR and provides information covering the period from January 1958 to March 2020. Placement of computational mesh hindcasting grid nodes selected to characterize wind and wave climate is indicated in Figure 2a. The eight points located in the outer edges of the archipelago are hereinafter cited as external points (EPX), where X indicates each specific point. Similarly, five hindcasting grid nodes used to explore wind and wave conditions in the southern flank of Tenerife are designated as internal points, denoted by IPX. The location of IPX points is more clearly depicted in Figure 2b, which also shows the position of two meteoceanic buoys measuring wind and wave conditions, one located northwest of Gran Canaria and other south of Tenerife, respectively referred to as BGN and BTS wave buoys, and belonging to the network of meteoceanic deep water buoys (REDEXT) of the Spanish Port Authority. These buoys are anchored in areas deeper than 200 m and are located at positions virtually coincident with two SIMAR points. Time series provided by both buoys have an hourly sampling rate and cover the periods from June 1997 to December 2019 (BGN) and from April 1998 to March 2020 (BTS), but directional sensors were not available until 2003. Regarding mean sea level fluctuations, measurements have been carried out with a tide gauge at the northeastern tip of Gran Canaria (TG), as shown in Figure 2b. The corresponding dataset includes hourly values of sea water level and cover the period from July 1992 to March 2020.

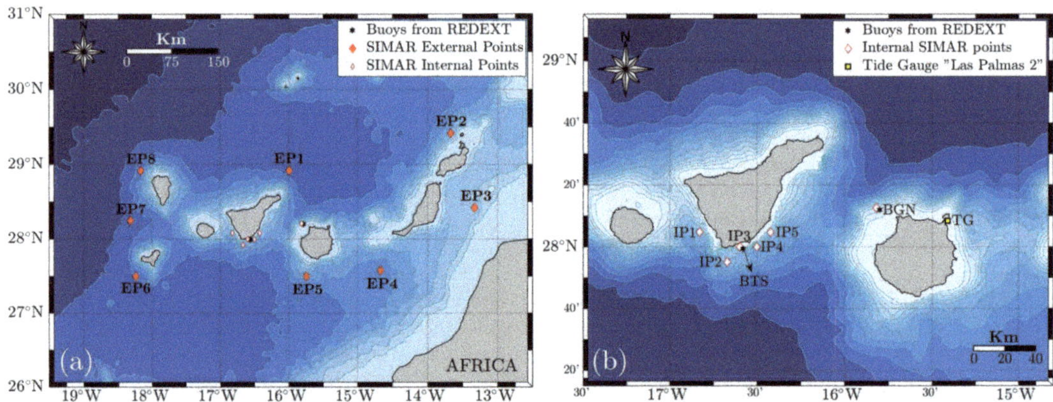

Figure 2. Map of the Canary archipelago showing location of external (**a**) and internal SIMAR points, as well as wave buoys and tide gauge (**b**).

Regarding the geometric configuration of the archipelago, it is important to visualize the existence of two deep channels between La Gomera and Tenerife, and between Gran Canaria and Tenerife (Figure 2b), hereinafter referred to as G-T and GC-T channels. The approximate average depths of both channels are 1600 m (G-T) and 2700 m (GC-T), while the minimum widths are 29 km (G-T) and 61 km (GC-T), approximately. It is also important to mention that these three central islands are substantially high, with altitudes exceeding 1500 m and reaching up to 3700 m, approximately.

Standard sea state parameters used to characterize wave climatology have been derived from spectral moments of the directional spectral density function, $S(f,\theta)$, which represents the energy contribution of any wave component to the measured wave field in terms of the frequency, f, and propagation direction, θ. The most important parameter for the characterization of a sea state is the significant wave height, H_{m0}, defined as four times the square root of the zero-th order spectral moment, m_0, which represents the total energy of the process. Therefore, H_{m0} is proportional to the energy content of the corresponding sea state and, consequently, it is the parameter used to represent its severity, as a general rule. In the case of wave periods, there are several optional characteristic periods that can be used in light of the objectives pursued. Two of the most widely employed in practice are used in this study. These are the average wave period, T_{02}, and the spectral peak period, T_p, which is the period associated with the most energetic spectral wave components. In this regard, it is important to note that, in contrast to T_{02}, T_p is not computed by means of spectral moments but considering a single spectral estimate and therefore presents considerable statistical uncertainty, or statistical variability [15]. In terms of wave direction propagation, the most common parameter used to characterize the directional properties of a wave field is the mean spectral direction, θ_m, which represents the mean approaching direction averaged over all the frequency bands in the directional wave spectrum. The analytical expression of the above-described characteristic parameters in terms of the directional spectrum can be found throughout the literature (e.g., [16]).

3. Methodology

The study has been performed at two different spatial levels. On one hand, wave climate in the outer flanks of the archipelago has been explored to identify the characteristics of wave fields reaching the zone, the second one regarding specifically the southern coastal zones of Tenerife island to examine wave climate and the characteristics of extreme wave storms in this coastal stretch in greater detail.

3.1. Wave Storm Concept and Definition

Wave storm is a concept that is intuitively easy to understand but difficult to formally define. To a large extent, this is because a storm is a relative term referring to a period of time during which the severity of wave conditions is significantly intense with respect to the conditions normally observed at that location. In this context, wave storms represent extreme events with low frequency of occurrence but potentially severe impacts.

In terms of the above, there is no universally accepted procedure for the identification of wave storms. However, from a practical perspective, a wave storm is commonly considered a sequence of sea states with significant wave height exceeding a given threshold (H_t) for a specified minimum time period selected, known as the minimum storm duration (D_{min}), so that when the duration (D) of an exceedance of H_{m0} is smaller than D_{min}, the event is not considered a storm. In addition, it is assumed that two consecutive events must be considered a single storm if the significant wave height in the time between these events does not drop below H_t during a period larger than a certain minimum time interval, usually named the minimum inter-storm duration (ISD_{min}), where the time interval between the end of one storm and the beginning of the next is called the inter-storm period, or duration (ISD) (e.g., [17]). These parameters are schematically illustrated in Figure 3.

The above definition is not completely rigorous and free of drawbacks. The main problem is the selection of a threshold level that satisfies the criterion of independence of events and at the same time allows the identification of a sufficiently large number of events for the sample to be statistically representative. For this reason, there is great variability in criteria to establish this parameter, depending mainly on the geographic location and the local average wave climate [18].

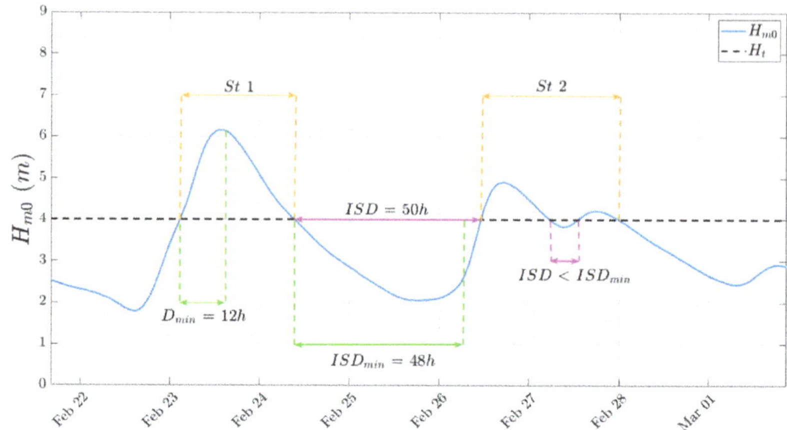

Figure 3. Illustrative sketch of parameters used to define and select independent wave storms.

3.2. Data Validation

The need for data providing information on long-term wave conditions over extensive areas makes it necessary to use alternative sources of information to that provided by measuring instruments, which are generally scarce, if any, and cover short periods of time. In this sense, the joint use of wind and wave numerical prediction models to obtain long-term databases on a spatial grid is currently a very common and useful tool. However, due to various reasons, wind and wave prediction models have some limitations in their ability to reproduce real conditions, mainly in areas with complex topography, such as islands (e.g., [19,20]). Consequently, whenever possible, it is necessary to validate the results against other sources of information, preferably from specific experimental measurement devices, such as meteoceanic buoys. Data validation has been achieved in this study by comparing data derived from models at two nodes, located at positions coinciding with that of BGN and BTS wave buoys (see Figure 2b), and data measured by these buoys, considering only the periods for which measured data were available.

Due to the interest in extreme wave storms and considering the difficulty of models to correctly reproduce extreme wave conditions [21,22], characteristic parameters associated with each selected wave storm were validated by linear regression and graphically represented for visually assessing the degree of agreement between simulations and experimental recordings. Moreover, joint validation of circular (wind and wave direction) and linear variables (wind speed, wave height and period) has been done by examining wind and wave roses for measured and simulated data.

3.3. Identification of Storms with Severe Impacts on Beaches

Digitized newspaper archives covering large time periods constitute a rich source of socioeconomic and environmental information. In particular, archives of regional newspapers are of great value for identifying the occurrence of non-recorded past and present extreme events of natural phenomena, as well as for obtaining a rough idea of their damage intensity and their socioeconomic impacts, which is of great importance for coastal managers [23,24]. Nevertheless, despite its considerable usefulness in this sense, this type of information should be considered with due caution [25]. With this in mind, a Boolean search with different combinations of keywords was developed in the JABLE database to detect dates on which the press reported wave storms causing damages on beaches, or nearby infrastructures in the study area, to be used as a source of evidence of the impact of severe damaging events.

3.4. Wave Storm Severity

Several parameters have been suggested in the literature to assess the severity of wave storms. The Storm Power Index [26] was introduced as the product of the squared maximum value of H_{m0} during the storm and its duration. Clearly, this parameter overestimates the storm severity by considering only the maximum of H_{m0} during the storm. Accordingly, an integral parameter to quantify the total wave power, *TWP*, has been proposed [27] and can be expressed as

$$WP = \int_{t_i}^{t_f} H_{m0}^2(t)dt \approx \Delta t \sum_{t_1}^{t_2} H_{m0}^2 \tag{1}$$

where t_i and t_f are the starting and ending times of the storm event, so that the storm duration, D, is given by $D = (t_f - t_i)$.

It is interesting to remark that *TWP* is a function of the storm duration and therefore does not allow comparison of the severity of storms with different lengths. To this end, it is appealing to standardize this value with respect to D to obtain the storm energy [28], denoted as E and given by

$$E = \frac{1}{D} \int_{t_i}^{t_f} H_{m0}^2(t)dt \approx \frac{\Delta t}{D} \sum_{t_1}^{t_2} H_{m0}^2 \tag{2}$$

3.5. Wave Storm Seasonality

To assess the existence of seasonal variations in the timing of wave storms throughout the annual period, the day of the calendar year on which the maximum significant wave height of a given storm occurred has been converted to an angular value, θ, assuming that the number of days per year is 365. Thus, in leap years, the data corresponding to 29 February have been removed [11].

To examine whether storms in a given region exhibit a seasonal pattern, it is necessary to know whether it is statistically possible to accept that the time of occurrence of wave storms throughout the year is uniformly distributed. The acceptance or rejection of the uniformity hypothesis is assessed in this study through the use of the Rayleigh and Kuiper tests, by considering the storm peaks' timing throughout the year as a circular variable. The selection of these two tests from among the multitude of existing alternatives is due to the fact that the Rayleigh test is powerful only when it is possible to assume that the population distribution has only one mode, while the Kuiper test is specifically indicated in the case of multimodal distributions. More detailed information on these and other uniformity tests can be found in [11,29] and references therein.

3.6. Wave Storm Directionality

Directional characteristics of wave conditions observed or simulated, both at outer and inner points of the archipelago, have been explored by elaborating wind and wave roses. This type of representation in polar coordinates allows an easy visualization of the directional distributions for characteristic wave heights and periods. In particular, the bivariate empirical distributions of the following pairs of characteristic wave parameters have been obtained, $\theta_m - H_{m0}$, $\theta_m - T_{02}$, and $\theta_m - T_p$, for both the total dataset and the values associated with selected storms, at each of the selected points.

4. Results and Discussion

4.1. Data Validation

Regression analysis during wave storm conditions reveals that values of the regression coefficient, r^2, for H_{m0} are quite good, both in the north and south points, although with slightly higher values in the north (0.66 for BGN) than in the south (0.62 for BTS). Figure 4 shows three examples of the significant height evolution, both measured and modeled, during storm conditions. It can be observed that, even with significantly high values of r^2, models may overestimate (panel a, r^2 = 0.82) or underestimate (panel c,

r^2 = 0.87) the experimental measures, although, often, the degree of correspondence is quite good or very good in some cases (panel b, r^2 = 0.92). In this sense, it is interesting to underline that although the general trend of the models is to underestimate experimental observations during extreme events (e.g., [16,17]), the results in this study include cases in which storms are underestimated, overestimated, as well as events considerably well reproduced, especially on the northern side of the islands. These results should not be surprising taking into account factors such as the altitude and complexity of the archipelago's topography, the irregularity of the coastline, and the islands' self and mutual shading effects, among others, which limit the ability of models to correctly reproduce wind and wave conditions in these areas.

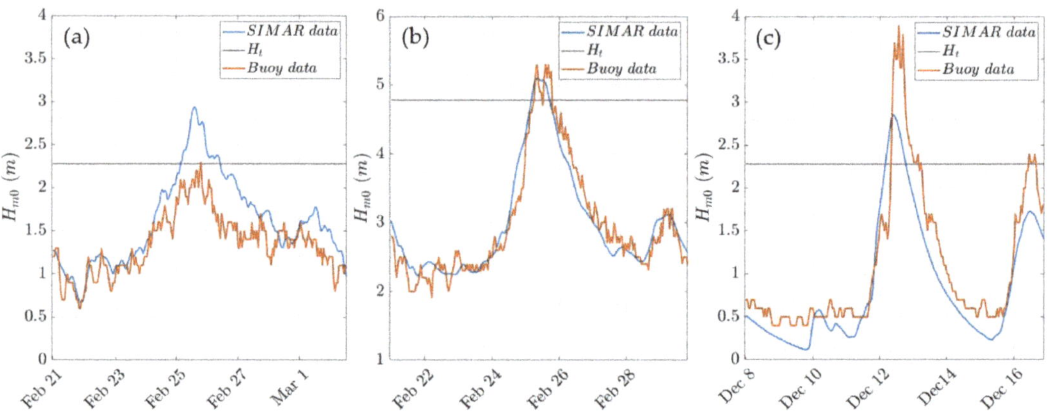

Figure 4. Examples of simulated and measured H_{m0} sequences during wave storm episodes measured and simulated at BTS (**a**,**c**) and BGN (**b**), with corresponding regression coefficients r^2 = 0.82 (**a**), r^2 = 0.92 (**b**), and r^2 = 0.87 (**c**).

Regarding the simultaneous evolution of wind and wave directions, the analysis of both parameters as obtained from the buoy and by the model, during storms shown in Figure 4a,b, reveals that there is a fairly good correlation between the experimental measurements and the simulations. A quantitative measure of the correlation between two circular variables can be obtained by means of the circular correlation coefficient, ρ_c [29]. The value of this coefficient for wind measured and simulated directions observed at BTS during the storm of Figure 4a is 0.29, while for the storm detected at BGN during the storm depicted in Figure 4b is 0.78. On the other side, the circular correlation coefficients for wave measured and simulated directions in these two cases are 0.59 and 0.88. In other words, there is a better degree of correlation between wave direction measurements and simulations than between wind direction measurements and simulations. Moreover, the correlation coefficient between both circular variables is higher in the north than in the south. These results can be explained by the lower directional variability of waves than that of wind, as well as the superior ability of the models to simulate wind and wave conditions to the north than to the south of the islands, due to the complexity of their orography. Unfortunately, there is no directional information for the latter case (Figure 4c) since it occurred prior to 2003, the date on which the directional sensors were incorporated into the buoys.

An overview of wind and wave directional variability, as well as their respective combined variability with wind speed, significant wave height, and peak period, can be qualitatively explored by means of wind and wave roses, as shown in Figure 5, for both measured and simulated data at point IP3, south of Tenerife. Panels on the left and in the middle show overall good agreement between average measured and simulated wave direction, although the models slightly overestimate directional dispersion around the mean. Regarding wave height (panels a and d) and period (panels b and e), it can be observed that, in general, the wave model tends to slightly overestimate the significant wave height while weakly underestimating the peak period.

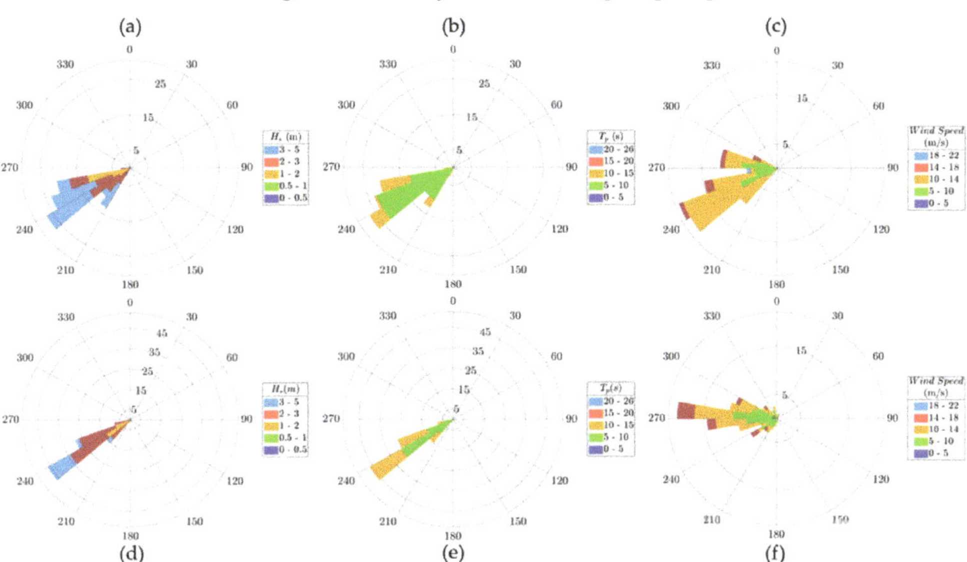

Figure 5. Directional distributions of H_{m0} (**a**,**d**), T_p (**b**,**e**) and wind speed (**c**,**f**) during wave storm periods for simulated (IP3, upper plots: (**a**–**c**)) and measured (BTS, lower plots (**d**–**f**) data.

The aforementioned increment in directional dispersion is more noticeable in the case of wind (panels c and f), revealing some weaknesses of the atmospheric model for reproducing wind conditions south of the archipelago during stormy conditions. However, despite the above differences, mostly due to large orographic complexity, the average wind direction and wind intensity are reasonably well reproduced. It is important to note that these effects are much less important in the north of the islands, and are substantially attenuated, both north and south, when all data, not only those of stormy scenarios, are considered. In conclusion, even though the models exhibit some weaknesses, there is overall fairly good agreement between wave measurements and simulations, so that the model results may be used to explore the spatiotemporal variability of wave storms in the study area, although not without some caution.

4.2. Wave Storm Parameter Selection

As discussed above, the selection of the optimal threshold is an open issue. Consequently, there are several procedures for selecting an appropriate threshold value. Among these, one of the most widely used in practice is the percentile method, which, like the others, incorporates a certain degree of subjectivity but has the advantage of simplicity. Furthermore, some authors attribute a fairly good degree of robustness to this procedure, depending on the specific use (e.g., [30,31]).

In line with previous comments, the approach used to select the threshold, H_t, has been conditioned by three main factors. The first is the specific interest of the study in storms capable of causing substantial damage on sandy beaches. The second is the knowledge of the prevailing wind and wave conditions in the archipelago, and the third is the observed general slight overestimation of significant wave height by the model, mainly during stormy conditions. Accordingly, after trying different quantiles ($Q_{95.0}$, $Q_{99.0}$, $Q_{99.5}$ and $Q_{99.9}$) to select a threshold considering these aspects, it has been observed that using percentiles lower than or equal to $Q_{99.5}$ led to the identification of a substantially high number of storms for an area with a rather moderate wave climate and subjected to strong sheltering effects against the most frequent storms, generally coming from the NW directional sector, as will be discussed in Section 4.5. Finally, the quantile $Q_{99.9}$ has been identified as the most appropriate threshold for the objectives of the study.

Considering the importance of the occurrence of wave storms coinciding with high tide conditions, the minimum allowable temporal distance between storms, ISD_{min}, was set at 48 h, which includes four low and high tide levels and is close to the average duration of atmospheric disturbances in the North Atlantic [11,17]. In the same vein, the minimum duration of the storm has been stated as the duration of a complete tidal cycle, D_{min} = 12 h.

4.3. Wave Storm Severity

The severity of wave storms has been evaluated by estimating the total wave power, *TWP*, as in Equation (1), and storm energy, *E*, as in Equation (2). Results obtained at the outer flanks of the Canary Islands for the period covered by the datasets are shown in Figure 6a. Numbers next to each point indicate the number of identified storms, while values in brackets stand for the threshold used to select the events, corresponding in each case to the associated $Q_{99.9}$. It can be observed that the highest values of *E* and *TWP* are located on the western (EP6, EP7, EP8) and northern (EP8, EP1, EP2) flanks of the archipelago, since these are the sides most exposed to harsh wave fields reaching the islands from directional sectors with north and/or west components. On the contrary, the average energy and total wave power of wave storms detected south or east of the archipelago are substantially lower due to sheltering effects against the direct action of storms approaching from any directional sector, except for those travelling from the SW and S sectors. Points EP4 and EP3 are special cases. The first is located at the south but in the middle of the channel formed by Gran Canaria and Fuerteventura, through which waves coming from the NNW-NNE can propagate. Point EP3 is placed at the eastern side of the archipelago, so that wave directions at this point are restricted to NE-SW. It should be noted that proximity to Africa imposes significant fetch restrictions for wave propagation from the east.

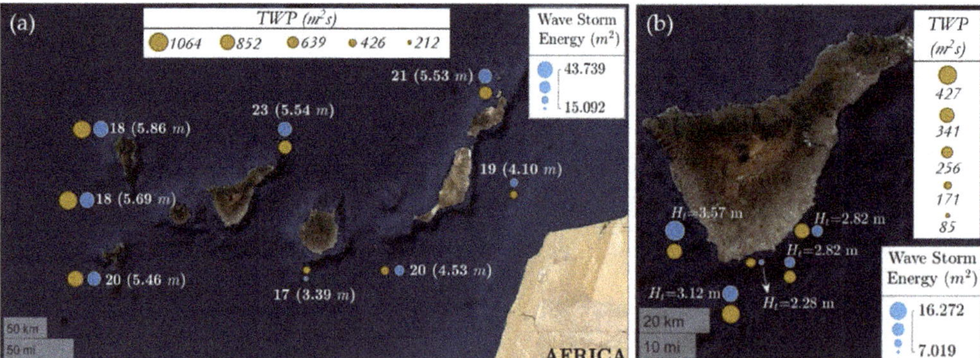

Figure 6. (**a**) Average wave storm energy, *E* and *TWP*, at the outer flanks of Canary Islands (1958–2020). Numbers next to each point indicate the number of identified storms and, in brackets, the threshold used to define such storms. (**b**) Average values of *E* and *TWP* at inner points south of Tenerife for the same period. Numbers next to each point indicate the threshold used to define wave storms.

It is worth mentioning that the thresholds established at each point to select the storms follow the same pattern as the average value of the storm energy. However, this is not the case for the number of wave storms, which is very similar for all the outer locations around the archipelago. Nevertheless, this result is totally consistent since wave storms are defined as a function of the threshold that changes from point to point.

Storminess conditions for inner points located at the southern coasts of Tenerife are shown in Figure 6b, which indicates both the storm energy and total wave power, as well as the selected storm threshold at each point. Results reveal that the average storm energy and total wave power are larger in the southwestern and southern strips, exposed to relatively severe wave storms approaching from the southwest sector, as well as to storms travelling through the G-TF channel from the northwest. It must be stressed that, on this edge of the island, in the vicinity of points IP1 and IP2, are located the most famous beaches of the island. Wave conditions change substantially in the southeastern strips because these areas are protected against wave storms approaching the islands with northern, western, and even southern components. Energy reaching these points comes almost exclusively from the NNE-NE direction, coinciding with the predominant direction of the relatively weak trade winds.

4.4. Wave Storm Seasonality

The polar plot shown in Figure 7 presents the date of occurrence for each storm at locations south of Tenerife, with the bubble size indicating storm energy, E. It is easy to observe the existence of two distinct climatic seasons, one remarkably mild, from April to October, and other relatively stormy, which extends from November to April, although the period of occurrence for more energetic storms is restricted to a shorter time span, from December to March, approximately. This is the period when severe storms are detected in the southernmost places, while extreme storms reach the southwestern locations from November to April, and the southeastern coastal strip is influenced by less severe wave storms from mid-November to mid-April. An interesting, but rare, feature is the detection of two relatively moderate energy storms at location IP5 during summer. These storms are associated with intense trade wind events, generating moderately severe wave conditions and arriving at this location traveling from the NNE-NE, through the GC-T channel.

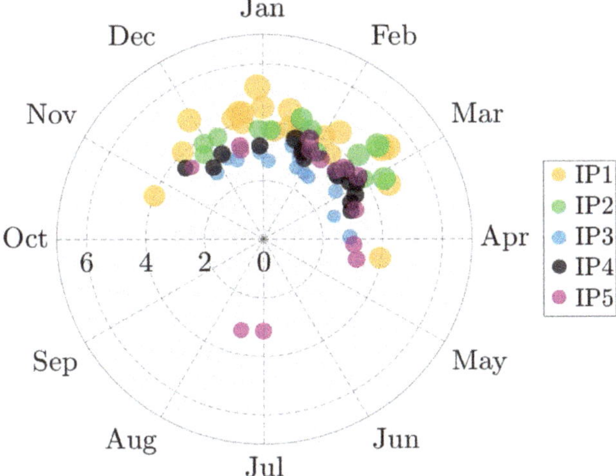

Figure 7. Polar plot of $(H_{m0})_{max}$ and the timing of its occurrence throughout the year at all inner points in the southern flank of Tenerife, for the period 1958–2020. Bubble size indicates storm energy.

Generally speaking, extreme storms occur mainly during winter, between December and March, with a substantially lower probability of presentation during spring and autumn, becoming practically null in summer. Accordingly, Rayleigh and Kuiper tests clearly reject the hypothesis of uniformity at any level of significance. In other words, extreme storms are not randomly distributed around the circle (i.e., along the year) but concentrate during late autumn, winter, and early spring, with the more intense conditions observed during winter. Henceforth, the used tests allow us to accept the existence of a seasonal pattern in the timing of extreme wave storms on a statistical basis. It is interesting to note that the tests indicated above have been applied to be rigorous in the analysis, although, in this particular case, the simple observation of Figure 7 provides qualitative confirmation of this fact.

4.5. Wave Storm Directionality

Wind and wave directionality in the external locations is shown in Figure 8. The left panels depict the directionality of wind and wave conditions considering the whole datasets, while the right panels illustrate directionality during selected storm events. The upper panels reveal that the predominant wind conditions in the outer flanks of the archipelago are associated with the prevalence of trade winds, giving rise to relatively mild wind conditions blowing principally from the NNE sector (panel a). However, during severe wave storms, the prevailing wind conditions change significantly, except for the eastern strip (EP3), where the average direction remains from NNE, with a very small directional spread, but the intensity increases notably (panel b). Wind at locations in the northern and western flanks become dominated by intense winds blowing from the WNW-NNW sector. It is interesting to note that winds at locations north of Lanzarote (EP2) display mixed conditions, receiving the influence of the trade winds, intensified during the summer, and the arrival of strong winds from the NW during winter. At the southern location (EP5), wind conditions during wave storm situations become dominated by W-SW intense wind conditions, while point EP4, located in the channel between GC and F, receives the influence of winds with western, northern, and northeastern components.

With respect to the directionality of wave fields reaching the islands at the outer edges and their severity (middle panels), average directional conditions depict a similar pattern to that of the wind, with the predominance of low and moderate wave conditions travelling mainly from the N-NE directional sector, but rolling towards NW when shifting away from the African coast along the northern side and especially on the western coast (panel c). This effect becomes very clear during stormy wave conditions (panel d). In these situations, severe wave storms travel from the NW sector, affecting predominantly the western and northern sides of the archipelago. Due to self-sheltering effects, locations at the eastern side receive much less energy, with waves travelling from the NNE, while southern locations are affected by mild or moderate storms, following a similar pattern to that of the winds, including its propagation through the channels formed between islands.

Regarding the bivariate distribution of wave direction and period (lower panels), the situation is almost similar to that of wave height and directionality. Thus, during stormy conditions (panel f), waves approaching northern and western locations have notably long periods, revealing a swell structure. However, locations sheltered from these conditions (EP3 and EP5) receive smaller and shorter waves, indicating the frequent presence of wind-driven seas travelling from the NNE (EP3) and from WSW (EP5).

In brief, it is worth noting that most frequent and severe storms arrive from the N-NW sector and affect mainly the western and northern sides of the archipelago. The eastern flank is subjected to a strong sheltering effect against these events, so that it is almost exclusively exposed to wave fields generated by the trade winds (N-NNE). The presence of channels between islands allows the propagation of NW-NE wave fields towards the eastern and western edges of the central islands' southern flanks. In the southern areas, during storm conditions, waves approaching from the S-SW sector predominate, although these are usually more moderate than on the rest of the flanks.

Even taking into account the above-mentioned uncertainty associated with the reproducibility of some parameters during extreme events with models, the differences are substantially large and consistent, showing meaningful and realistic changes between the characteristics of the wave fields under general conditions and during extreme episodes.

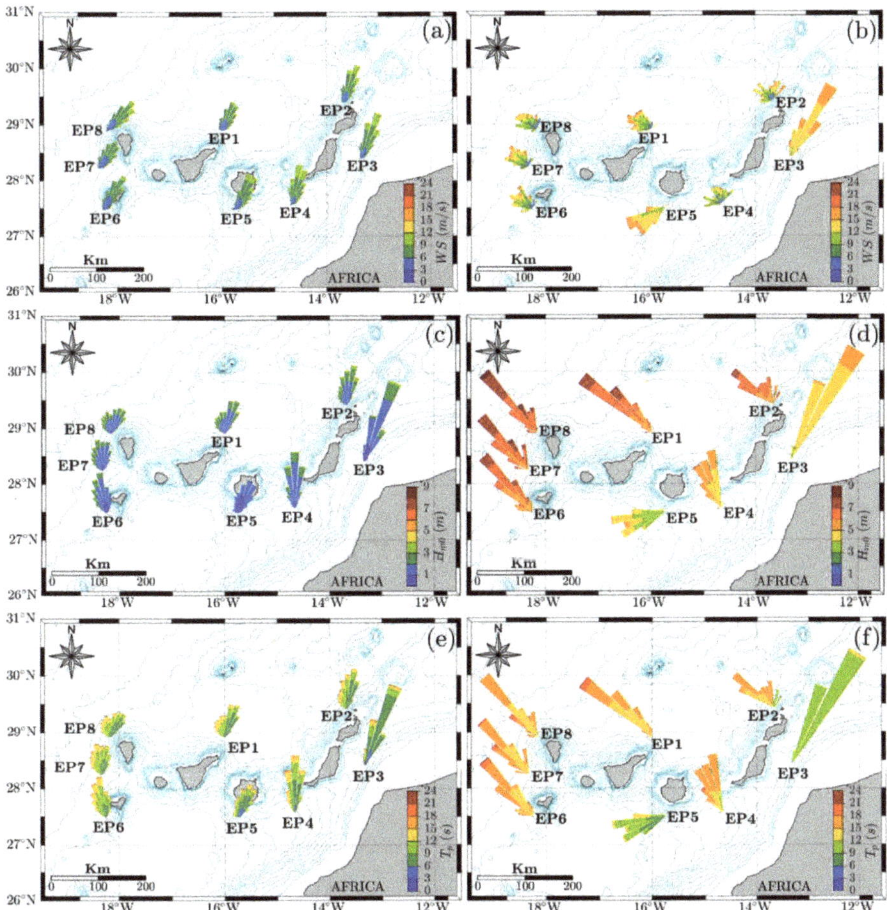

Figure 8. Wind and wave roses at outer locations. Wind roses for the whole dataset (**a**), during stormy conditions (**b**). Wave roses for H_{m0} during all the studied period (**c**) and under wave storm conditions (**d**). Wave roses for peak wave period during the whole study interval (**e**) and during storms (**f**).

Results from the analysis of wind and wave directionality at the inner points, located along the southwest, south, and southeast areas of Tenerife, in terms of wind speed, H_{m0}, and T_p, are shown in Figure 9. Variations in each pair of values are examined under two different types of situation: on the one hand, under average conditions, or average regime, extracted by using the whole dataset, and, on the other hand, during the extreme events selected in each location. Figure 9a,d show that, in general, the dominant direction of the wind regime is NE. Wind speed reaches relatively low values very often (panel a). In contrast, wind speeds observed during wave storms are substantially higher and directionali-ty exhibit a more complex spatiotemporal pattern. In the southwest locations of the island (IP1), the dominant wind direction during wave storms (panel d) shows a considerable dispersion in the NW-NNW sector, although the intensity of the wind is notable only during periods when the wind flows from the NNW, passing through the G-T channel. At points located on the southern flank, the intensity is notably higher than the global average, but the direction in which the wind flows is the opposite, highlighting the presence of strong wind events from the SW. On the other hand, in the southeast locations (IP4, IP5), the predominant direction remains around the NW, but with strong winds blowing along the GC-T channel.

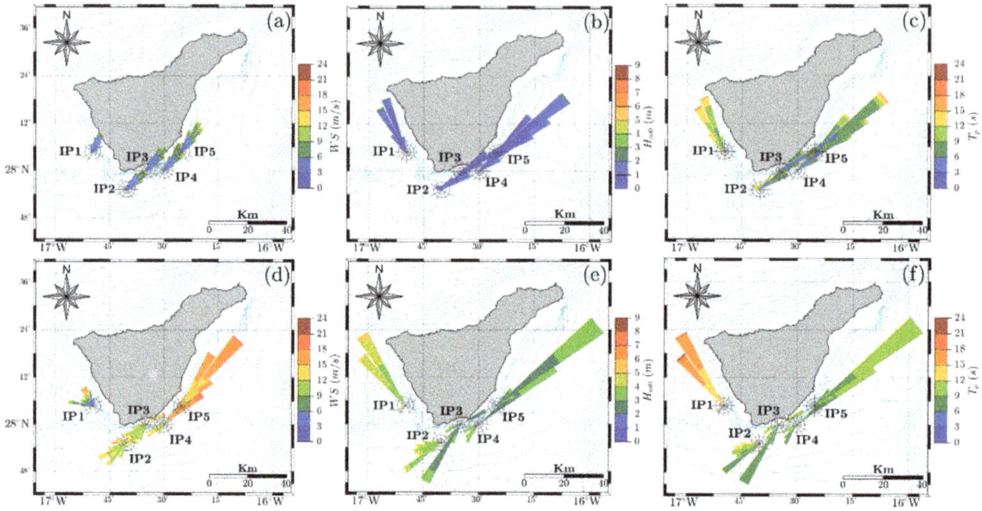

Figure 9. Wind and wave roses at inner locations. Wind roses for the whole dataset (**a**), during stormy conditions (**d**). Wave roses for H_{m0} during all the studied period (**b**) and under wave storm conditions (**e**). Wave roses for peak period during the whole study interval (**c**) and during storms (**f**).

With regard to wave conditions south of Tenerife, it should be noted that, under global conditions (panel b), the southern coasts rarely experience wave conditions with H_{m0} values above 2 m. The most western point (IP1) receives waves from the NNW, while waves reaching the southern (IP2, IP3, IP4) and eastern points (IP5) travel predominantly from the NE. Under severe conditions (panel e), the general pattern of wave directionality is practically the same as for wind in these conditions (panel d). However, note that at the westernmost point (IP1), the directional dispersion observed for the wind disappears, with waves travelling almost exclusively from the NNW, through the G-T channel.

The joint distribution of wave direction and peak period reflects an average behavior (panel c) very similar to that observed for wave direction and height. However, it is interesting to note that T_p values associated with storms observed at south and southeast areas (panel f) correspond mainly to locally wind-driven seas or young swell waves. In particular, waves reaching point IP5 and IP4 are associated with dominant trade winds and are funneled through the GC-T channel. Wave storms detected in the southwest points (IP1) are often due to wave fields reaching these coasts from distant storms located in the northwest area of the North Atlantic, through the channels between G-T, while those located in the southern border (IP2, IP3) show a much more complex pattern generated by the alternation of storms arriving through G-T and GC-T channels, as well as less frequently from the SSW-S directional sector.

4.6. Illustrative Wave Storm Event with Severe Impacts

With the aim of characterizing in more detail wind and wave conditions during storm events on the southern edge of Tenerife, the variability of wind and wave conditions during the selected extreme events has been examined considering the evolution of wind speed and direction, as well as wave height, period, and direction. Furthermore, mean water wave elevation data measured at the tidal gauge north of GC (TG) have been used to explore the contribution of this phenomenon to the impacts of selected extreme wave storms. Although the analysis was carried out for each of the storms identified south of Tenerife, the evolution of the parameters during one selected storm is briefly described below as an example. The selection of this storm has been made on the basis of several storm characteristics (storm energy, maximum value of H_{m0} during the event, storm duration, wave direction during the storm) and the existence of evidence of its impact on tourist beaches extracted from the local press historical archives.

The selected storm occurred during late February and early March 2018, and it has been selected principally because of its significantly long duration (around 90 h). The maximum significant wave height during the storm was 4.1 m and the associated return period 12.3 years. This storm, named Emma, began to develop around 23 and 24 February to the NW of the Canary Islands, and progressively increased in intensity to reach its maximum between 26 February and 01 March, while it remained more or less stationary over the archipelago. Then, it continued its trajectory towards Northern Europe, progressively moving away from the islands. This low-pressure system resulted in strong winds blowing from the W-SW sector towards the islands and generating a wave storm that affected almost all the islands and, in particular, the southern and southwestern areas of Tenerife, with a large impact on the tourist beaches located on this coastal stretch.

Figure 10 shows the wave storm evolution as observed at point IP2. Figure 10c shows that before 23 February, the wind direction presents a remarkable variability, until 23 February, when it was established in the S-SW sector, as with the wave direction. On the other hand, Figure 10a reveals a rather positive relationship between the temporal evolutions of wind speed and wave height, indicating that it was a locally generated wave storm. This is also evidenced in Figure 10b, in which it can be seen that both the average and peak periods adopt quite low values during the episode. Finally, in Figure 10d, it can be seen that the storm's peak coincided with a fairly large high tide, with a tidal range of around 2 m. Regarding the effect of tidal level on the wave storm impacts, the evaluation of tidal ranges coincident with the timing of storm occurrence, during the period when tidal information is available (1992–2020), reveals that the tidal range during the storms identified as damaging events was, in all cases, higher than the average value (1.5 m), pointing out the well-known significant contribution of this factor to the impact of wave storms on the coastline. In this particular example, the combination of both phenomena caused important negative impacts, including damage to coastal structures and buildings near the coastline, as well as erosion problems on the beaches of Arona (Southwestern Tenerife), which were reported by the local press, such as observed in Figure 11, which shows the lack of sand and the severe damage caused on the promenade of Los Tarajeles beach, in Los Cristianos (Arona), which had to be closed during the period of their repair. Furthermore, the "Francisco Andrade Fumero" promenade, on Las Américas beach, was closed to public use due to flooding and damage. During the stormy period, the red flag was present on all the beaches of the highly tourism-dependent municipality of Arona.

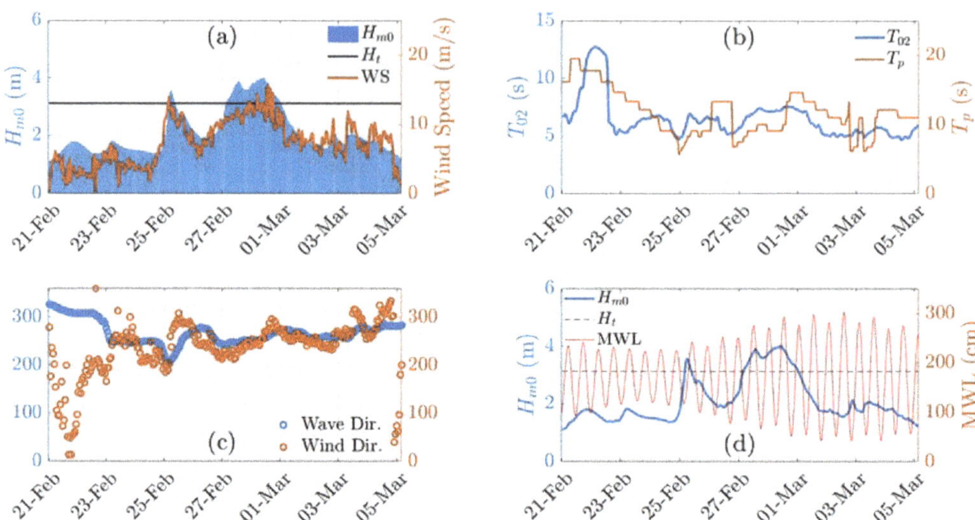

Figure 10. Significant wave height, wind speed (**a**), mean and peak periods (**b**), wind and wave directions (**c**), H_{m0}, and mean water level (**d**) evolution at point IP2 during a severe storm affecting southern strips of Tenerife island.

Figure 11. Photograph illustrating the effects of the storm Emma on Los Tarajales beach and promenade, municipality of Arona, Tenerife (Diario de Avisos, 4 March 2018).

5. Conclusions

Results of the long-term wind and wave datasets' analysis reveal a complex spatial pattern of wave storminess around the islands and in the southern flank of Tenerife, due to the intricacy of the coastline geometry, the presence of deep channels between islands, the high altitude and complex topography of the islands, and the sheltering effects exerted by each island over the others, depending on the directionality of the incident wave fields.

The most energetic events are detected on the western and northern flanks of the archipelago, while wave storms detected south or east of the archipelago are comparatively much milder, mainly due to sheltering effects against the direct action of storms approaching from any directional sector, except from those travelling from the SW and S sectors. In this sense, the points in the middle of the channels formed between the islands are special cases because they can receive waves from the NNW-NNE sector.

South of Tenerife, the severity of wave storms shows substantial spatial variability, with larger values in the southwestern and southern strips, exposed to relatively severe wave storms approaching from the SW, as well as to storms travelling through the G-TF channel from the NW.

The timing of extreme wave storms throughout the year in this area exhibits a marked seasonal character, occurring mainly during winter, between December and March, and becoming practically null in summer.

The directionality of wave storms at the southern flank of the island shows considerable spatial variability but reduced directional dispersion, mainly during stormy conditions, the period during which the channels on both sides of the island play a major role, with waves travelling through both channels but mainly from the NNW, through the G-T channel.

With regard to the role of sea water level during stormy conditions, it has been observed that the tidal range during the storms identified as damaging events was, in all cases, higher than the average value, highlighting the significant contribution of the tide to the impact of wave storms on the coastline.

In brief, the results evidence the importance of sheltering effects and the role of the G-T and GC-T channels, allowing NW and NE wave fields to reach the eastern and western edges of Southern Tenerife, with special relevance of the G-T channel through which waves generated by N-NW storms can reach tourist beaches located in the southwest coastal strip. Moderate storms from the W-SW-S sector predominate in the south-central area, while relatively weak storms propagating through the GC-T channel and associated with N-NE wind conditions dominate in the eastern coastal stretch.

Finally, the detailed analysis of specific severe storms highlights the vulnerability of tourist beaches on the southern and southwestern strips of Tenerife to unusual wave storms approaching from the S-SW sector, or from the NNW through the G-T channel, and the consequent strong socioeconomic impact of such events on this strategic beach tourism destination.

Author Contributions: Conceptualization, methodology and supervision, G.R.; software and validation, formal analysis, resources, and data curation, D.G.-M.; writing—original draft preparation, G.R.; writing—review and editing, G.R. All authors have read and agreed to the published version of the manuscript.

Funding: This research received no external funding.

Institutional Review Board Statement: Not applicable.

Informed Consent Statement: Not applicable.

Data Availability Statement: Datasets used in the study can be obtained from Puertos del Estado (http://www.puertos.es/es-es/oceanografia/Paginas/portus.aspx (accessed on 27 May 2021)).

Acknowledgments: The authors would like to thank Puertos del Estado for providing the meteoceanic data used in this work.

Conflicts of Interest: The authors declare no conflict of interest.

References

1. Belibassakis, K.A.; Karathanasi, F.E. Modelling nearshore hydrodynamics and circulation under the impact of high waves at the coast of Varkizain Saronic-Athens Gulf. *Oceanologia* **2017**, *59*, 350–364. [CrossRef]
2. Oreja Rodríguez, J.R.; Parra-López, E.; Yanes-Estévez, V. The sustainability of island destinations: Tourism area life cycle and teleological perspectives. The case of Tenerife. *Tour. Manag.* **2008**, *29*, 53–65. [CrossRef]
3. ISTAC. Instituto Canario de Estadística. Available online: http://www.gobiernodecanarias.org/istac/ (accessed on 28 May 2021).
4. Domínguez, L.M. The tourism model of the Canary Islands (in Spanish). *Etudes Caribeennes* **2008**, *III*, 81–98.
5. Lusito, L.; Francone, A.; Strafella, D.; Leone, E.; D'Alessandro, F.; Saponieri, A.; De Bartolo, S.; Tomasicchio, G.R. Analysis of the sea storm of 23rd–24th October 2017 offshore Bari (Italy). *Aquat. Ecosyst. Health Manag.* **2020**, *23*, 445–452. [CrossRef]

6. Rodríguez, G.; Pacheco, M.; González, J. Wind-wave energy assessment around Canary Islands. In *Proceedings of MAREC-2002, International Conference on Marine Renewable Energy*; Institute of Marine Engineering, Science, and Technology: Newcastle, UK, 2002; pp. 183–190. Available online: https://accedacris.ulpgc.es/handle/10553/51860 (accessed on 28 May 2021).
7. Iglesias, G.; Carballo, R. Wave power for la Isla Bonita. *Energy* **2010**, *35*, 5013–5021. [CrossRef]
8. Pérez, N.; Rodríguez, G.; Pacheco, J.M. Atmospheric recirculation on the east coast of Gran Canaria Island. *WIT Trans. Ecol. Environ.* **2014**, *183*, 15–25.
9. Chiri, H.; Pacheco, M.; Rodríguez, G. Spatial variability of wave energy resources around the Canary Islands. *WIT Trans. Ecol. Environ.* **2013**, *169*, 15–26.
10. Di Paola, G.; Aucelli, P.P.C.; Benassi, G.; Iglesias, J.; Rodríguez, G.; Rosskopf, C.M. The assessment of the coastal vulnerability and exposure degree of Gran Canaria Island (Spain) with a focus on the coastal risk of Las Canteras Beach in Las Palmas de Gran Canaria. *J. Coast. Conserv.* **2018**, *22*, 1001–1015. [CrossRef]
11. Vega, J.L.; González, J.; Rodríguez, G. Statistical assessment of annual patterns in coastal extreme wave conditions. *WIT Trans. Ecol. Environ.* **2013**, *169*, 39–49.
12. Di Paola, G.; Iglesias, J.; Rodríguez, G.; Benassi, G.; Aucelli, P.P.C.; Pappone, G. Estimating coastal vulnerability in a meso-tidal beach by means of quantitative and semi-quantitative methodologies. *J. Coast. Res.* **2011**, *61*, 303–308. [CrossRef]
13. Puertos del Estado. REDMAR (Tide Gauges Network of the Spanish Port Authority) Resumen de Parámetros Puerto de Las Palmas. 2019. Available online: https://bancodatos.puertos.es/BD/informes/globales/GLOB_2_3_3450.pdf (accessed on 28 May 2021).
14. JABLE Digital Press Archive, University of Las Palmas de Gran Canaria. Available online: https://jable.ulpgc.es/ (accessed on 28 May 2021).
15. Rodríguez, G.; Guedes Soares, C.; Machado, U. Uncertainty of the sea state parameters resulting from the methods of spectral estimation. *Ocean Eng.* **1999**, *26*, 991–1002. [CrossRef]
16. Holthuijsen, L.H. *Waves in Oceanic and Coastal Waters*; Cambridge University Press: Cambridge, UK, 2007.
17. Salvadori, G.; Tomasicchio, G.R.; D'Alessandro, F.; Lusito, L.; Francone, A. Multivariate sea storm hindcasting and design: The isotropic buoy-ungauged generator procedure. *Sci. Rep.* **2020**, *10*, 1–12. [CrossRef] [PubMed]
18. Di Paola, G.; Rodríguez, G.; Rosskopf, C.M. Short- to mid-term shoreline changes along the southeastern coast of Gran Canaria Island (Spain). *Rend. Fis. Acc. Lincei* **2020**, *31*, 89–102. [CrossRef]
19. Abdolali, A.; van der Westhuysen, A.; Ma, Z.; Mehra, A.; Moghimi, S. Evaluation the accuracy and uncertainty of atmospheric and wave model hindcast during severe events using model ensembles. *Ocean Dyn.* **2021**, *71*, 217–235. [CrossRef]
20. Tuomi, L.; Pettersson, H.; Fortelius, C.; Tikka, K.; Björkqvist, J.V.; Kahma, K.K. Wave modelling in archipelagos. *Coast. Eng.* **2014**, *83*, 205–220. [CrossRef]
21. Bidlot, J.; Holmes, D.J.; Wittmann, P.A.; Roop Lalbeharry, R.; Chen, H.S. Intercomparison of the performance of operational ocean wave forecasting systems with buoy data. *Weather Forecast.* **2002**, *17*, 287–310. [CrossRef]
22. Cavaleri, L. Wave modelling—Missing the peaks. *J. Phys. Oceanogr.* **2009**, *39*, 2757–2778. [CrossRef]
23. Guzzetti, F.; Tonelli, G. Information system on hydrological and geomorphological catastrophes in Italy (SICI): A tool for managing landslide and flood hazards. *Nat. Hazards Earth Syst. Sci.* **2004**, *4*, 213–232. [CrossRef]
24. Garnier, E.; Ciavola, P.; Spencer, T.; Ferreira, O.; Armaroli, C.; McIvor, A. Historical analysis of storm events: Case studies in France, England, Portugal and Italy. *Coast. Eng.* **2018**, *134*, 10–23. [CrossRef]
25. Boholm, M. Risk and Casuality in Newspaper Reporting. *Risk Anal.* **2009**, *29*, 1566–1577. [CrossRef]
26. Dolan, R.; Davis, R.E. An intensity scale for Atlantic coast northeast storms. *J. Coast. Res.* **1992**, *8*, 840–853.
27. Mendoza, E.; Jiménez, J.A. Storm-induced beach erosion potential on the Catalonian Coast. *J. Coast. Res.* **2006**, *48*, 81–88.
28. Lin-Ye, J.; García-León, M.; Gràcia, V.; Ortego, M.I.; Stanica, A.; Sánchez-Arcilla, A. Multivariate hybrid modelling of future wave-storms at the northwestern Black Sea. *Water* **2018**, *10*, 221. [CrossRef]
29. Jammalamadaka, S.R.; SenGupta, A. *Topics in Circular Statistics*; World Scientific Press: Singapore, 2001; p. 176.
30. Davies, G.; Callaghan, D.P.; Gravois, U.; Jiang, W.; Hanslow, D.; Nichol, S.; Baldock, T. Improved treatment of non- stationary conditions and uncertainties in probabilistic models of storm wave climate. *Coast. Eng.* **2017**, *127*, 1–19. [CrossRef]
31. Arns, A.; Wahl, T.; Haigh, I.D.; Jensen, J.; Pattiaratchi, C. Estimating extreme water level probabilities: A comparison of the direct methods and recommendations for best practise. *Coast. Eng.* **2013**, *81*, 51–66. [CrossRef]

Article

Geomorphological Approach to Cliff Instability in Volcanic Slopes: A Case Study from the Gulf of Naples (Southern Italy)

Giuseppe Di Crescenzo [1], Nicoletta Santangelo [2], Antonio Santo [1] and Ettore Valente [2,*]

[1] Department of Civil, Constructional and Environmental Engineering, University of Naples Federico II, Piazzale Tecchio 80, 80125 Naples, Italy; giuseppe.dicrescenzo@unina.it (G.D.C.); antonio.santo@unina.it (A.S.)

[2] Department of Earth, Environmental and Resources Sciences, DISTAR, University of Naples Federico II, Via Cinthia 21, 80126 Naples, Italy; nicsanta@unina.it

* Correspondence: ettore.valente@unina.it

Abstract: This paper deals with the problem of cliff stability and proposes a geomorphological zonation of a cliff using a sector of the Posillipo promontory (named the Coroglio-Trentaremi sea cliff, Italy), in the Campi Flegrei coastal area, as a case study. A detailed geological and geomorphological analysis was carried out, by combining field work with analysis of detailed scale topographic maps, orthophoto, and stratigraphical data from deep boreholes. Field and borehole data, together with structural data collected in seven different stations along the cliff, allowed us to derive six geological cross-sections and to reconstruct the complex stratigraphical and structural setting of the cliff. Geomorphological analysis focused on the detection of the main geomorphological factors predisposing to cliff instability. We selected the most significant factors and divided them into two groups: factors influencing landslide intensity and factors influencing cliff instability. Then, by means of a heuristic approach, we constructed a matrix that was used to derive a map showing the geomorphological zonation of the sea cliff. This map may enable to development of a reliable scenario of cliff instability and consequent retreat, which may be useful either to plan intervention works in the most critical areas or to organize prevention plans aimed at risk mitigation.

Keywords: volcanic coast; cliff instability; rock falls; Campi Flegrei; southern Italy

Citation: Di Crescenzo, G.; Santangelo, N.; Santo, A.; Valente, E. Geomorphological Approach to Cliff Instability in Volcanic Slopes: A Case Study from the Gulf of Naples (Southern Italy). *Geosciences* **2021**, *11*, 289. https://doi.org/10.3390/geosciences11070289

Academic Editors: Gianluigi Di Paola, Germán Rodríguez, Carmen M. Rosskopf and Jesus Martinez-Frias

Received: 29 May 2021
Accepted: 7 July 2021
Published: 12 July 2021

1. Introduction

Studying the evolution of coastal areas is a crucial issue because coastal areas have experienced human frequentation since historical and even pre-historical times [1,2] and host many socio-economic activities [3]. Coastal areas are very dynamic environments that are subject to modifications even at short timescales. Such modifications depend on several factors, such as climate change, wave action, isostasy, geology, tectonics, and anthropic actions [4]. This is true both for sandy and rocky coasts. Moreover, among the above-mentioned factors, geology and tectonics play a crucial role in the evolution of rocky coasts. In fact, sea cliffs carved in soft sediments evolve more rapidly than sea cliffs carved in hard rock-types [5]. For example, Brooks and Spencer [6] estimated a retreat rate of 0.9–3.5 m/y since 1883 for a sector of the East Anglia sea cliff carved in soft sediments, e.g., mainly unconsolidated sands and clays. Young [7] estimated an average cliff retreat rate of 0.25 m/y since the early 1990s for a sector of the California coastline in the USA, carved in lithified Cenozoic units (e.g., mudstone, shale, sandstone, and siltstone). Epifano et al. [8] estimated an average retreat rate of 0.044 m/y in a 60-year-long period for a rocky coast carved in Jurassic marls and sandstones in Portugal. Furthermore, retreat rates of rocky coasts are also dependent on the amount and density of discontinuities, e.g., bedding, fractures, and faults, even when they affect shore platforms. Naylor and Stephenson [9] analyzed discontinuities within shore platform in Wales and Australia and derived that the higher the density of discontinuities the more rapidly the shore platform will erode, with consequences also for sea cliffs behind shore platforms.

In this paper we have carried out a geological and geomorphological analysis of a portion of the Campi Flegrei coastline, in the Gulf of Naples (southern Italy). The Campi Flegrei is a volcanic area located along the Tyrrhenian flank of the Southern Apennines (Figure 1). Morphoevolution of the Campi Flegrei has been characterized by the interaction of volcanism and relative sea-level variation due to climate and bradyseism ([10] and reference therein). As a result, both sandy beaches and rocky coasts are present along the coast of the Campi Flegrei. Regarding rocky coasts, they are all carved in volcanic units and are affected by mainly NE-SW and NW-SE trending faults and fractures [11]. These volcanic units include both consolidated tuffaceous deposits, e.g., the Neapolitan Yellow Tuff, hereinafter NYT, aged 15 ka [12], and unconsolidated pre- and post-NYT pyroclastics.

Several studies have been carried out on specific sectors of the Campi Flegrei coast, e.g., the Monte di Procida sea cliff [13], the Miseno Cape [14], the gulf of Pozzuoli [15], and the Posillipo promontory [16]. These studies highlighted the role of ground motions, e.g., bradyseism, besides rock-types, volcano-tectonics, and climate, in shaping the coastline. Furthermore, researches along the Posillipo promontory have been focused on very specific areas, such as the Nisida island [17], the Coroglio sea cliff [18], and the Marechiaro area [19]. Researches on the Posillipo promontory have also pointed out that this amazing portion of the Campi Flegrei coastline has attracted human frequentation since historical times [17,19]. Evidence of this includes the Roman ruins of Palazzo degli Spiriti, the archaeological area of Pausyllipon and the Roman villae of Villa Diana. By the way, despite these very detailed scale analyses, research along the Posillipo coastline has never been performed along a wider portion. To tackle this issue, we analyzed the westernmost portion of the Posillipo promontory, which extends from the Coroglio sea cliff to the NW, to the La Gaiola islet, to the SE (Figure 1).

The main goal of our paper is to provide robust geological and geomorphological field data that may allow us to characterize this sector of the Campi Flegrei coastline with the aim of identifying the areas more susceptible to cliff retreat. The assessment of cliff retreat susceptibility is not a simple task and researchers can adopt various approaches (heuristic, statistic, deterministic ([20] and references therein)) at different scales (from local to regional) [3,20–26]. Our case study is presented at a local scale and is representative of cliffs in volcanic rocks made up of alternating soft and hard pyroclastic deposits. We adopted a heuristic criterion and analyzed the main predisposing factors occurring in the study area. Our aim is not to propose a new method to assess susceptibility but only to present a case study that may be a useful example and source of data for further studies on the stability of similar volcanic coastal cliffs [13,16,18,27].

2. Geological and Geomorphological Setting of the Campi Flegrei

The Campi Flegrei is a volcanic area placed at the border of the Campana Plain, a wide tectonic depression occurring along the Tyrrhenian, inner, flank of the Southern Apennines ([28] and reference therein). The Southern Apennines are a NE-vergent fold and thrust belt formed by the collision of the Eurasian and African plates [29,30]. Tectonic depressions along the inner sector of the mountain belt include large peri-Tyrrhenian grabens formed because of the extensional tectonics due to the opening of the Tyrrhenian back-arc basin [31,32]. Among these grabens, the Campania Plain is the largest one. The sedimentary filling of the Campania plain consists of ~3000 m of marine, transitional, and continental deposits, with abundant volcaniclastic deposits produced by both Vesuvius and the Campi Flegrei [28,33,34]. The study area falls within the Campi Flegrei (Figure 1), which is a volcanic field placed in a resurgent caldera (Figure 2).

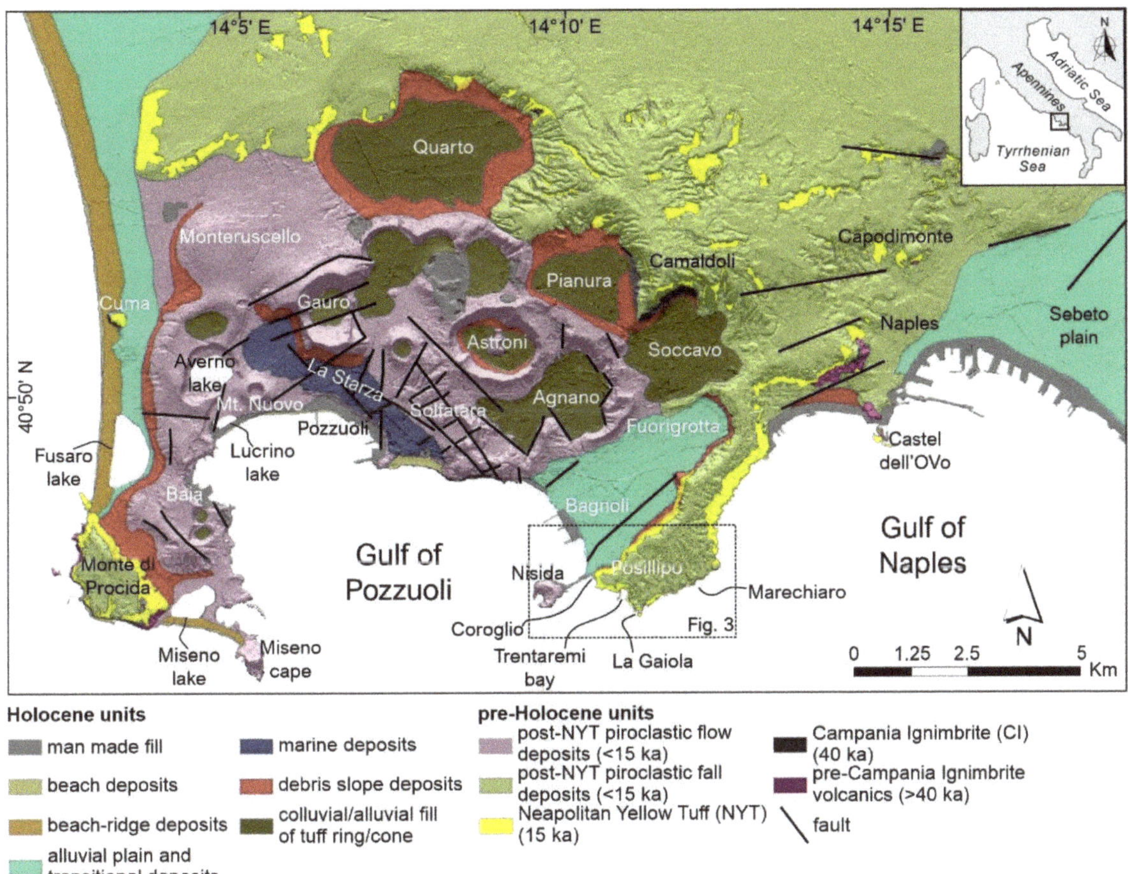

Figure 1. Geolithological map of the Campi Flegrei (modified from [10]). Faults are derived from the Geological Sheet 447—Napoli of the Geological Map of Italy at scale 1:50,000, CARG Project [35]. Dashed rectangle indicates location of the Coroglio-Trentaremi sea cliff shown in Figure 3.

The time at which volcanic activity in the Campi Flegrei began is unknown and the oldest outcropping volcanic units are ~60 ka old [36]. The caldera, which is quasi-circular and whose diameter is ~8 km long (Figure 2), formed because of collapses during the two strongest eruptions, which are related to the Campanian Ignimbrite (hereinafter CI, ~39 ka old [37–39]) and the Neapolitan Yellow Tuff (hereinafter NYT, ~15 ka old [12]). The CI eruption emplaced ~300 km^3 of pyroclastic fall and flow deposits [40], which have been covered by younger volcanoclastic units. The CI outcrops only in a few sectors of the Campi Flegrei, e.g., at the base of the Camaldoli hill, at Cuma and at the base of the Monte di Procida hill (Figure 1). The NYT eruption extruded at least 40 km^3 of pyroclastic fall and flow deposits that accumulated at the caldera boundaries, thus forming the slopes that limit the entire Campi Flegrei (Figure 1) [10]. Moreover, the hills bounding the Campi Flegrei caldera are mainly formed by the NYT; the thickness of these may exceed several tens of meters, they are blanketed by younger pyroclastic fall deposits, and their age is constrained at 15 ka [35]. Volcanic activity after the NYT eruption has been confined within the caldera and consisted of hydromagmatic phenomena with occasional Plinian and effusive events (which are evidenced by the diffuse presence of post-NYT fall deposits and by local outcrops of lavas). Consequently, several monogenic vents formed, including tuff rings, tuff cones, cinder cones and lava domes (Figure 2) [41–45]. The last volcanic event in the Campi Flegrei occurred in 1538 AD with the formation of Monte Nuovo volcano [46,47] (Figures 1 and 2).

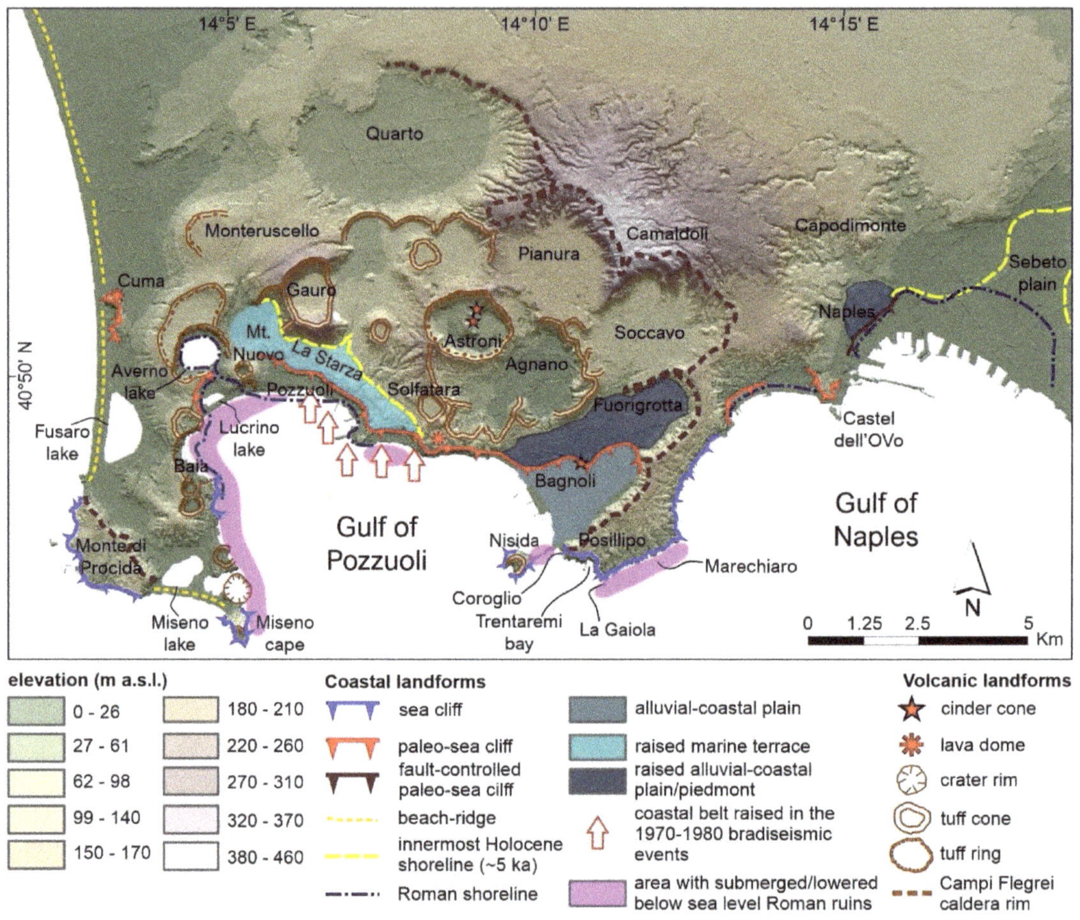

Figure 2. Geomorphological map of the Campi Flegrei (modified from [10]).

The Campi Flegrei's history is also characterized, besides volcanic activity, by ground motions in the form of episodes of uplift and subsidence during bradyseismic crises [48,49]. These episodes range from several meters to several tens of meters and continue nowadays (Figure 2) [50–53]. Ground motions are testified to, above all, by the raised marine terrace of La Starza (Figure 2), near Pozzuoli [43,50,51,54], and by submerged archaeological ruins near Baia, Nisida, La Gaiola and Castel dell'Ovo [19,51,52,55,56] (Figure 2).

The complex volcanic history of the Campi Flegrei has influenced its geomorphological setting, which results in a hilly landscape mainly modelled by slope, volcanic and coastal processes. Elevation within the Campi Flegrei spans from 0 m a.s.l. to 460 m a.s.l., with the highest peak corresponding with the Camaldoli hill (Figure 2). Slope processes include landslides affecting the volcanic slopes that often cause severe damage to people and buildings [57]. Volcanic landforms are clustered within the caldera's inner slope and consist of tuff rings, tuff cones, cinder cones and lava domes formed by volcanic events in the last 15 ka ([10] and reference therein).

Coastal landforms consist of both sandy and rocky coasts. Sandy coasts occur near the main alluvial plains, e.g., the Sebeto and the Fuorigrotta-Bagnoli plain (Figure 2) and suffered man-made intervention and strong urbanization. Sandy beaches are also present to the west of the Campi Flegrei where they still preserve remnants of beach-ridge and pass, towards the inner side, to coastal lakes (e.g., Fusaro lake, Miseno lake and Lucrino lake; Figure 2). Rocky coasts form high and steep sea cliffs occurring near Miseno Cape and around Posillipo. Sea cliffs are cut in the pre-NYT pyroclastics, in the NYT and in the post-NYT pyroclastics. In addition, sea cliffs are dissected by a dense network of faults and fractures, mainly NE-SW and NW-SE trending [11], which may act as predisposing factors for rock falls. Sea cliff instability is also enhanced by diffuse extraction of tuff deposits, since Roman times, which results in anthropic cavities at the base of most of the sea cliffs [10]. Rocky coasts include our study area, which is a sector of the Posillipo coastline, between the Coroglio sea cliff, to the west, and the Trentaremi bay and La Gaiola islet, to the east (Figures 2 and 3).

3. Materials and Methods

The study was based on the geological and geomorphological analysis of the Coroglio-Trentaremi sea cliff by combining field work with analysis of detailed scale topographic maps and orthophotos, and stratigraphical data from deep boreholes.

Geomorphological analysis has been carried out by means of a detailed scale topographic map (Technical Map of the Regione Campania at scale 1:5000), Google Earth images and field work.

This analysis allowed us to reconstruct the main geomorphological features of the sea cliff, and to map the main landform indicators of slope instability. We gave particular attention to detachment niches and landslide bodies and to the presence of notches and cavities at the base of the sea cliff. The latter are indicators of undercutting by waves, which is undoubtedly the most important factor in causing coastal retreat [4,58]. The rate of undercutting is controlled by the complex and wide-ranging behavior of geological materials and by the great variability in their geotechnical properties. According to Budetta [24], the increasing depth of the notch into fractured rock mass, because of wave erosion, is responsible for the spreading of the shear stresses towards the top of the cliff and for the destabilization of the overlying slope. We also considered the presence of ancient quarries (mainly roman in age) for tuff extraction which, accordingly to Ruberti et al. [27], may also be considered as instability factors, especially if involved in the shear stress propagation caused by waves undercutting.

Geological analysis was conducted by means of field work, both inland and by the sea using small boats that allowed to reach some areas not accessible by foot (e.g., most of the sea cliff base), and borehole data [59]. The combination of field and borehole data allowed us to derive 6 geological cross-sections representative of the subsurface stratigraphical setting of the Coroglio-Trentaremi sea cliff. Furthermore, field work also allowed us to define the structural setting of some portions of the sea cliff by establishing 7 structural stations. Structural stations have been established directly on the cliff by climber geologists along 10 vertical rope descents [59]. These structural data have been combined with field data collected at the base of the sea cliffs. As a result, we have been able to establish the relationships between bedding, sea cliff trend, and fractures by defining dip and dip direction and to propose failure mechanisms for each structural station.

By comparing geological and geomorphological collected data, we used a heuristic approach to identify and select two main groups of predisposing factors which are listed in Table 1. The first group includes factors that may influence landslide intensity (e.g., sea cliff height, I1; volume of landslide body, I2; volume of detachment niches, I3; volume of blocks or projecting sectors, I4), and factors that may increase the weakness of a sea cliff (e.g., fracture spacing, W1; persistence of beating fractures, W2; volume of caves at the sea cliff base, W3; distance between the shoreline and the base of active sea cliff, W4). The volume of the caves has been derived by expeditive surveys with a laser pointer. The distance between the present shoreline and the base off the cliff considers the presence/absence of a debris body that may act as protection from the wave action. Each factor includes four group values, which have been weighted according to their attitude to instability.

Table 1. Instability (I) and weakness (W) factors adopted to evaluate the geomorphological zonation to landslide along a sea cliff, with relative weight.

Weight	I1 Sea Cliff Height (m)	I2 Volume of Landslide Body (m^3)	I3 Volume of Detachment Niche (m^3)	I4 Volume of Blocks or Projecting Sectors (m^3)	W1 Fracture Spacing (m)	W2 Persistence of Beating Fractures (m)	W3 Volume of Caves at the Sea Cliff Base (m^3)	W4 Distance between the Shoreline and the Base of Active Sea Cliff (m)
1	0–20	0–10	0–100	0–5	>20	0–5	0	>20
2	21–100	11–30	101–300	6–10	10–19	6–10	1–20	11–20
3	100–200	31–50	301–1000	11–30	5–9	11–30	21–100	2–10
4	>200	>50	>1000	>30	0–4	>30	101–1000	0–1

To evaluate the susceptibility to landslide of a sea cliff, we have crossed the weights of the Intensity (I) and the Weakness (W) factors (Table 2). This matrix suggests that high magnitude (M) landslides may occur when high weight I and W factors occur. As an example, either a very high sea cliff or a sea cliff with a high volume of blocks or projecting sectors coupled with large caves at the sea cliff base may contribute to high magnitude landslides (M3 class in Table 2). On the other hand, a low height sea cliff carved in poorly fractured rocks and with no caves at the sea cliff base may contribute to low magnitude landslides (M1 class in Table 2).

Table 2. Matrix for the evaluation of landslide magnitude (M) by crossing weights of both the Intensity (I) and Weakness (W) factors. Red, orange and yellow colors indicate high (M3), moderate (M2) and low (M1) magnitude classes.

Weight of Weakness Factor \ Weight of Intensity Factor	1	2	3	4
1	M3	M3	M2	M2
2	M3	M3	M2	M1
3	M2	M2	M2	M1
4	M1	M1	M1	M1

4. Results

4.1. Geological Setting of the Coroglio-Trentaremi Sea Cliff

The geological setting of the Coroglio-Trentaremi sea cliff is reported in Figure 3, whereas the stratigraphical relationships between the outcropping rock-types are reported in the stratigraphical columns of Figure 4 and in the geological cross-sections of Figure 5. The sea cliff is mainly cut in tuffaceous deposits of the NYT and in older volcanics, that are mantled by unconsolidated post-NYT pyroclastics.

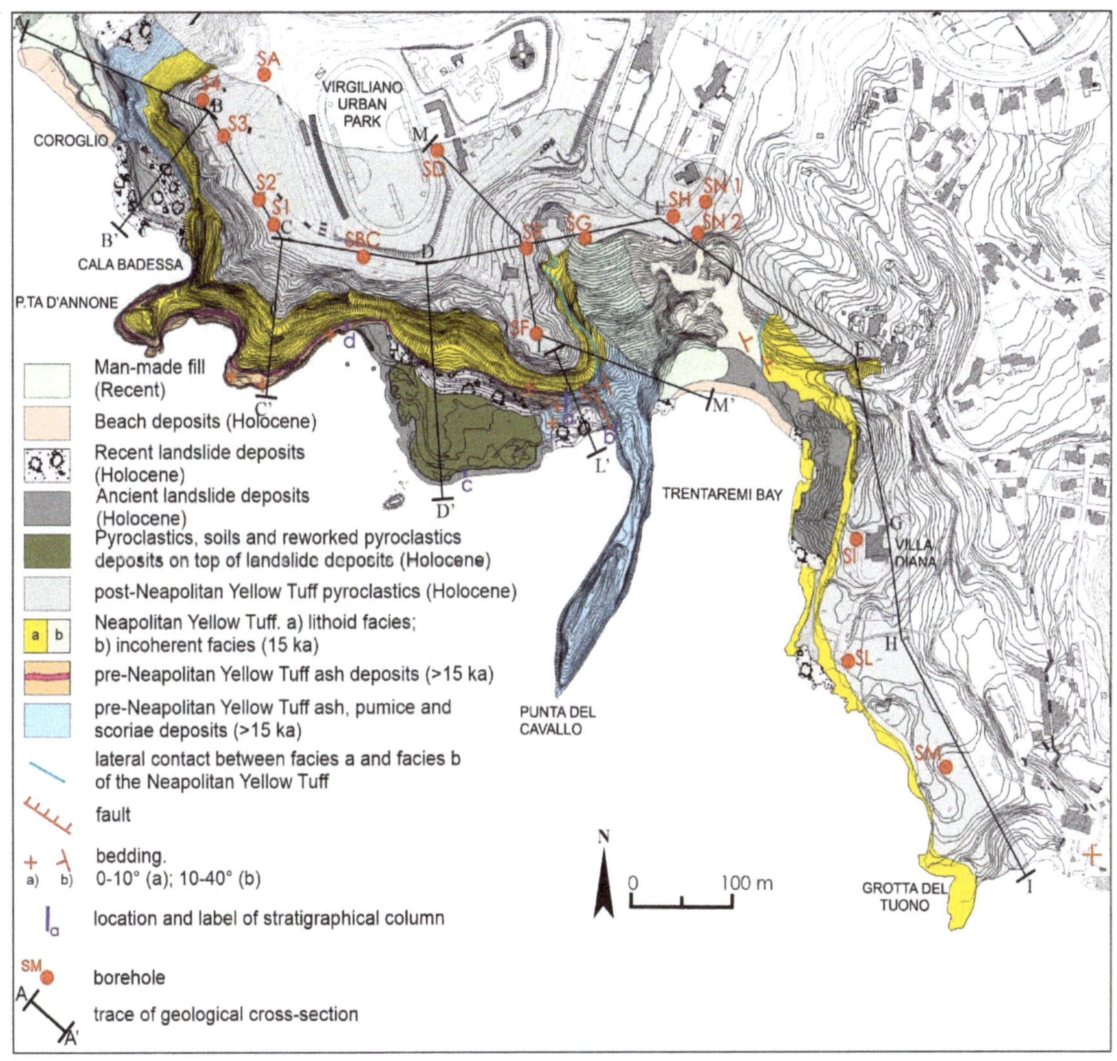

Figure 3. Geological map of the Coroglio-Trentaremi area.

The oldest volcanic unit consists of whitish and stratified tuff deposits, made of a pre-NYT ash, pumice and scoriae (Figure 3), which outcrops in the Coroglio area, to the west, and at Punta Cavallo and Trentaremi bay, to the east (stratigraphical columns A and B in Figure 4). Geological cross-sections A-A′, B-B′, L-L′ and M-M′ in Figure 5 show that this unit forms a morphostructural low where the more recent pre-NYT ashy unit and the NYT accumulated. This unit is strongly inclined in the surroundings of the Trentaremi bay (dip around 45°) and its bottom is made of boulders, scoriae and large pumices, suggesting that this area is pretty close to the possible source area, as has been already hypothesized by Cole et al. [60]. The unit continues upward with greyish-whitish, pisolithic ashes and pumices, with interbedded large pumice-rich layers, whose bedding ranges from plan-parallel to gently dipping and slightly undulating. The base of this unit is not recognizable, whereas its top surface has a probably erosional origin. The age of this unit ranges between 37 ka and 15 ka and is constrained at 20.9 ka by Scarpati et al. [61] (Figure 3).

The pre-NYT ash, pumice and scoriae deposits pass upward to a mainly ashy unit, which has been named pre-NYT ash deposits in Figure 3. This unit outcrops to the west of the study area, near Punta d'Annone (geological cross-sections A-A', C-C' and D-D' of Figure 5) and its thickness ranges between 20 and 30 m (stratigraphical column D in Figure 4). It consists of an alternation of horizontally stratified ashy layers and undulated scoriae rich and ashy layers probably of fall origin. This unit is either not reported in the literature or it is grouped with the NYT [35], its age ranges between 37 ka and 15 ka, and it is tentatively correlable with the Whitish Tuff of Scarpati et al. [62], which is dated at 19.7 ka (Figure 3).

The pre-NYT ash deposits pass upward to the NYT, which outcrops in the entire study area, with thickness ranging from 30 m (to the east, near the Trentaremi bay) to 80–100 m (to the west, near the Coroglio sea cliff) (Figures 3 and 5). It consists of lithified pyroclastic deposits made of pumices, lithics and cm-sized scoriae, yellowish to greenish, in a yellowish ashy matrix [63]. The unit is massive (as a typical pyroclastic flow deposit) being just locally thinly layered and with abundant sub-vertical fractures. The top surface of the NYT is a low-dipping erosional surface and its age is 15 ka [12].

The NYT passes upward to fall deposits related to Phlegrean eruption which occurred in the last 15 ka [43,60], which have been named post-NYT pyroclastics in Figure 3. This unit consists of an alternation of decimeter thick sandy and ashy layers and centimeter thick layers rich in whitish pumices with interbedded thin paleosols that testify to a period of quiescence of the eruptive phases. A paleosol also occurs between the top surface of the NYT and the base surface of the post-NYT pyroclastics. The top surface of the post-NYT pyroclastics is almost flat and corresponds with the topographic surface (Figures 3 and 5). The thickness of this unit is not uniform (Figure 5), ranging from 10–15 m (to the east, near the Trentaremi bay) to 40 m (to the west, near the Coroglio sea cliff), and its age is younger than 15 ka (Figure 3).

The sea cliff foothill, from the Coroglio area to the Trentaremi bay, is mantled by slope debris, landslide bodies and beach deposits, which are mainly located near the Trentaremi bay and whose maximum thickness is around 20 m (Figure 3 and stratigraphical column C in Figure 4). Moreover, landslide deposits consist of reworked ashes with NYT blocks, whose size may exceed some cubic meters, and are due to rock falls affecting the sea cliff, some of which are very recent as suggested by the lack of vegetation and the low erosion by sea waves.

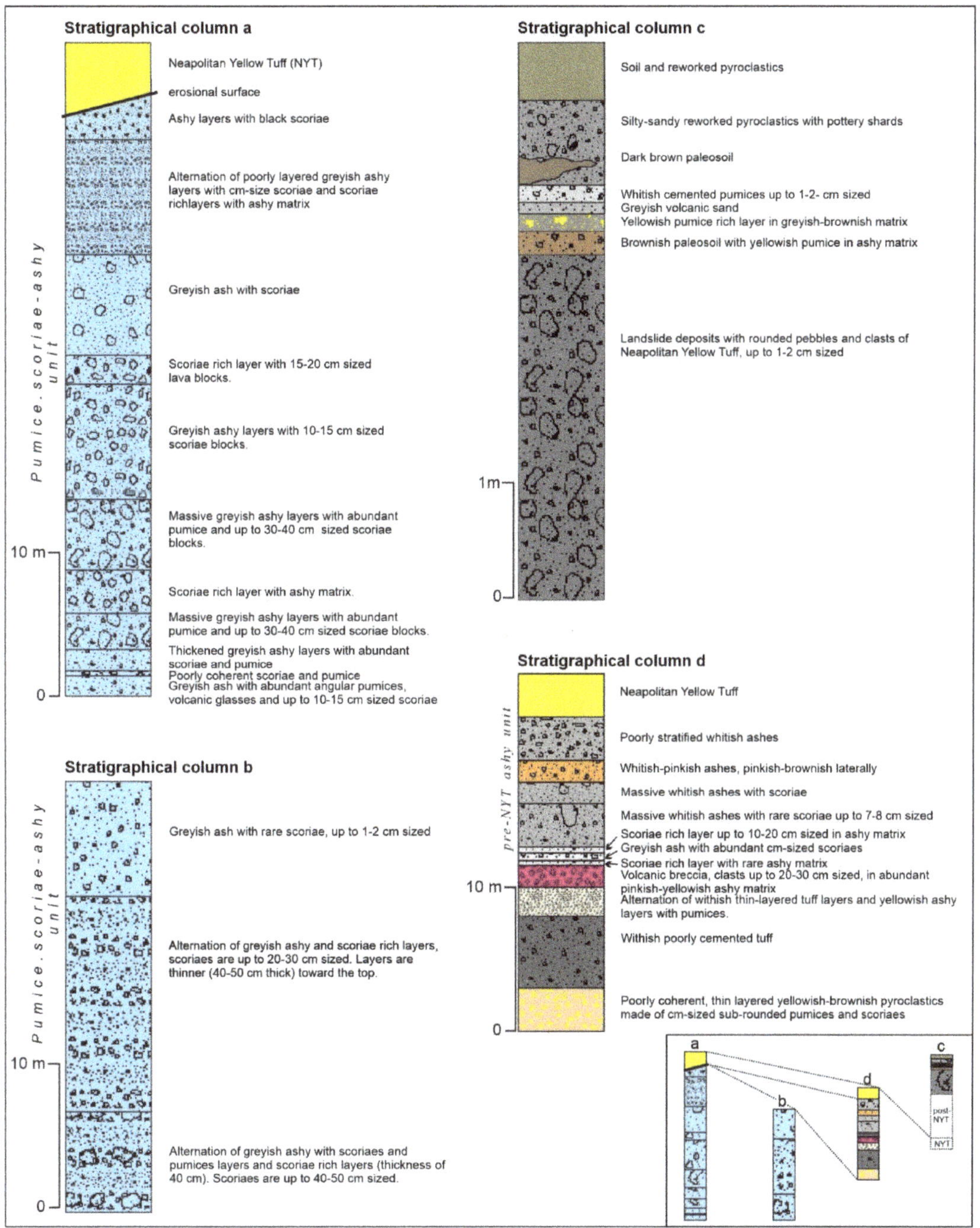

Figure 4. Stratigraphical columns representative of the geological setting of the Coroglio-Trentaremi sea cliff base. Location of stratigraphical column is reported in Figure 3. Inset scheme in the lower right corner of the figure shows stratigraphical correlation between stratigraphical columns.

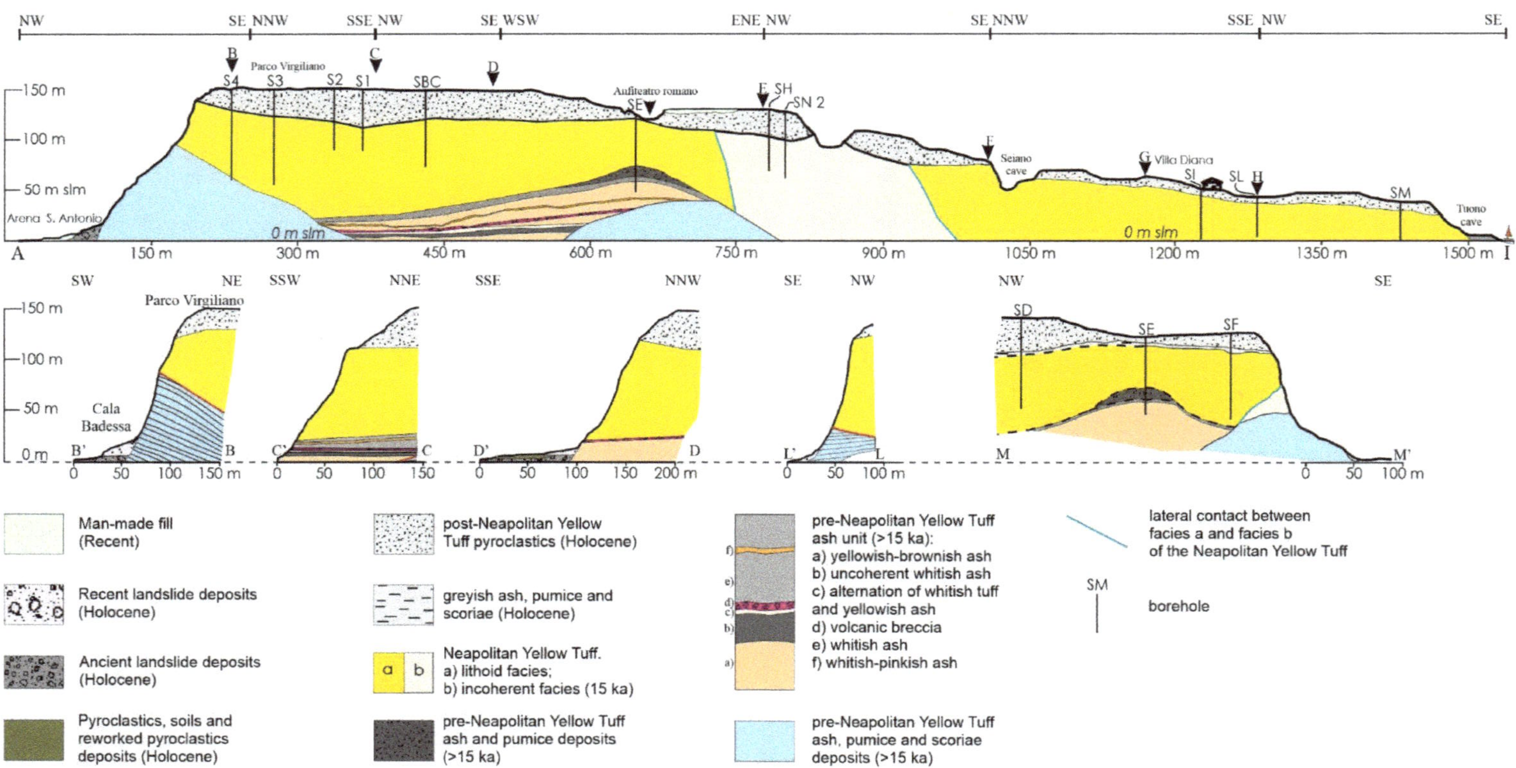

Figure 5. Geological cross-section along the Coroglio-Trentaremi sea-cliff. Cross-section location is reported in Figure 3.

4.2. Geomorphological Setting of the Coroglio-Trentaremi Sea Cliff

Geomorphological analysis has been addressed to the recognition of the predisposing factors that may cause landslides and cliff retreat, and the results are reported in the geomorphological map of Figure 6. The study area is characterized by high sea cliffs that are prone to retreat because of both their stratigraphical setting (alternation of poorly cemented pyroclastics and tuffs) and the occurrence of a dense net of sub-vertical fractures in the tuffaceous deposits.

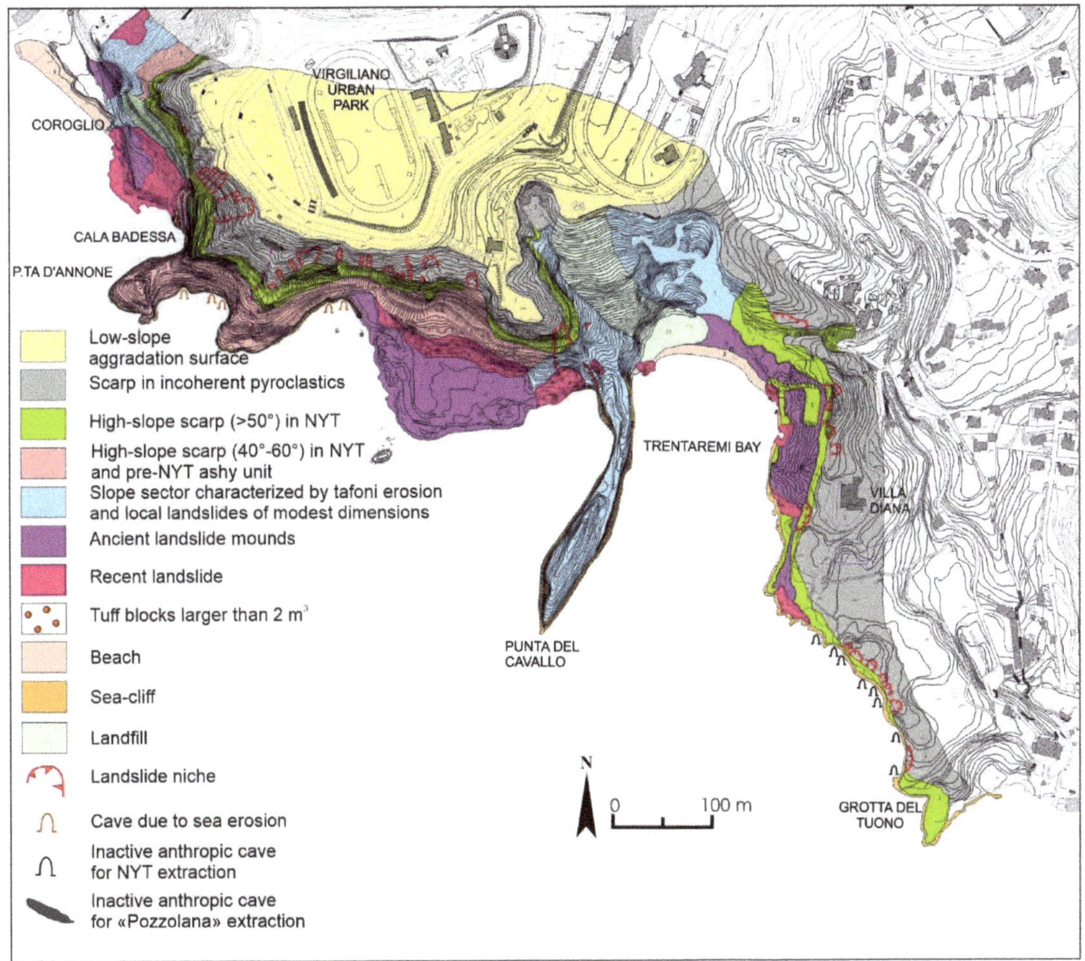

Figure 6. Geomorphological map of the Coroglio-Trentaremi area.

The geomorphological map of Figure 6 highlights the diffuse presence of landslides, which have also been reported on Google Earth 3D view (Figure 7), where they have been classified according to the type of movement and the type of material involved. Moreover, flow-like landslides (yellow dots in Figure 7) affect the sectors of the Coroglio-Trentaremi sea cliff where post-NYT pyroclastics outcrop and with slope angle between 45° and 70°, whereas rock-falls (red dots in Figure 7) occur in the NYT with slope angle between 70° and 90°. We also mapped in Figure 7 the caves at the base of the sea cliff, which are mainly due to anthropic activities for the NYT extraction. The collected data clearly suggest that the main factors influencing the types of landslide are lithology and slope degree (Figure 8). Flow-like landslides remain confined to the upper part of the cliff, where soft unconsolidated deposits rest on top of the NYT formation. The presence of hard/soft rock-types contacts along the sea cliff may also increase the possibility of detachment of rock falls because of selective erosion. Also, the presence of caves (both natural and artificial) plays a crucial role in the evolution of the sea cliff by identifying sectors more prone to instability because of the presence of vacuums that may enhance sea erosion at the base of the cliffs.

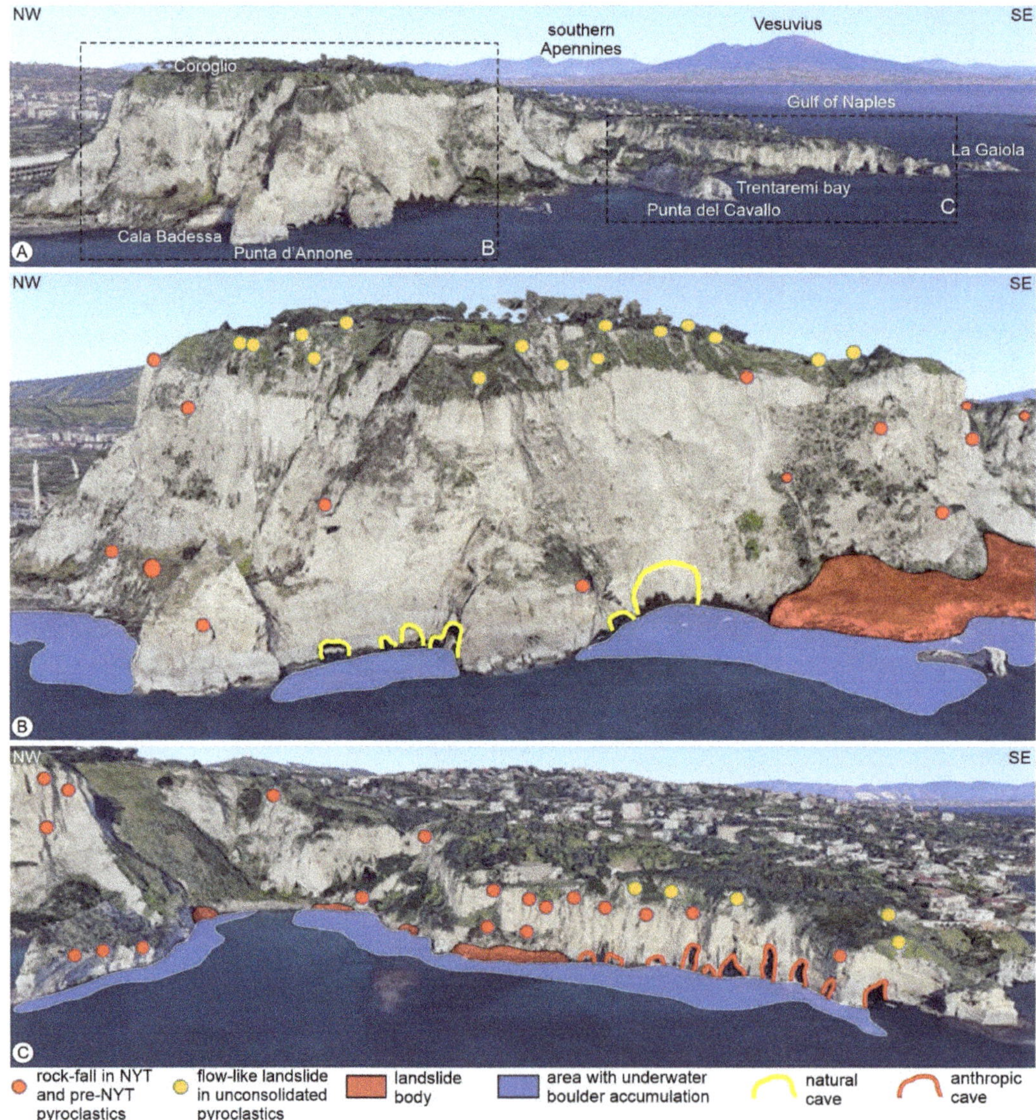

Figure 7. (**A**) Google Earth 3D views of the Coroglio-Trentaremi coastal sector. (**B**) Detail of the Coroglio sea cliff. (**C**) Detail of the Trentaremi sea cliff.

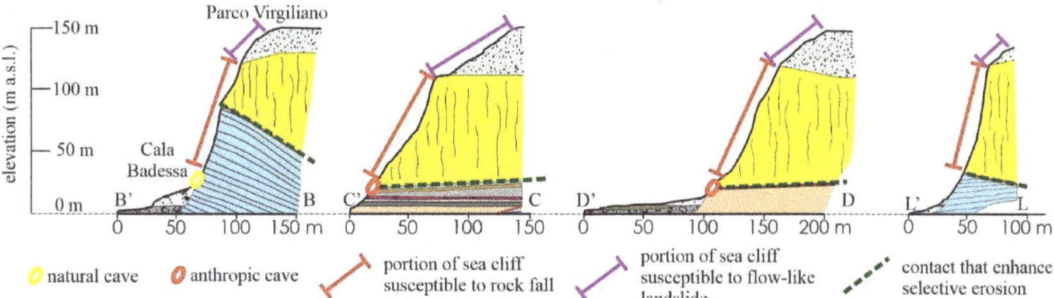

Figure 8. Main predisposing factors controlling cliff stability along the Posillipo promontory. Geological cross-sections are modified from Figure 5. Vertical lines in the NYT indicate sub-vertical fractures affecting this unit.

In Figure 9 we show some photos of the sea cliff that allow us to appreciate the size of the landslide bodies, which are up to some hundreds of cubic meters. Landslide bodies of different ages may also be appreciated because of the lack of vegetation, which testify for the high frequency of rock falls along this sea cliff. The tuffaceous deposits of the NYT are also locally affected by a very dense fractures network that shapes the sea cliff profile. Fractures are in many cases open, thus suggesting the possible occurrence of a landslide event in the near future (Figure 10).

4.3. *Morphostructural Analysis and Failure Mechanism*

Faults and fractures trends have been analyzed by combining a morphostructural analysis of Google Earth orthophotos and detailed scale topographic maps with field surveys aimed at collecting structural data in seven geo-mechanical stations.

Morphostructural lineaments along the Posillipo coast mainly trend N20, N45 and N135 (Figure 11A). Moreover, the more diffuse trends are the N20 and N45, which result to be parallel to the fault that bound the Posillipo promontory towards the NW, i.e., towards the Bagnoli-Fuorigrotta plain [35] (Figure 1).

In addition, NW-SE trending faults and fractures have also been recognized in most of the structural stations along the Posillipo sea cliff, in the area between the Coroglio promontory and the Trentaremi bay (structural stations A to E in Figure 11B). The only exceptions are the structural stations F and G, near the Trentaremi bay, which also record some N-S trend. Faults and fractures measured in all the structural stations have high dip angles that range between 75° and 90°.

The diffuse presence of both NE-SW and NW-SE trending faults and fractures is consistent with data from Vitale and Isaia [11], who performed detailed structural analysis in the Campi Flegrei caldera, and suggested that these trends are the prevalent ones.

Overall data from morphostructural analysis and from the structural stations point out the occurrence of a pervasive system of sub-vertical, mainly NE-SW and NW-SE trending faults and fractures that affect the tuff deposits. These systems also cause the formation of rupture surfaces from which rock-falls originate, thus conditioning the stability of the Coroglio-Trentaremi sea cliff.

Figure 9. Landsliding along the Coroglio-Trentaremi sea cliff. Red arrows indicate landslide bodies, whereas yellow arrows indicate detachment niches.

Figure 10. Beating fractures and rock-falls niches along the Coroglio-Trentaremi sea cliff. Unstable fractures, dihedrons and pinnacles set on high angle systems are well evident.

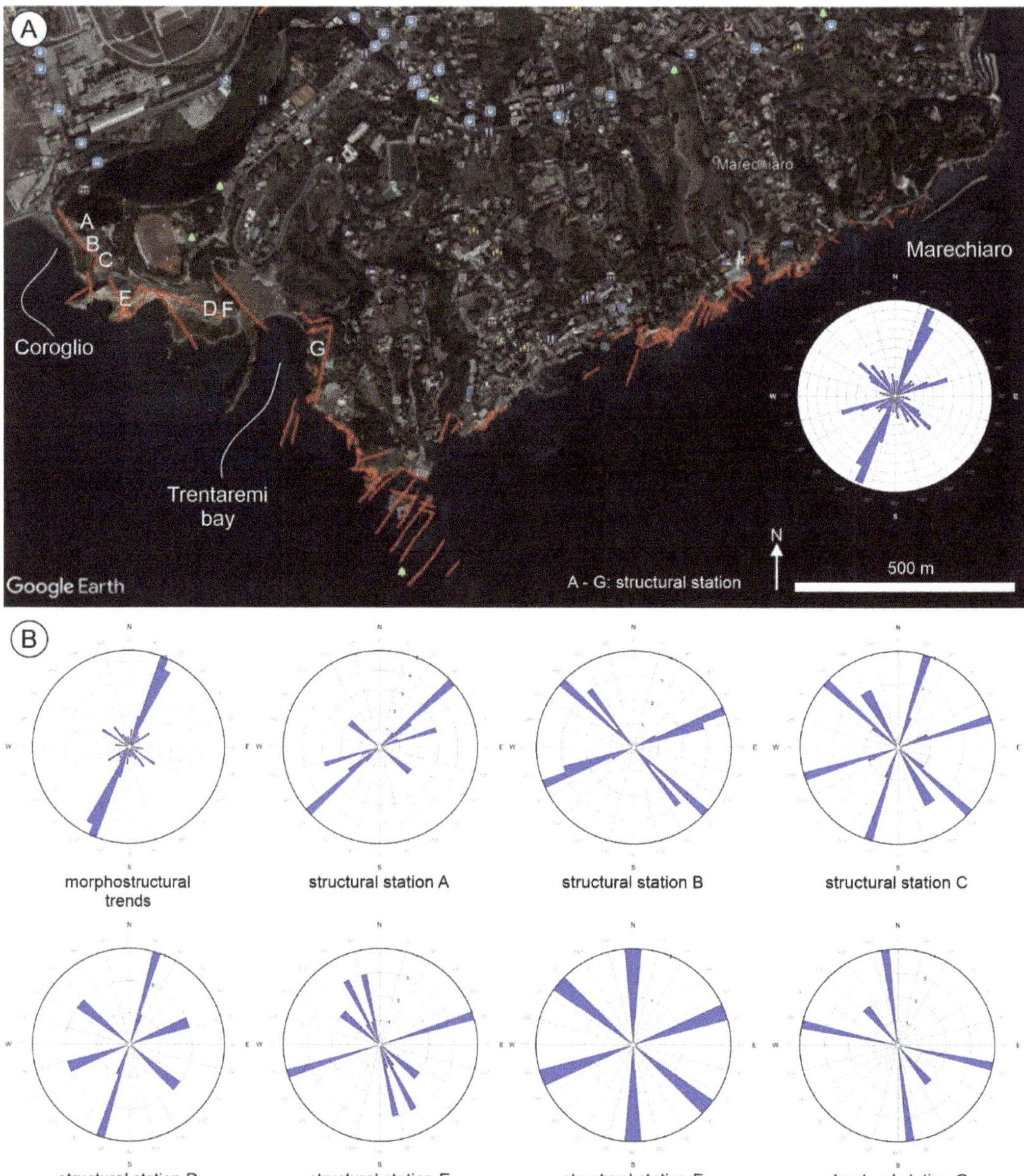

Figure 11. (**A**) Morphostructural lineaments (red lines) along the Coroglio-Trentaremi sea cliff plotted on Google Earth orthophoto. Labels A to G indicate the location of the structural stations. The rose diagram to the right of panel A shows the direction of all the morphostructural lineaments, the faults and fractures shown in the rose diagrams of panel B. (**B**) Rose diagrams showing the direction of the morphostructural lineaments and the direction of the faults and fractures measured in the single structural stations along the Coroglio-Trentaremi sea cliff. Roses represent the absolute number of data.

To define the failure mechanism of the tuffaceous cliff we analyzed field data from seven structural stations, whose locations are shown in Figure 11. All the structural stations highlight the occurrence of a dense net of fractures that, in many cases, allow us to identify several low-equilibrium sectors in the sea cliff. More specifically, we recognized five systems of fractures and bedding whose dip-direction and dip are listed in Table 3.

Table 3. List of fracture systems (K1 to K5) along the Coroglio-Trentaremi sea cliff with indication of the dip-direction (α) and dip (β). Bedding (Ks) and sea cliff front trend are also reported. Structural stations' locations are reported in Figure 11.

Structural Station A	α	β	**Structural Station B**	α	β
K1	310	85	K1	310	85
K2	250	85	K2	250	85
K3	45	85	K4	140	70
Ks	200	20	Ks	200	20
sea cliff front	230	85	sea cliff front	245	85
Structural Station C	α	β	**Structural Station D**	α	β
K1	310	85	K1	310	85
K2	250	85	K2	250	85
K5	330	70	Ks	200	20
Ks	330	70	sea cliff front	200	80
sea cliff front	200	20			
Structural Station E	α	β	**Structural Station F**	α	β
K1	310	85	K1	310	85
K2	250	85	K2	250	85
K5	330	70	Ks	200	20
Ks	200	20	sea cliff front	180	85
sea cliff front	165	80			
Structural Station G	α	β			
K4	140	70			
K5	175	65			
Ks	200	20			
sea cliff front	280	85			

The intersection between fracture systems and bedding allows us to hypothesize different failure mechanisms along the Coroglio-Trentaremi sea cliff, which are listed in Table 4. To synthesize, structural analysis of the sea cliff points out a high susceptibility to landsliding of several tuffaceous blocks by means of rock-fall and toppling. Systems of fractures are 5 to 7 m spaced and intercept the bedding, thus isolating large rocky dihedrons and pinnacles whose volume may exceed some hundreds of cubic meters (Figure 10). This volume estimation is confirmed by measurements of landslide niches' dimensions, that may reach values of 500 m^3, and of large blocks within the landslide bodies.

Table 4. Failure mechanism proposed for each structural station.

Structural Station	Fractures' System	Proposed Failure Mechanism
A	K1-K2-K3-Ks	Toppling
B	K1-K2-K4-Ks	Toppling
C	K1-K2-K5-Ks	Wedge breaks and planar slide
D	K1-k2-Ks	Rock-fall
E	K1-K2-K5-Ks	Rock-fall and toppling
F	K1-K2-Ks	Rock-fall and toppling
G	K4-K5-Ks	Wedge break

4.4. Geomorphological Zonation of Instability along the Coroglio-Trentaremi Sea Cliff

Overall geological, structural, and geomorphological data highlight the diffuse presence of landslides affecting the sea cliff. We propose, in Figure 12, a geomorphological zonation of the Coroglio-Trentaremi sea cliff, which highlights the location of areas more susceptible to instability. This geomorphological zonation is based on the evaluation of I and W factors (Table 1), whose weights have been crossed according to the matrix reported in Table 2.

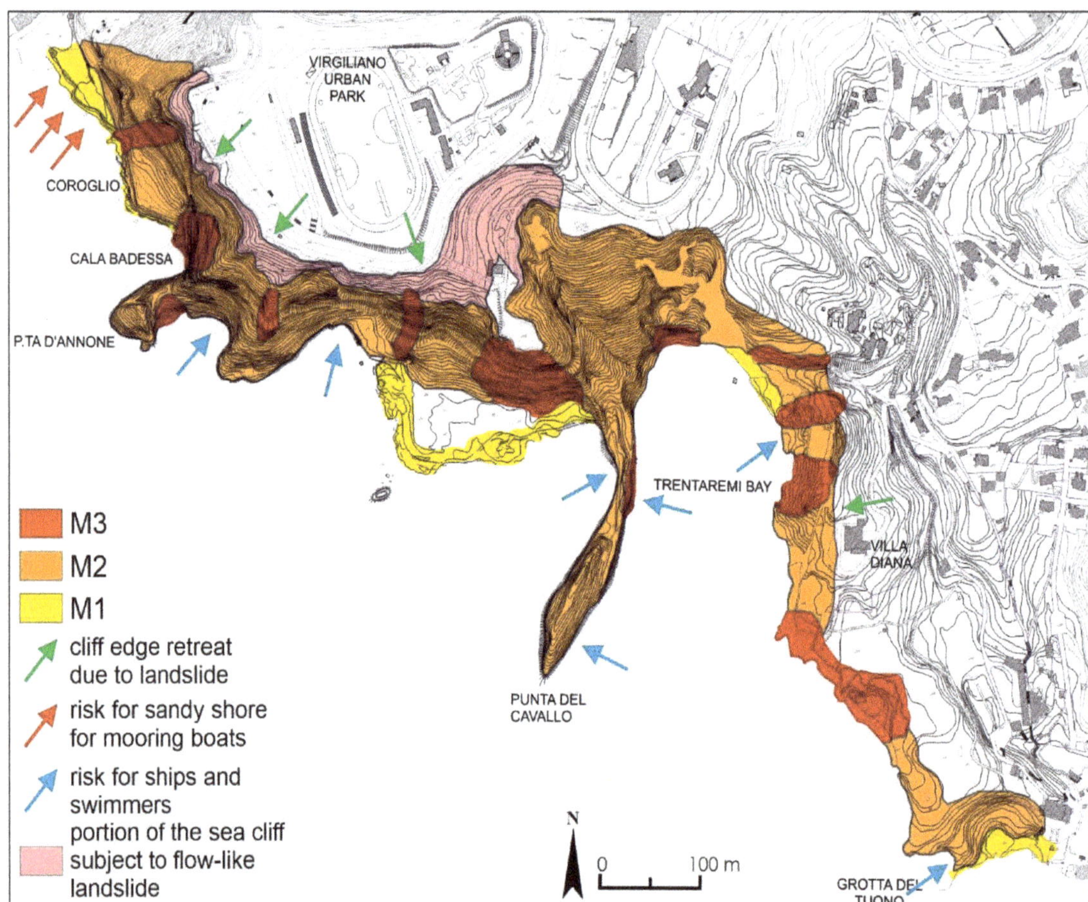

Figure 12. Geomorphological zonation of landslide magnitude along the Coroglio-Trentaremi sea cliff. M3: high magnitude landslide (hundreds to thousands of cubic meters involved); M2: moderate magnitude landslide (tens of cubic meters involved); M1: low magnitude landslide (few cubic meters involved). Arrows indicate areas subject to risk due to cliff edge retreat (green arrows), to the presence of sandy shore for mooring boats (red arrows) and to the presence of ships and swimmers (blue arrows).

The map shows that the more diffuse magnitude class is M2 (e.g., moderate magnitude). High magnitude landslides (class M3 in Figure 12) are limited to few sectors of the Coroglio-Trentaremi sea cliff. These include portions of the sea cliff with high sea cliff height values, presence of a dense network of fractures, some of which are beating fractures, and with large caves at the base of the sea cliff. Low magnitude landslides (class M1 in Figure 12) include few sectors at the base of the sea cliff, occurring near Coroglio, the Trentaremi bay and the La Gaiola islet. In this area, the sea cliff height is low to very low and landslide bodies locally protect the sea cliff base from wave action.

Sectors of the sea cliff where the presence of human activities may enhance landslide risk are limited to few portions of the Coroglio-Trentaremi sea cliff. These include both areas placed on top of the sea cliff edge and areas placed at the base of the sea cliff. On top of the sea cliff, the area with sport facilities at the Virgiliano urban park, to the west, and the archaeological roman villae of Diana, to the east, are present (Figure 12). In both cases, cliff edge retreat due to landslide may cause several problems to both human activities and archaeological finds. At the base of the sea cliff, sandy shores for mooring boats are present, which may increase landslide risk especially in summer due to the increasing number of boats that attend this sector of the Campi Flegrei coastline.

5. Discussion and Conclusions

Overall data highlight the complex geological, geomorphological and morphostructural setting of the Coroglio-Trentaremi sea cliff, which is carved in volcanic deposits whose ages range between 20 ka and the present. The main volcanic units are made up of both loose (mainly ashy and sandy) and lithified deposits (generally tuffs). The contact between the different volcanic units is characterized by volcano-tectonic collapses and erosional surfaces locally marked by paleosols, which also control the bedding of strata. All these conditions represent significant predisposing factors for selective erosion and, consequently, for cliff instability.

The lithified deposits are affected by a pervasive system of sub-vertical faults and fractures, mainly NE-SW and NW-SE trending, which may favor the occurrence of landslides. We recognized five systems of fractures, a finding that is consistent with geostructural analysis along the Coroglio cliff by Matano et al. [16]. The interaction between bedding and fracture' systems determine the failure mechanism, which mainly consists of rock fall and toppling that may involve up to some hundreds or even thousands of cubic meters of material.

Landslide type and related volumes are consistent with detailed topographic data produced by Caputo et al. [64] along the Coroglio sea cliff, who recognized rock fall, debris fall, earth flow and soil slip that caused, in the 2013–2015 time span, the accumulation of 210 m^3 of material. Furthermore, Caputo et al. [64] also estimated average cliff retreat in the 2013–2015 time span to be of 0.07 m/yr.

The cliff zonation that we propose in this study is based on an empiric geomorphologic approach and considers a high quantity of field data. We know that it is not exhaustive and that to better define the landslide risk, further investigations are necessary. Nonetheless, it may enable the development of a reliable scenario of cliff instability, which may be useful either to plan intervention works in the most critical areas or to organize prevention plans aimed at risk mitigation.

The studied coastal segments surely have high naturalistic and historical value, including the beautiful bays of Coroglio, Trentaremi and, at the top of promontory, the Pausyllipon- La Gaiola archaeological site and the Virgiliano urban park.

Our study pointed out that, at some points, landslide niches are quite close either to some important archaeological remains, such as the Roman villae of Diana near the Trentaremi bay, or to touristic areas, such as the Virgiliano Urban Park on top of the Coroglio cliff, also showing evidence of cliff retreat (Figure 12). These data suggest that these sectors deserve mitigation intervention to prevent damage to people and to man-made structures, and to preserve this unique and peculiar site for future generations.

Author Contributions: Conceptualization, A.S. and G.D.C.; methodology, A.S. and G.D.C.; software, E.V.; validation, A.S. and N.S.; formal analysis, E.V.; investigation, A.S., G.D.C., N.S. and E.V.; writing—original draft preparation, A.S., N.S. and E.V.; writing—review and editing, A.S., N.S. and E.V.; visualization, G.D.C.; supervision, A.S. and N.S. All authors have read and agreed to the published version of the manuscript.

Funding: This research received no external funding.

Acknowledgments: The authors wish to thank three anonymous reviewers whose suggestions helped us to improve the manuscript.

Conflicts of Interest: The authors declare no conflict of interest.

References

1. Walsh, K.; Berger, J.-F.; Roberts, C.N.; Vannière, B.; Ghilardi, M.; Brown, A.G.; Woodbridge, J.; Lespez, L.; Estrany, J.; Glais, A.; et al. Holocene demographic fluctuations, climate and erosion in the Mediterranean: A meta data-analysis. *Holocene* **2019**, *29*, 864–885. [CrossRef]
2. Amato, V.; Cicala, L.; Valente, E.; Ruello, M.R.; Esposito, N.; Ermolli, E.R. Anthropogenic amplification of geomorphic processes along the Mediterranean coasts: A case-study from the Graeco-Roman town of Elea-Velia (Campania, Italy). *Geomorphology* **2021**, *383*, 107694. [CrossRef]
3. Del Río, L.; Gracia, F.J. Erosion risk assessment of active coastal cliffs in temperate environments. *Geomorpholohy* **2009**, *112*, 82–95. [CrossRef]
4. Castedo, R.; Murphy, W.; Lawrence, J.; Paredes, C. A new process—Response coastal recession model of soft rock cliffs. *Geomorphology* **2012**, *177–178*, 128–143. [CrossRef]
5. French, P.W. *Coastal Defences*; Routledge: London, UK, 2002.
6. Brooks, S.; Spencer, T. Temporal and spatial variations in recession rates and sediment release from soft rock cliffs, Suffolk coast, UK. *Geomorphology* **2010**, *124*, 26–41. [CrossRef]
7. Young, A.P. Decadal-scale coastal cliff retreat in southern and central California. *Geomorphology* **2018**, *300*, 164–175. [CrossRef]
8. Epifânio, B.; Zêzere, J.L.; Neves, M. Identification of hazardous zones combining cliff retreat rates with landslide susceptibility assessment. *J. Coast. Res.* **2013**, *165*, 1681–1686. [CrossRef]
9. Naylor, L.; Stephenson, W. On the role of discontinuities in mediating shore platform erosion. *Geomorphology* **2010**, *114*, 89–100. [CrossRef]
10. Ascione, A.; Aucelli, P.P.; Cinque, A.; Di Paola, G.; Mattei, G.; Ruello, M.; Ermolli, E.R.; Santangelo, N.; Valente, E. Geomorphology of Naples and the Campi Flegrei: Human and natural landscapes in a restless land. *J. Maps* **2020**, 1–11. [CrossRef]
11. Vitale, S.; Isaia, R. Fractures and faults in volcanic rocks (Campi Flegrei, southern Italy): Insight into volcano-tectonic processes. *Acta Diabetol.* **2013**, *103*, 801–819. [CrossRef]
12. Deino, A.L.; Orsi, G.; de Vita, S.; Piochi, M. The age of the Neapolitan Yellow Tuff caldera-forming eruption (Campi Flegrei caldera—Italy) assessed by 40Ar/39Ar dating method. *J. Volcanol. Geotherm. Res.* **2004**, *133*, 157–170. [CrossRef]
13. Esposito, G.; Salvini, R.; Matano, F.; Sacchi, M.; Danzi, M.; Somma, R.; Troise, C. Multitemporal monitoring of a coastal landslide through SfM-derived point cloud comparison. *Photogramm. Rec.* **2017**, *32*, 459–479. [CrossRef]
14. Aucelli, P.P.; Mattei, G.; Caporizzo, C.; Cinque, A.; Amato, L.; Stefanile, M.; Pappone, G. Multi-proxy analysis of relative sea-level and paleoshoreline changes during the last 2300 years in the Campi Flegrei caldera, Southern Italy. *Quat. Int.* **2021**. [CrossRef]
15. Aucelli, P.P.C.; Mattei, G.; Caporizzo, C.; Cinque, A.; Troisi, S.; Peluso, F.; Stefanile, M.; Pappone, G. Ancient Coastal Changes Due to Ground Movements and Human Interventions in the Roman Portus Julius (Pozzuoli Gulf, Italy): Results from Photogrammetric and Direct Surveys. *Water* **2020**, *12*, 658. [CrossRef]
16. Matano, F.; Iuliano, S.; Somma, R.; Marino, E.; Del Vecchio, U.; Esposito, G.; Molisso, F.; Scepi, G.; Grimaldi, G.M.; Pignalosa, A.; et al. Geostructure of Coroglio tuff cliff, Naples (Italy) derived from terrestrial laser scanner data. *J. Maps* **2015**, *12*, 407–421. [CrossRef]
17. Mattei, G.; Troisi, S.; Aucelli, P.P.C.; Pappone, G.; Peluso, F.; Stefanile, M. Sensing the Submerged Landscape of Nisida Roman Harbour in the Gulf of Naples from Integrated Measurements on a USV. *Water* **2018**, *10*, 1686. [CrossRef]
18. Matano, F.; Caccavale, M.; Esposito, G.; Fortelli, A.; Scepi, G.; Spano, M.; Sacchi, M. Integrated dataset of deformation measurements in fractured volcanic tuff and meteorological data (Coroglio coastal cliff, Naples, Italy). *Earth Syst. Sci. Data* **2020**, *12*, 321–344. [CrossRef]
19. Aucelli, P.; Cinque, A.; Mattei, G.; Pappone, G.; Stefanile, M. Coastal landscape evolution of Naples (Southern Italy) since the Roman period from archaeological and geomorphological data at Palazzo degli Spiriti site. *Quat. Int.* **2018**, *483*, 23–38. [CrossRef]
20. Fall, M.; Azzam, R.; Noubactep, C. A multi-method approach to study the stability of natural slopes and landslide susceptibility mapping. *Eng. Geol.* **2006**, *82*, 241–263. [CrossRef]
21. Forte, G.; De Falco, M.; Santangelo, N.; Santo, A. Slope Stability in a Multi-Hazard Eruption Scenario (Santorini, Greece). *Geosciences* **2019**, *9*, 412. [CrossRef]
22. Nunes, M.; Ferreira, Ó.; Schaefer, M.; Clifton, J.; Baily, B.; Moura, D.; Loureiro, C. Hazard assessment in rock cliffs at Central Algarve (Portugal): A tool for coastal management. *Ocean Coast. Manag.* **2009**, *52*, 506–515. [CrossRef]
23. Budetta, P.; Santo, A.; Vivenzio, F. Landslide hazard mapping along the coastline of the Cilento region (Italy) by means of a GIS-based parameter rating approach. *Geomorphology* **2008**, *94*, 340–352. [CrossRef]
24. Budetta, P. Stability of an undercut sea-cliff along a Cilento coastal stretch (Campania, Southern Italy). *Nat. Hazards* **2010**, *56*, 233–250. [CrossRef]
25. Budetta, P.; De Luca, C. Wedge failure hazard assessment by means of a probabilistic approach for an unstable sea-cliff. *Nat. Hazards* **2014**, *76*, 1219–1239. [CrossRef]
26. Marques, F.M.S.F.; Matildes, R.; Redweik, P. Sea cliff instability susceptibility at regional scale: A statistically based assessment in the southern Algarve, Portugal. *Nat. Hazards Earth Syst. Sci.* **2013**, *13*, 3185–3203. [CrossRef]
27. Ruberti, D.; Marino, E.; Pignalosa, A.; Romano, P.; Vigliotti, M. Assessment of Tuff Sea Cliff Stability Integrating Geological Surveys and Remote Sensing. Case History from Ventotene Island (Southern Italy). *Remote. Sens.* **2020**, *12*, 2006. [CrossRef]

28. Santangelo, N.; Romano, P.; Ascione, A.; Ermolli, E.R. Quaternary evolution of the Southern Apennines coastal plains: A review. *Geol. Carpathica* **2017**, *68*, 43–56. [CrossRef]
29. Cello, G.; Mazzoli, S. Apennine tectonics in southern Italy: A review. *J. Geodyn.* **1998**, *27*, 191–211. [CrossRef]
30. Turco, E.; Macchiavelli, C.; Mazzoli, S.; Schettino, A.; Pierantoni, P.P. Kinematic evolution of Alpine Corsica in the framework of Mediterranean mountain belts. *Tectonophysics* **2012**, *579*, 193–206. [CrossRef]
31. Doglioni, C.; Innocenti, F.; Morellato, C.; Procaccianti, D.; Scrocca, D. On the Tyrrhenian sea opening. *Mem. Descr. Carta Geol. d'Italia* **2004**, *64*, 147–164.
32. Cinque, A.; Patacca, E.; Scandone, P.; Tozzi, M. Quaternary kinematic evolution of the Southern Appennines. relationships between surface geological features and deep lithospheric structures. *Ann. Geophys.* **1993**, *36*, 249–259. [CrossRef]
33. Brancaccio, L.; Cinque, A.; Romano, P.; Rosskopf, C.; Russo, F.; Santangelo, N.; Santo, A. Geomorphology and neotectonic evolution of a sector of the Tyrrhenian flank of the Southern Apennines (Region of Naples, Italy). *Z. Geomorphol. Supp.* **1991**, *82*, 47–58.
34. Caiazzo, C.; Ascione, A.; Cinque, A. Late Tertiary–Quaternary tectonics of the Southern Apennines (Italy): New evidences from the Tyrrhenian slope. *Tectonophysics* **2006**, *421*, 23–51. [CrossRef]
35. Geological Sheet 447–Napoli of the Geological Map of Italy at scale 1:50000, CARG Project. Available online: https://www.isprambiente.gov.it/Media/carg/447_NAPOLI/Foglio.html (accessed on 20 April 2021).
36. Pappalardo, L.; Civetta, L.; D'Antonio, M.; Deino, A.; Di Vito, M.; Orsi, G.; Carandente, A.; de Vita, S.; Isaia, R.; Piochi, M. Chemical and Sr-isotopical evolution of the Phlegraean magmatic system before the Campanian Ignimbrite and the Neapolitan Yellow Tuff eruptions. *J. Volcanol. Geotherm. Res.* **1999**, *91*, 141–166. [CrossRef]
37. De Vivo, B.; Rolandi, G.; Gans, P.B.; Calvert, A.; Bohrson, W.; Spera, F.J.; Belkin, H. New constraints on the pyroclastic eruptive history of the Campanian volcanic Plain (Italy). *Miner. Pet.* **2001**, *73*, 47–65. [CrossRef]
38. Giaccio, B.; Hajdas, I.; Isaia, R.; Deino, A.; Nomade, S. High-precision 14C and 40Ar/39Ar dating of the Campanian Ignimbrite (Y-5) reconciles the time-scales of climatic-cultural processes at 40 ka. *Sci. Rep.* **2017**, *7*, srep45940. [CrossRef]
39. Valente, E.; Buscher, J.T.; Jourdan, F.; Petrosino, P.; Reddy, S.; Tavani, S.; Corradetti, A.; Ascione, A. Constraining mountain front tectonic activity in extensional setting from geomorphology and Quaternary stratigraphy: A case study from the Matese ridge, southern Apennines. *Quat. Sci. Rev.* **2019**, *219*, 47–67. [CrossRef]
40. Fedele, F.G.; Giaccio, B.; Isaia, R.; Orsi, G. The Campanian Ignimbrite Eruption, Heinrich Event 4, and palaeolithic change in Europe: A high-resolution investigation. *Large Igneous Prov.* **2003**, 301–325. [CrossRef]
41. de Vita, S.; Orsi, G.; Civetta, L.; Carandente, A.; D'Antonio, M.; Deino, A.; di Cesare, T.; Di Vito, M.; Fisher, R.; Isaia, R.; et al. The Agnano–Monte Spina eruption (4100 years BP) in the restless Campi Flegrei caldera (Italy). *J. Volcanol. Geotherm. Res.* **1999**, *91*, 269–301. [CrossRef]
42. Di Renzo, V.; Arienzo, I.; Civetta, L.; D'Antonio, M.; Tonarini, S.; Di Vito, M.; Orsi, G. The magmatic feeding system of the Campi Flegrei caldera: Architecture and temporal evolution. *Chem. Geol.* **2011**, *281*, 227–241. [CrossRef]
43. Di Vito, M.; Isaia, R.; Orsi, G.; Southon, J.; de Vita, S.; D'Antonio, M.; Pappalardo, L.; Piochi, M. Volcanism and deformation since 12,000 years at the Campi Flegrei caldera (Italy). *J. Volcanol. Geotherm. Res.* **1999**, *91*, 221–246. [CrossRef]
44. Smith, V.; Isaia, R.; Pearce, N. Tephrostratigraphy and glass compositions of post-15 kyr Campi Flegrei eruptions: Implications for eruption history and chronostratigraphic markers. *Quat. Sci. Rev.* **2011**, *30*, 3638–3660. [CrossRef]
45. Isaia, R.; Vitale, S.; DI Giuseppe, M.G.; Iannuzzi, E.; Tramparulo, F.D.; Troiano, A. Stratigraphy, structure, and volcano-tectonic evolution of Solfatara maar-diatreme (Campi Flegrei, Italy). *GSA Bull.* **2015**, *127*, 1485–1504. [CrossRef]
46. Liedl, A.; Buono, G.; Lanzafame, G.; Dabagov, S.; Della Ventura, G.; Hampai, D.; Mancini, L.; Marcelli, A.; Pappalardo, L. A 3D imaging textural characterization of pyroclastic products from the 1538 AD Monte Nuovo eruption (Campi Flegrei, Italy). *Lithos* **2019**, *340–341*, 316–331. [CrossRef]
47. Arzilli, F.; Piochi, M.; Mormone, A.; Agostini, C.; Carroll, M.R. Constraining pre-eruptive magma conditions and unrest timescales during the Monte Nuovo eruption (1538 ad; Campi Flegrei, Southern Italy): Integrating textural and CSD results from experimental and natural trachy-phonolites. *Bull. Volcanol.* **2016**, *78*, 72. [CrossRef]
48. Lima, A.; De Vivo, B.; Spera, F.J.; Bodnar, R.J.; Milia, A.; Nunziata, C.; Belkin, H.E.; Cannatelli, C. Thermodynamic model for uplift and deflation episodes (bradyseism) associated with magmatic–hydrothermal activity at the Campi Flegrei (Italy). *Earth Sci. Rev.* **2009**, *97*, 44–58. [CrossRef]
49. Cannatelli, C.; Spera, F.J.; Bodnar, R.J.; Lima, A.; De Vivo, B. Ground movement (bradyseism) in the Campi Flegrei volcanic area. *Vesuvius Campi Flegrei Camp. Volcanism* **2020**, 407–433. [CrossRef]
50. Cinque, A.; Rolandi, G.; Zamparelli, V. L'estensione dei depositi marini olocenici nei Campi Flegrei in relazione alla vulcano-tettonica. *Boll. Soc. Geol. Ital.* **1985**, *104*, 327–348.
51. Cinque, A.; Aucelli, P.P.C.; Brancaccio, L.; Mele, R.; Milia, A.; Robustelli, G.; Romano, P.; Russo, F.; Santangelo, N.; Sgambati, D. Volcanism, tectonics and recent geomorphological change in the bay of Napoli. *Geogr. Fis. Din. Quat.* **1997**, *3*, 123–141.
52. Morhange, C.; Marriner, N.; Laborel, J.; Todesco, M.; Oberlin, C. Rapid sea-level movements and noneruptive crustal deformations in the Phlegrean Fields caldera, Italy. *Geology* **2006**, *34*, 93. [CrossRef]
53. Del Gaudio, C.; Aquino, I.; Ricciardi, G.; Ricco, C.; Scandone, R. Unrest episodes at Campi Flegrei: A reconstruction of vertical ground movements during 1905–2009. *J. Volcanol. Geotherm. Res.* **2010**, *195*, 48–56. [CrossRef]

54. Isaia, R.; Vitale, S.; Marturano, A.; Aiello, G.; Barra, D.; Ciarcia, S.; Iannuzzi, E.; Tramparulo, F.D. High-resolution geological investigations to reconstruct the long-term ground movements in the last 15 kyr at Campi Flegrei caldera (southern Italy). *J. Volcanol. Geotherm. Res.* **2019**, *385*, 143–158. [CrossRef]
55. Aucelli, P.; Cinque, A.; Mattei, G.; Pappone, G.; Rizzo, A. Studying relative sea level change and correlative adaptation of coastal structures on submerged Roman time ruins nearby Naples (southern Italy). *Quat. Int.* **2019**, *501*, 328–348. [CrossRef]
56. Pappone, G.; Aucelli, P.P.; Mattei, G.; Peluso, F.; Stefanile, M.; Carola, A. A Detailed Reconstruction of the Roman Landscape and the Submerged Archaeological Structure at "Castel dell'Ovo islet" (Naples, Southern Italy). *Geosciences* **2019**, *9*, 170. [CrossRef]
57. Di Martire, D.; De Rosa, M.; Pesce, V.; Santangelo, M.A.; Calcaterra, D. Landslide hazard and land management in high-density urban areas of Campania region, Italy. *Nat. Hazards Earth Syst. Sci.* **2012**, *12*, 905–926. [CrossRef]
58. Wolters, G.; Müller, G. Effect of Cliff Shape on Internal Stresses and Rock Slope Stability. *J. Coast. Res.* **2008**, *241*, 43–50. [CrossRef]
59. Di Crescenzo, G.; Santo, A. *Collina di Posillipo e Collina di Camaldoli: Studio Geologico, Geomorfologico e Strutturale*; Convenzione di Ricerca CUGRI–Comune di Napoli: Naples, Italy, 2001.
60. Cole, P.D.; Perrotta, A.; Scarpati, C. The volcanic history of the southwestern part of the city of Naples. *Geol. Mag.* **1994**, *131*, 785–799. [CrossRef]
61. Scarpati, C.; Perrotta, A.; Lepore, S.; Calvert, A. Eruptive history of Neapolitan volcanoes: Constraints from 40Ar–39Ar dating. *Geol. Mag.* **2012**, *150*, 412–425. [CrossRef]
62. Scarpati, C.; Perrotta, A.; Sparice, D. Volcanism in the city of Naples. *Rend. Online Soc. Geol. Ital.* **2015**, *33*, 88–91. [CrossRef]
63. Scarpati, C.; Cole, P.; Perrotta, A. The Neapolitan Yellow Tuff? A large volume multiphase eruption from Campi Flegrei, Southern Italy. *Bull. Volcanol.* **1993**, *55*, 343–356. [CrossRef]
64. Caputo, T.; Marino, E.; Matano, F.; Somma, R.; Troise, C.; De Natale, G. Terrestrial Laser Scanning (TLS) data for the analysis of coastal tuff cliff retreat: Application to Coroglio cliff, Naples, Italy. *Ann. Geophys.* **2018**, *61*, 110. [CrossRef]

Article

Shoreline Evolution and Erosion Vulnerability Assessment along the Central Adriatic Coast with the Contribution of UAV Beach Monitoring

Gianluigi Di Paola [1,2], Antonio Minervino Amodio [3,*], Grazia Dilauro [1], Germàn Rodriguez [4,5] and Carmen M. Rosskopf [1]

1 Department of Biosciences and Territory, University of Molise, Contrada Fonte Lappone, I-86090 Pesche, Italy
2 Department of Biological, Geological, and Environmental Sciences, University of Bologna "Alma Mater Studiorum", Via Zamboni 67, I-40126 Bologna, Italy
3 National Research Council (CNR), Institute of Heritage Science (ISPC), I-00185 Tito, Italy
4 Departamento de Física, Universidad de Las Palmas de Gran Canaria, 35001 Las Palmas de Gran Canaria, Spain
5 Institute of Environmental Studies and Natural Resources (iUNAT), Universidad de Las Palmas de Gran Canaria, 35001 Las Palmas de Gran Canaria, Spain
* Correspondence: antonio.minervinoamodio@ispc.cnr.it; Tel.: +39-0971-427-309

Abstract: Coastal erosion and its impacts on the involved communities is a topic of great scientific interest that also reflects the need for modern as well as cost and time-effective methodologies to be integrated into or even to substitute traditional investigation methods. The present study is based on an integrated approach that involves the use of data derived from UAV (Unmanned Aerial Vehicle) surveys. The study illustrates the long- to short-term shoreline evolution of the Molise coast (southern Italy) and then focuses on two selected beach stretches (Petacciato and Campomarino beaches), for which annual UAV surveys were performed from 2019 to 2021, to assess their most recent shoreline and morpho-topographical changes and related effects on their coastal vulnerability. UAV data were processed using the Structure from Motion (SfM) image processing tool. Along the beach profiles derived from the produced DEMs, the coastal vulnerability of the selected beach stretches was evaluated by using the Coastal Vulnerability Assessment (CVA) approach. The results obtained highlight some significant worsening of CVA indexes from 2019 to 2021, especially for the Campomarino beach, confirming the importance of the periodic updating of previous data. In conclusion, the easy use of the UAV technology and the good quality of the derived data make it an excellent approach for integration into traditional methodologies for the assessment of short-term shoreline and beach changes as well as for monitoring coastal vulnerability.

Keywords: shoreline evolution; beach erosion; UAV data elaboration; coastal vulnerability assessment; CVA approach; Molise coast; Italy

Citation: Di Paola, G.; Minervino Amodio, A.; Dilauro, G.; Rodriguez, G.; Rosskopf, C.M. Shoreline Evolution and Erosion Vulnerability Assessment along the Central Adriatic Coast with the Contribution of UAV Beach Monitoring. *Geosciences* **2022**, *12*, 353. https://doi.org/10.3390/geosciences12100353

Academic Editors: Patrick Seyler and Jesus Martinez-Frias

Received: 23 July 2022
Accepted: 16 September 2022
Published: 22 September 2022

1. Introduction

Coasts are dynamic systems that change in form at different space and time scales in response to geomorphological and marine forces [1–3]. Coastal landforms, affected by short-term perturbations such as storms, frequently return to their pre-disturbance morphology, thus reflecting a basic, morphodynamic equilibrium [4]. Most coasts are constantly adjusting towards a dynamic equilibrium, frequently adopting different 'states' in response to variable wave energy and sediment supply [5]. Coasts exhibit natural variability in response to changes in environmental factors, which can make it difficult to identify the impact of climate change. Thus, for instance, most beaches across the world show evidence of recent erosion, but ongoing sea-level rise is not always considered to be the major cause. Erosion can result from several other factors, both natural and anthropogenic, such as reduced fluvial sediment input [6], offshore bathymetric changes [7], altered wind and wave patterns [8,9], or a mix of factors [10].

The assessment of coastal erosion and flooding is an important issue for the scientific community, considering that a significant and increasing share of the world population currently lives in coastal areas. In fact, the United Nations, during the Ocean Conference 2017 [11], highlighted that more than 10% of the world population (600 million people) live along coastal areas that are less than 10 m above sea level, while about 40% of the world population (2.4 billion people) live within 100 km from the coastline. Moreover, the possible increase of seal level could force the displacement of up to 187 million people globally during this century [12]. As for the Italian peninsula, which has approximately 7500 km of coastline, the data released by the Ministry of Environment, Land, and Sea Protection [13] for the period of 1960–2012 show that 23% (1534 km) of this coastline has experienced erosion, resulting in a total land loss of 35.5 km^2.

Assessing coastal evolution and recent erosion trends along with related natural and/or anthropogenic causes is crucial for the correct definition of near-future planning and interventions aimed at combatting erosion and ensuring the sustainable development of coastal zones.

To this end, an appropriate experimental observation of coastal zones with their complex and variable dynamics and configurations is of paramount importance. In this regard, traditional methods suitable for measuring, mapping, and monitoring beach–dune systems and shoreline changes are commonly applied, such as the interpretation of topographic maps and aerial images. Related activities, however, take a long time and immense effort to complete, and they may not provide the necessary spatial scale and local details and/or comprehensively cover the most recent time intervals [14]. Fortunately, such limitations in spatio-temporal resolution are now being overcome, or at least alleviated, thanks to the access to large satellite databases and the development of various technologies appropriate for this purpose. In particular, the progress and easy accessibility of Unmanned Aerial Vehicle (UAV) technology has enabled the development of an alternative coastal monitoring technique that efficiently captures spatial and temporal requirements across a wide range of environmental applications [15–19], with the important added advantage of making the task much less tedious.

This study presents an integrated approach involving traditional methods and the innovative UAV methodology to study the shoreline evolution and morphodynamic behaviour of the Molise coast, a well-known coastal system that has been under continuous study for some decades and one of the most vulnerable coastal regions in Italy (e.g., see [20–22] and references therein).

The major aims of this study are: (i) to assess the spatial-temporal distribution of shoreline changes that occurred along the Molise coast from 1954 to 2016 using remotely sensed data and topographic maps; (ii) to assess long- to short-term shoreline changes of two selected test areas and, by means of UAV-derived data, their most recent shoreline and morpho-topographic beach–dune changes, and (iii) to assess the possible influence of such changes on the vulnerability of the test areas to coastal erosion and inundation.

2. The Molise Coast

2.1. Physiographic and Geomorphological Features

The coastline of the Molise region (Figure 1) is approximately 36 km long and part of the Central Italian Adriatic coast. Its northern and southern boundaries are respectively the Formale del Molino channel and the Saccione Stream (Figure 1). A major part of the Molise coastline (22 km) consists in a low sedimentary coast with sandy beaches, while its central sector (ca. 14 km long) consists in a high rocky coast. The latter was formed by Plio-Pleistocene sedimentary successions composed of clayey-sandy marine and sandy-conglomeratic fluvial sediments (Montesecco Clay, Serracapriola Sand, and Campomarino Conglomerate formations) [23]. The sea cliffs are cut into the Plio-Pleistocene bedrock and are mostly inactive and set back from the shoreline, with the only exception of the short but artificially protected cliff of the Termoli promontory [21]. Consequently, the beaches are practically continuous along the entire Molise coast, leaving aside the interruptions owing to the major river mouths (Trigno and Biferno rivers and Sinarca and Saccione streams, Figure 1) and the three harbours present along the coast (Figure 1). The beaches are largely made up of fine to medium-grained sands and characterized by very variable widths ranging between a few meters and 93 m [21]. The dominant longshore transport occurs from NW to SE [10,21].

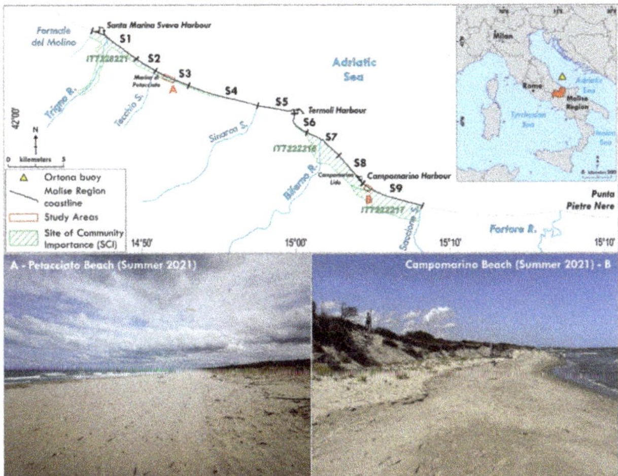

Figure 1. The Molise coast. Location and panoramic views of the two study areas, A and B.

The Termoli promontory divides the Molise coast into two nearly independent sectors (physiographic sub-units) [21]. These sectors, hereinafter named the northern and southern Molise coasts, are oriented in the WNW and NW-SE directions, respectively.

The Molise coast has an important ecological value thanks to the presence of 18 habitats of interest for the European Community [24,25]. In particular, there are three Sites of Community Importance (Figure 1), which include the three major river mouths (Trigno, Biferno, and Saccione) from north to south: IT7228221 (Foce Trigno-Marina di Petacciato), where test area A is located, IT7222216 (Foce Biferno-Litorale di Campomarino), which includes test area B, and IT7222217 (Foce Saccione-Bonifica Ramitelli). Habitats of community interest with a high naturalistic value cover an area of more than 340 hectares [25], highlighting the fact that the Molise coastline is one of the most important natural coastal stretches along the Italian Adriatic coast.

Concerning the reconstruction of the evolution of the Molise shoreline, in agreement with previous studies (e.g., see [21] and references therein), the subdivision in nine coastal segments (S1–S9, Figure 1) has been maintained to allow for an easy comparison of the shoreline change data calculated in the present study and those calculated previously.

Literature data (e.g., [21] and references therein) show that erosion strongly controlled the long-term evolution of the Molise coast. The coastal segments that include the Trigno and Biferno river mouths (S1 and S7) suffered important shoreline retreat (Figure 2), totalling average annual rates of −2.7 m/y and −2.9 m/y, respectively, during the period of 1954–2014 [21]. On the other hand, segments S2, S4, S5, S8, and S9 remained substantially stable (0.1 m/y–0.3 m/y), while S3 and S6 even experienced some slight advance (0.8 m/y and 1.0 m/y, respectively). Shoreline erosion was driven in particular by the decrease of fluvial sediment input to the coast, mostly resulting from in-channel mining and hydraulic interventions at the basin and from the fluvial reach scale from the 1950s onwards (e.g., [10,21] and references therein). To face the ongoing coastal erosion, numerous defence structures were built from the 1980s onwards, mostly under emergency conditions, allowing for the partial balancing out of previous erosion trends (for example, see S5 in [21]). In fact, hard defence structures (adherent breakwaters, emerged and submerged detached breakwaters, revetments, groins, and jetties) cover about 62% of the Molise coast [21]. Despite a relatively high degree of coastal protection as early as 1998, erosion continued to significantly affect the Molise coast [21]. Especially during very recent years (the period of 2011–2014), shoreline erosion not only involved segments S1 and S7 but also extended to segments S2, S3, and S9 (−4.9 m/y, −3.3 m/y, and −3.6 m/y, respectively), suggesting an ongoing process of progressive shoreline destabilization [21].

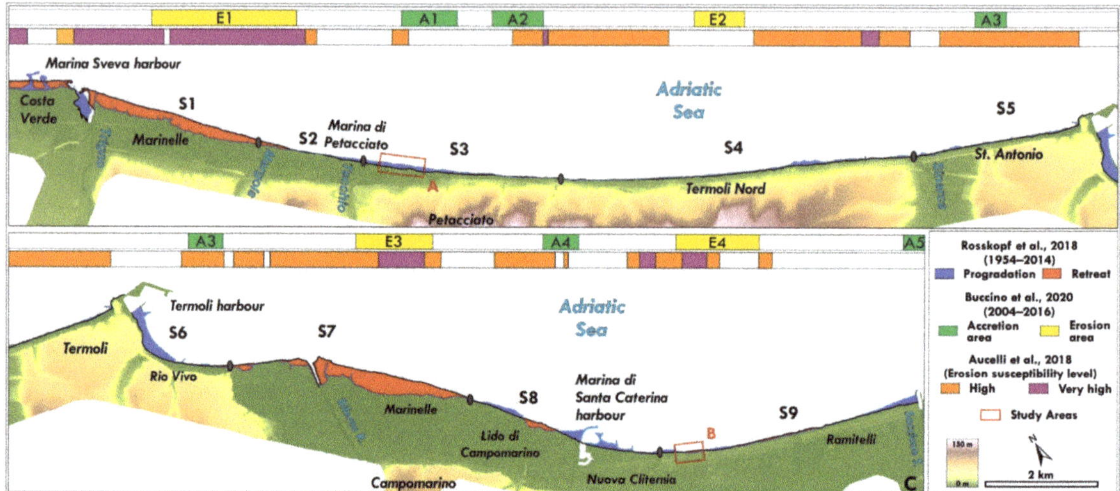

Figure 2. Shoreline progradation and retreat rates from 1954 to 2014 [21], location of areas with high to very high levels of erosion susceptibility [22], and location of mid-term (2004–2016) major shoreline erosion/accretion areas E1–E4 and A1–A5 [10].

Recent data provided by Buccino et al. 2020 [10] on the evolution of the Molise shoreline from 2004 to 2016 (Figure 2) highlight the presence of two major erosion areas located south of the Trigno and Biferno river mouths (E1 and E3), and of two others (E2 and E4) falling within coastal segments S4 and S8, respectively. Furthermore, the study shows the presence of several major accretion areas (A1–A5), which are characterized by a positive sediment balance.

Finally, the data reported by Aucelli et al. (2018) [22] show that most of the Molise shoreline is characterized by a high level of erosion susceptibility, while nearly the entire segment S1 and smaller portions of S4, S7, S8, and S9 are characterized by a very high susceptibility level (Figure 2).

2.2. Meteomarine Features of the Molise Coast

Wind, wind-generated waves, and astronomically and meteorologically induced sea level variations affecting the Molise coast are framed in the meteo-marine context that characterizes the Adriatic Sea.

Typical strong winds above the Adriatic Sea are generally of two types, Bora and Sirocco. Apart from these, any other class of windstorms has negligible importance regarding wave storms, both in frequency and intensity [26]. Sirocco is a humid, warm, and steady wind blowing over the Adriatic basin from SE-SSE. This wind is strongly influenced by the orography as it is channelled along the major axis of the basin by the Apennines and the Dynaric Alps. It blows on most of the length of the basin, usually not very strong, but is able to produce storms in the Central and Northern Adriatic areas [26]. On the other hand, Bora is a cold and gusty, strong, low-level, downslope wind blowing mainly from

NE, across the mountain barrier of the Dinaric Alps. Strong bora events may give rise to structured jets and multiple jet systems flowing through orographic gaps, with strong sub-basin scale spatial gradients across the Adriatic [27,28].

Because of the meteorological scenario above, wave climate in the Adriatic Sea is usually mild or moderate most of the year. However, the severity of wave climate varies throughout the basin depending on the relative importance of the dominant and prevailing winds, Bora and Sirocco, as well as on the fetch available for generating waves. In general, Bora is more intense than Sirocco and is fetch-limited, blowing mainly along the minor axis of the Adriatic, although it can suddenly attain very high speeds. On the contrary, Sirocco may blow over much larger fetches, along the major axis of the basin, and may grow slowly. Furthermore, Sirocco generally reaches the highest speeds in the eastern Adriatic regions and decreases while proceeding to the western coasts [27]. In particular, wind fields in the Central Adriatic Sea are strongly conditioned by the bordering orography, with the wind channelled along the main axis and with the presence of a reduced but still important effect of the transverse component.

In line with these, coastal topography and geometry also significantly influence wave generation by Bora and Sirocco in the Central Adriatic Sea. In particular, wave climate in the Central Western Adriatic Sea is conditioned by the presence of the Conero and Gargano headlands at its northern and southern limits, which significantly restrict the fetch in the longitudinal direction. Accordingly, potential fetches for the study area are larger along sectors oriented around the N and E directions and decrease towards the NE.

Given these wind conditions, waves approaching the Molise coast exhibit a bimodal directional distribution, with two prevailing directional sectors. One of these sectors is roughly represented by the directional quadrant between the NW and the NE, hereinafter referred to as the main wave direction, while the second directional sector extends from the NE to the SE, and is referred to as the secondary wave direction (Table 1). Wave fields approaching from the NE-SE sector are characterized by low and moderate significant wave heights, $H_s < 3$ m, and with the largest percentage of observations corresponding to events with $H_s < 1$ m. The contribution of low and moderate sea states also prevails in the NW-NE directional sector. However, in this case, the sea states may be more severe, with H_s values of more than 3 m and up to 6 m, although the frequency of these events is notably low. Nevertheless, the bimodal character of the bivariate distribution of H_s and wave direction becomes clearly unimodal and oriented mainly around the NNE sub-sector for sea states with $H_s > 3$ m [21]. In this sense, an analysis of the annual maximum significant wave height at the Ortona buoy (Sea Wave Measurement Network, RON) in the period of 1990–2006 reveals that this parameter ranges from 3 to 6 m, approaching from the N-NE directional sector, although the most severe events arrive from the N-NNE sub-sector (Figure 3A). The occurrence of these extreme episodes takes place during the late autumn, winter, and early spring months (Figure 3B). The main wave climate features of the Molise coast, including the Average Significant Wave Height (H_s) and the Average Significant Wave Height of sea states exceeding 2 m (H_t), as well as their associated periods, are reported in Table 1.

Table 1. Main wave climate features estimated for the Molise coast. Average Significant Wave Height (H_s) and the associated period (T_s); Average Significant Wave Height of events exceeding 2 m (H_t) and the corresponding period (T_t).

Recording Period (Ortona Buoy)	Main Wave Direction (°N)	Secondary Wave Direction (°N)	Effective Fetch (km)	H_s (m)	T_s (s)	H_t (m)	T_t (s)
1990–2006	340–10	70–100	476	0.7	3.5	3.5	6.6

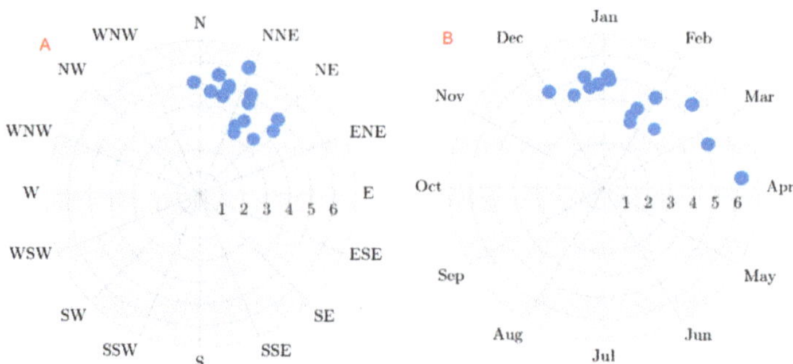

Figure 3. Annual maximum wave storms, directionality (**A**) and timing during the year (**B**) at the Ortona buoy (1999–2006). For the location of the Ortona buoy, see Figure 1.

In correspondence with the directions of the prevailing and dominant waves and the shoreline orientation of the Molise coast, longshore drift occurs in the north–south direction [21,29,30].

Regarding astronomical tides, the study area is microtidal and experiences ordinary tidal excursions of 30–40 cm [21], although the maximum high tide recorded in the tide tables is 0.6 m and the minimum height is –0.2 m, which are referenced to mean lower low water [31]. Sea-level oscillations induced by meteorological conditions (storm surges or meteorological residues) are obtained as the difference between the measured sea level and the predicted astronomical tide. Analyses of meteorological residues obtained from sea level measurements at the Ortona tidal gauge [32] for the period between 1979 and 2019 indicate that storm surge values range approximately between −0.15 m and 0.45 m. Nevertheless, more than 95% of the observations lie in the range of −0.07 m–0.27 m, and the probability of the occurrence of values greater than 0.3 m are less than 0.01.

2.3. The Test Areas

To assess the most recent shoreline changes with the use of UAV-derived data, we selected two test areas (A and B, Figures 1 and 2). These test areas allowed us to investigate two different situations: one concerning conditions of essential stability for both the shoreline and the beach–dune system, and the other relating to evident erosion and destabilization, either recent or perhaps ongoing, which involve previously stable coastal segments.

Test area A falls under segment S3 (northern Molise coast, Figure 2) and is part of the Petacciato coastline. It consists in a 400 m long stretch of coast located 500 m south of Marina di Petacciato. This area is characterized by a sandy beach about 30 m wide, supported by a well-preserved dune system that is wider than 10 m and up to 4 m high.

Test area B is located along the southern Molise coast in the northernmost part of segment 9 (Figure 2), at approximately 1.5 km south of the Campomarino touristic port, and is part of erosion area E4, indicated by Buccino et al. (2020) [10]. It consists in an approximately 200 m long coastal stretch that is characterized by a 6–25 m wide sandy beach and an up to 4 m high dune system, whose original dune front has been replaced by a sub-vertical erosion scarp up to 2.5 m high that clearly separates the dune from the backshore. In front of test area B, 160–200 m from the shoreline, there are three breakwaters. They belong to a row of nine (originally) emerged and detached breakwaters approximately 700 m long, oriented slightly obliquely to the coast, and placed between 2001 and 2005. Furthermore, in front of the central part of area B, a detached groin (placed in 2011–2014) is located approximately 30 m from the shoreline (for further details, see Section 4.2).

3. Materials and Methods

This section describes the materials and methods used to investigate the long to short-term evolution of the Molise coast, the short-term morpho-topographical changes of the test areas from 2019 to 2021, and the possible impact of such changes on the coastal vulnerability assessment.

3.1. Evaluation of Shoreline Changes through the DSAS Tool

The digitization of the Molise shorelines was carried out in the ArcGIS environment using aerial photos from 1954, orthophoto maps from 2004, and images taken by Google Earth in 2016 (Table 2).

Table 2. Data sources used to calculate the shoreline variations in the test areas. RMSE = root mean square error.

Date	Data Source	Scale	RMSE (m)
1954	Aerial photo	1:36,000	5
2004	Orthophoto map	1:2500	3
2016	Google Earth image	1:500	1

Regarding the shorelines of the test areas in 2019, 2020, and 2021, we used the orthophoto images realized by the UAV surveys performed during this study, as shown below. As the test areas are part of a microtidal environment, the shoreline positions were defined as the water line at the time of the photo [33], but a maximum uncertainty of ± 1.6 m was assumed for the daily water line position, as it was not possible to reconstruct the tidal conditions for each image.

To evaluate shoreline changes, we used the Digital Shoreline Analysis System (DSAS), a freely available extension to ESRI's ArcGIS [34]. This tool, which automatically creates regularly spaced transects, provides two parameters. First, a distance parameter (Net Shoreline Movement—NSM) that measures the distance between the oldest and the youngest shoreline considered for each transect, and second, the Linear Regression Rate (LRR), which represents the average rate of accretion or erosion obtained after fitting a least-square straight line to each shoreline section, for each considered period. In detail, shoreline variations were determined for the Molise coast by using 352 transects placed at an equidistance of 100 m. As we needed a lot more details for the test areas, we fixed an equidistance value of 5 m.

3.2. Monitoring of Morpho-Topographic Changes in the Test Areas Based on UAV-Derived Data

The monitoring of the short-term evolution of the test areas was based on annual UAV surveys. To date, three UAV survey campaigns have been carried out—in July 2019, 2020, and 2021; a Phantom 3 Standard (Quadcopter) was used, which is a drone developed by Da-Jiang Innovations (DJI, Shenzhen, China). This drone was mounted with a stabilized camera that compensated for involuntary movements of the UAV due to wind, thus ensuring the correct orientation of the photos with respect to the ground. The coordinates for each photo were then registered by an internal GPS, allowing the drone to follow the points (waypoints) previously fixed with the GPS on the flight plan. An average flight altitude of 40 m was used for all surveys, which allowed for a ground resolution of 1.5 cm/px. Using the default camera of the Phantom 3 Standard, about 250 images with a longitudinal overlap (flight direction) of 85% and a flight strip overlap of 60% were obtained during each flight.

For all flights, six targets, 40×40 cm in size and easily visible from above, were placed to record the position of the Ground Control Points (GCPs) and consequently to orient the model in space. A GNSS receiver in static nRTK mode was used to acquire the GCPs.

To evaluate the variations in the plano-altimetric features of the test areas and the related volumetric changes, 3D models were generated. To obtain these models, the (2D) photos acquired with the drone were processed using the Structure from Motion (SfM) algorithm, which internally implements the photogrammetry and computer vision methods. A functional correlation between the 3D object points and the 2D image points via the collinearity condition is the core concept behind photogrammetric image data processing [35]. The examination of two photos and related orientation parameters allowed for the identification of the common points and the determination of the related 3D coordinates.

To obtain the final outputs (Dense Cloud Points, DEMs, and orthophotos), we proceeded as follows: (i) generation of the flight plan; (ii) field work consisting in the positioning of the targets, measurement of the position with the GNSS receiver, and acquisition of photos with the UAV; (iii) aerial data processing (as described by Snavely et al. (2008)) [36], error checking of the model against GCPs, and export of point clouds, DEMs, and orthophotos.

Minervino Amodio et al. (2022) [14] already highlighted that there is a strong correlation between the points surveyed on the beach with the GNSS and the UAV techniques. This correlation allowed us to directly use the DEMs 2019, 2020, and 2021 obtained from the UAV surveys for extracting the beach profiles needed for our investigation. Overall, 12 beach profiles, extending from the top of the dune ridge up to the water line (see below), were extracted.

3.3. Coastal Vulnerability Assessment

The vulnerability to coastal erosion and inundation was evaluated for the test areas by means of the Coastal Vulnerability Assessment (CVA), a method developed by Di Paola et al. (2014) [37], starting from the preliminary approach used for the coastal vulnerability assessment of several coastal areas [37–40]. For the vulnerability evaluation, the CVA method allowed us to consider both the beach retreat due to storm surges (using wave climate and geomorphological data such as bathymetry, beach sedimentology, and beach width) and the coastal inundation due to run-up on the beach. To this aim, morpho-sedimentary beach features, wave climate, and multi-temporal series of aerial photographs and topographic maps were analysed.

In detail, the CVA method evaluates the coastal vulnerability for each considered period according to the following equation:

$$CVA = I_{Ru} + I_R + E + T_i \tag{1}$$

where I_{Ru} is the wave run-up height index, which is given by the run-up level divided by the beach foreshore slope [41]; I_R is the short-term erosion index, which provides a measurement of the maximum beach recession due to storms, normalized with the beach width [42]; E is the beach erosion rate in m/y, and T_i measures the horizontal distance travelled by the tidal range. In this work, $T_i = 0$, because the Molise coast is microtidal and experiences ordinary tidal excursions of 30–40 cm [21].

The I_{Ru} index provides the measurement of the potential inundation capacity that characterizes natural beaches during wave storms. According to Stockdon et al. [41], the wave run-up height is provided by $Ru_{2\%}$, i.e., the wave run-up level exceeded by 2% of the number of incoming waves, which is measured vertically from the still water line. This value is projected along the beach through the calculation of $X_{Ru2\%}$, which corresponds to the horizontal distance travelled by the wave in the run-up process. Therefore, I_{Ru} takes on values that depend on the percentage associated with the maximum horizontal run-up distance of the wave on the beach ($X_{Ru2\%}$), which is normalized with respect to the width of the emerged beach (L). In this way, the I_{Ru} index can be customarily clustered into four discrete levels, as shown in Table 3.

Table 3. Coastal vulnerability assessment (CVA) classification scheme (according to [37]).

Variable	1	2	3	4
I_R (%)	≤15	16–30	31–50	>50
I_{Ru} (%)	≤40	41–60	61–80	>80
E (m/y)	≥−0.5	−0.6–−1.0	−1.1–−2.0	<−2.0
	Low	**Medium**	**High**	**Very High**
CVA	**≤6**	**7–9**	**10–12**	**≥13**

The I_R index instead provides the measurement of the potential beach retreat and is used for the dynamical calculation of the shoreline retreat based on the convolution method of Kriebel and Dean (1993) [42]. I_R values depend on the percentage associated with the maximum beach retreat (R_{max}), normalized with the beach width L. As a result, I_R values depend on the percentage associated with R_{max}, normalized with respect to the width of L. In addition, the I_{Ru} and I_R indexes can be customarily clustered into four discrete levels (Table 3).

Considering that the evolution of a beach is not only linked to the effects produced by coastal inundation during extreme events (H_t = 3.5 m, Table 1) but also to those caused by ordinary wave dynamics (H_s = 0.7 m, Table 1), the parameters I_{Ru} and I_R were calibrated considering both conditions, using Formula (2) and (3), respectively.

$$I_{Ru} = \left(I_{Ru\,(H_s)} + 2{\cdot}I_{Ru\,(H_t)}\right)/3 \tag{2}$$

$$I_R = \left(I_{R\,(H_s)} + 2{\cdot}I_{R\,(H_t)}\right)/3 \tag{3}$$

Regarding the E index, the evolution of the shoreline in the mid-term (periods 1954–2004 (E_1) and 2004–2016 (E_2)) and the short term (periods 2016–2019 (E_{2019}), 2016–2020 (E_{2020}), and 2016–2021 (E_{2021})) were considered. To this aim, we used the indications proposed by Cromwell [43], following Equation (4).

$$E = (E_1 + 2{\cdot}E_2 + 3{\cdot}E_{2019/2020/2021})/6 \tag{4}$$

The E index can be customarily clustered into four discrete levels (Table 3).

4. Results

4.1. Long- and Mid-Term Shoreline Changes along the Molise Coast from 1954 to 2016

The application of the DSAS method allowed for the calculation of the long and mid-term shoreline changes of the Molise coast (see northern and southern Molise coast, Figures 4 and 5) for the periods of 1954–2004, 2004–2016, and 1954–2016 (Table 4).

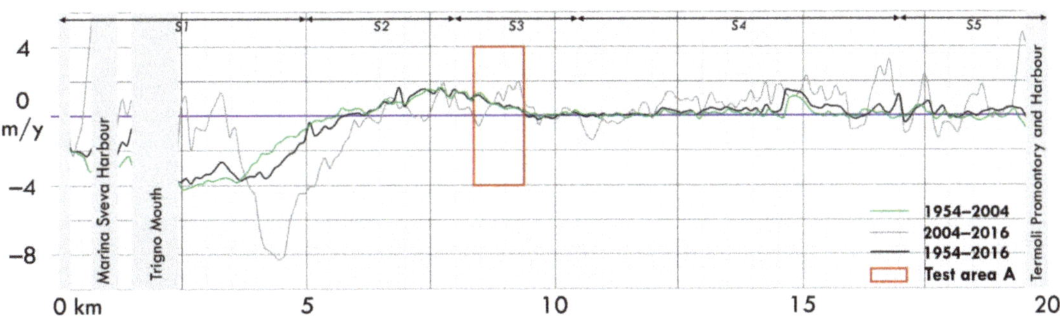

Figure 4. Long- to mid-term annual shoreline changes along the northern Molise coast (segments S1–S5) during the periods 1954–2016, 1954–2004, and 2004–2016.

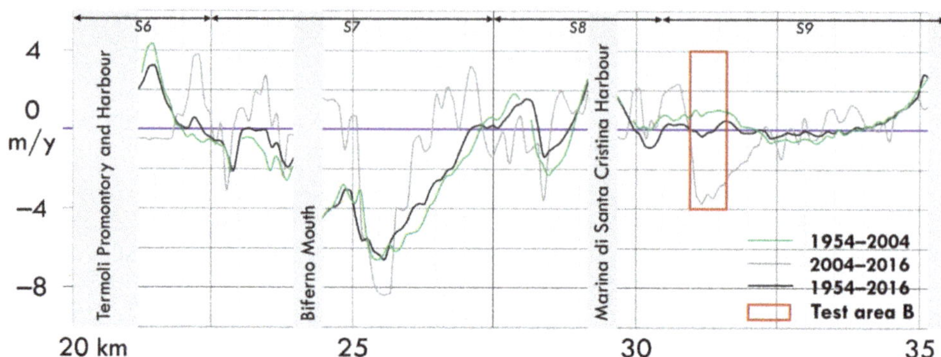

Figure 5. Long to mid-term annual shoreline changes along the southern Molise coast (segments S6–S9) during the periods 1954–2016, 1954–2004, and 2004–2016.

Table 4. Long and mid-term shoreline changes along the Molise coast.

Segment	1954–2016		1954–2004		2004–2016	
	NSM (m)	LRR (m/y)	NSM (m)	LRR (m/y)	NSM (m)	LRR (m/y)
S1	−132.5	−2.2	−142.1	−2.6	−26.1	−2.1
S2	−6.9	0.1	6.2	0.3	−13.1	−1.2
S3	33.9	0.7	27.5	0.7	6.4	0.5
S4	15.0	0.3	7.2	0.1	7.9	0.7
S5	4.8	0.2	−1.9	−0.1	6.7	0.7
S6	60.9	1.1	49.9	1.0	9.1	0.8
S7	−172.0	−2.7	−169.8	−3.4	−14.0	−1.2
S8	23.4	0.3	26.7	0.5	3.7	0.3
S9	9.4	0.3	9.8	0.2	−5.7	−0.5

Regarding the long-term evolution of the Molise coastline (time interval, 1954–2016), the results obtained (Figures 4 and 5, Table 4) confirm that erosion mainly affected segments S1 and S7, which underwent an overall average shoreline retreat of −132.5 m and −172.0 m, respectively (Table 4). Minor erosion areas are present in the northern part of S2, in the central and southernmost parts of S8, as well as in the northern portion of S9. However, segments S3, S4, S6, and S8 show an overall positive balance, and segments S2, S5, and S9 demonstrate substantial stability.

Shoreline changes calculated for the period of 1954–2004 obviously largely confirm this general trend, but these indicate major erosion rates for segments S1 and S7 (LRR of −2.6 m/y and −3.4 m/y, respectively; Table 4) with respect to the period of 1954–2016, thus highlighting a more intense shoreline erosion in these segments during the first 50 years of the considered time interval.

With reference to shoreline changes from 2004 to 2016, data highlight some trend inversions with respect to previous shoreline changes, from negative to positive and vice versa, which affect, for example, the northern part of S1, some parts of segments S6 and S7, as well as the two test areas. Shoreline retreat persists in segment S1, similar to the previous period (LRR of −2.1 m/y, Table 4), while the annual rates of S7 evidence a clear trend towards conditions of minor erosion (LRR of −1.2 m/y, Table 4). Furthermore, segments S2 and S9 show an overall trend towards moderate and slight shoreline retreat, respectively (LRR of −1.2 m/y and −0.5 m/y, respectively; Table 4), while the other coastal segments—and therefore also S3, where test area A is located—show an overall stability.

4.2. *Morpho-Topographic Changes of the Beach–Dune Systems in the Test Areas from 2019 to 2021*

Based on the data acquired during the UAV flights in 2019, 2020, and 2021, we produced the orthophotos and the related DEMs for test areas A and B, illustrated in Figures 6 and 7.

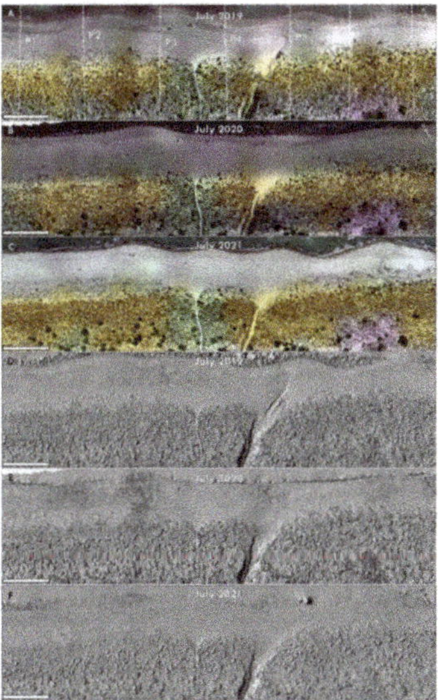

Figure 6. Orthophotos and DEM hillshades produced with the data acquired during the UAV flights in 2019 (**A**,**D**), 2020 (**B**,**E**), and 2021 (**C**,**F**) along Petacciato beach. (**A**) shows the locations of beach profiles P1–P7 extracted from the DEMs.

Figure 7. Orthophotos and DEM hillshades produced with the data acquired during the UAV flights in 2019 (**A**,**D**), 2020 (**B**,**E**), and 2021 (**C**,**F**) along Campomarino beach. (A) shows the locations of beach profiles C1–C5 extracted from the DEMs.

A comparison of the orthophotos and DEMs produced for test area A (Petacciato beach, Figure 6) gives evidence of the substantial stability of the dune vegetation cover and the variable coverage of vegetation debris on the beach. Slight shoreline movements are evidenced by small shoreline undulations varying from year to year, suggesting the probable role of longshore drift in local beach nourishment and erosion.

Comparing the orthophotos and DEMs obtained for test area B (Campomarino beach, Figure 7) highlights significant differences as to the morpho-topographic and vegetational features of its beach–dune system, allowing for the identification of a northern, southern, and central portion. The latter, in particular, is characterized by a larger beach and a dune front localised in a more internal position, approximately 15 m with respect to the adjacent beach portions. The comparison of recent Google Earth pictures (Figure 8) shows that the entire shoreline of test area B has undergone a more or less consistent retreat between June 2016 and July 2019 (up to approximately 30 m) despite the presence of the three breakwaters in front (but almost completely submerged, at least since 2014; Figure 8) and a groin, which was definitely detached in this period.

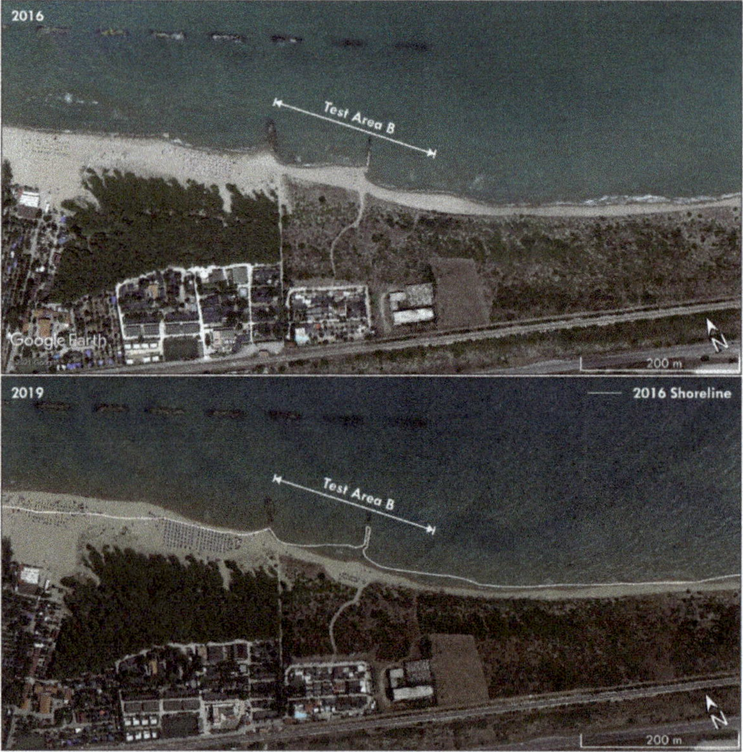

Figure 8. Evidence of shoreline and beach changes from 2016 to 2019 (Google Earth images dating back to 20 June 2016 and 20 July 2019) along the coastal stretch that includes test area B. On the Google Earth image from 2019, the white line indicates the shoreline of 2016.

Despite the fact that the entire shoreline of test area B has retreated from 2016 and 2019, the northern and southern sectors have not undergone a similar dune front retreat. This consideration, together with other observations (see below), led us to believe that the dune retreat in the central sector was not due to natural erosion but to human interventions aimed at enlarging the beach and also recovering, in this way, a part of the beach lost due to the shoreline retreat between June 2016 and July 2019 (Figure 8). In fact, this beach sector is used by the Marinelle campsite for bathing, as shown in several google earth images by the presence of beach umbrellas and pedestrian access that cuts through the dunes, connecting the campsite directly with the beach. The latter appears clean, without vegetation debris, while the presence of several excavator footprints in the images of 2020 (Figure 7B,E), along with direct observations in the field, confirm periodic operations of cleaning, levelling, and enlargement of the beach in the pre-summer period.

For all beach profiles surveyed in the test areas in 2019, 2020, and 2021 (for location, see Figures 6A and 7A), the major morphometric features were determined.

Table 5 illustrates the major morphometric features of the beach profiles surveyed in 2019 and 2021.

Table 5. Comparison of main morphometric features of beach profiles in 2019 and 2021.

Profiles	Backshore Width—L (m)		Backshore Slope—β_b (%)		Foreshore Slope—β_f (%)		Total Slope—m_o (%)		Berm—B (m)		Dune Front Retreat (m)	D_{50} (mm)
	2019	2021	2019	2021	2019	2021	2019	2021	2019	2021	-	-
P1	34.1	30.7	3.7	1.1	5.7	6.0	4.0	2.9	0.3	0.5	-	
P2	28.8	21.9	2.7	2.2	25.4	8.4	6.2	3.6	1.4	0.4	-	
P3	27.9	20.4	2.8	2.4	21.8	10.8	6.5	4.8	1.4	0.8	-	
P4	28.5	24.4	6.2	4.1	14.2	10.0	8.1	5.4	1.1	0.8	-	0.47
P5	23.9	20.5	4.7	5.3	8.0	6.4	5.7	5.6	0.8	0.5	-	
P6	22.5	21.2	6.7	4.5	17.5	8.7	10.3	5.9	1.1	0.6	-	
P7	18.9	16.5	4.0	5.6	13.1	6.8	6.3	6.0	0.7	0.5	-	
C1	10.5	8.4	9.8	15.6	8.6	18.2	9.3	22.2	0.6	0.3	0.4	
C2	9.2	7.5	9.4	22.8	10.9	10.1	10.0	17.5	0.5	0.6	1.2	
C3	24.1	20.0	4.2	4.8	10.0	5.3	5.1	5.0	0.5	0.6	0.4	0.50
C4	6.8	6.1	8.9	19.8	14.9	5.3	10.6	11.3	0.5	0.5	2.1	
C5	4.9	5.9	15.8	6.0	3.5	10.3	9.5	8.2	0.2	0.6	1.5	

Test Area A

Comparing data obtained for profiles P1–P7 (Table 5) for 2019 and 2021 highlights the following trends: A slight restriction of the backshore width, and aside from a few exceptions, the decrease of backshore, foreshore, and total slopes as well as a slight decrease in the height of the ordinary berm. However, no dune front retreat occurred along these profiles from 2019 to 2021.

Test Area B

Comparing data obtained for profiles C1–C5 highlights the following trends: A slight restriction of the backshore as well as partial increases and decreases of the backshore and foreshore slopes, which resulted in an overall increase of the total beach slopes along C1 and C2 from 2019 to 2021 and their substantial stability along C3–C5. In addition, berm heights alternately increased or decreased. Finally, all profiles showed some dune front retreat, between 0.4 and 2.1 m.

Figure 9 illustrates the short-term evolution of beach profiles P6 and C4 from 2019 to 2021. These profiles, indicated by Rosskopf et al. (2018) [21] as profiles P1 and P2, respectively, have already been under study for two decades and were surveyed in 2001, 2010, and 2016. Hence, they are taken as a reference for the overall evolution of the beach–dune systems in the test areas in the last two decades.

In particular, profile P6 well illustrates the persistence of the substantial stability of the beach–dune system in test area A, without any setback of the dune front and only slight morpho-topographic changes in the backshore; it is therefore consistent with its previous evolution.

Profile C4, on the other hand, well represents the overall evolution of test area B, which has been affected by progressive shoreline and dune front retreat, while the backshore width has remained relatively stable (Table 5), consistent with what we observed from 2001 onwards.

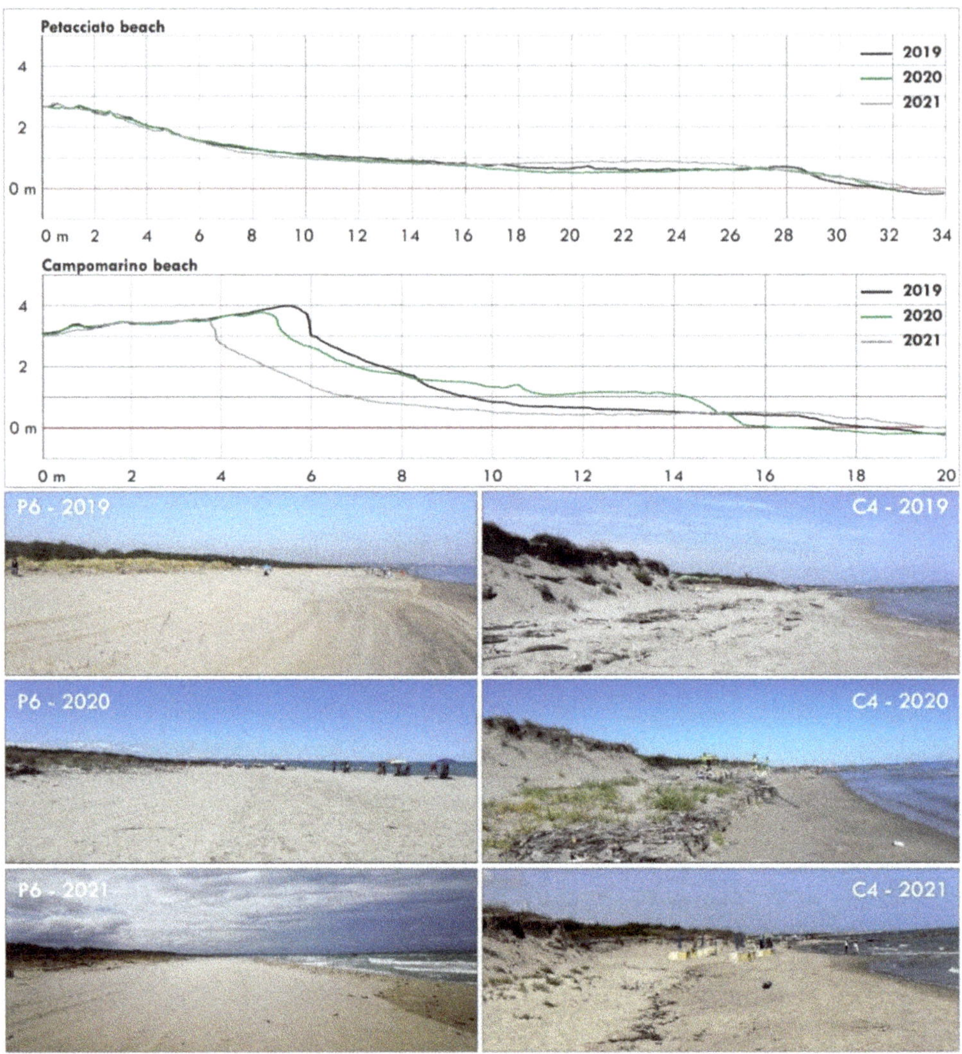

Figure 9. Representative beach profiles (C4 and P6) and panoramic views of test areas A and B.

4.3. Long- to Short-Term Shoreline Changes in the Test Areas and Related Erosion Indexes

The long- to short-term shoreline changes calculated for the beach profiles surveyed in the test areas, which were based on shoreline data referring to 1954, 2004, 2016, 2019, 2020, and 2021, provide the following data (Figure 10 and Tables 6 and 7).

Concerning the long-term evolution of test area A (the period of 1954–2016, Figure 10), the balance is clearly positive, as also evidenced by the average annual shoreline rates at around 1 m/y reconstructed for profiles P1–P7 (Table 6), and which is consistent with the evolution of S3 (Table 4). A minimum value of 1 has been attributed to the erosion indexes E_1 of profiles P1–P7.

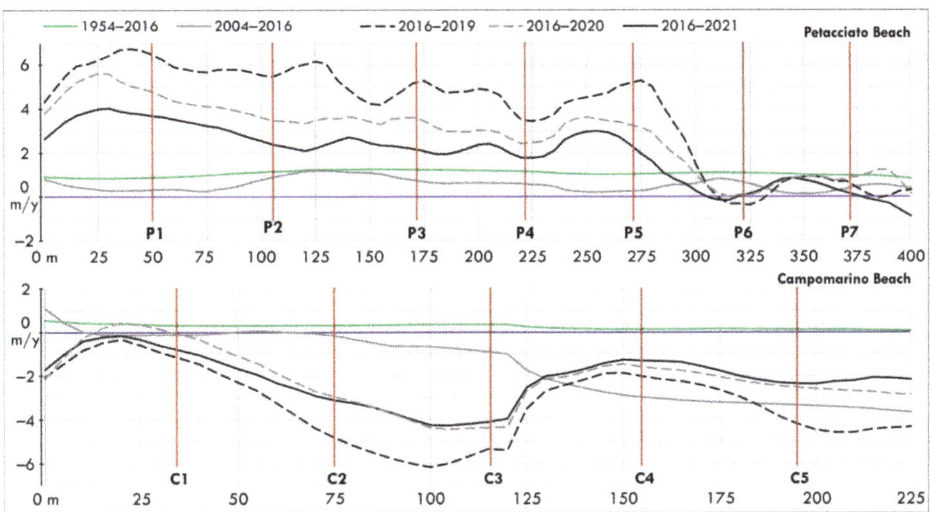

Figure 10. Annual shoreline change rates calculated for the beach profiles surveyed in the study areas for the periods 1954–2016, 2004–2016, 2016–2019, 2016–2020, and 2016–2021.

Table 6. Shoreline changes in m (NSM) and annual shoreline change rates in m/y (LRR) calculated for the beach profiles in study areas A and B for the periods 1954–2016 and 2004–2016, and an evaluation of the related erosion indexes (E_1 and E_2).

Beach Profiles		1954–2016			2004–2016		
		NSM (m)	LRR (m/y)	E_1	NSM (m)	LRR (m/y)	E_2
Petacciato beach	P1	48.0	0.9	**1**	−3.3	−0.1	**1**
	P2	67.1	1.2	**1**	6.8	0.6	**1**
	P3	71.3	1.3	**1**	4.7	0.4	**1**
	P4	67.7	1.2	**1**	3.3	0.3	**1**
	P5	60.0	1.0	**1**	1.2	0.1	**1**
	P6	68.0	1.1	**1**	9.2	0.8	**1**
	P7	61.3	1.0	**1**	5.2	0.5	**1**
Campomarino beach	C1	11.0	0.3	**1**	−7.8	−0.7	**2**
	C2	15.0	0.3	**1**	−5.1	−0.4	**1**
	C3	18.1	0.4	**1**	−11.3	−1.0	**2**
	C4	0.4	0.1	**1**	−34.6	−3.1	**4**
	C5	−0.3	0.1	**1**	−37.7	−3.2	**4**

Regarding the long-term evolution of test area B (Figure 10), the obtained data document its substantial stability (C4-C5) as well as a slight trend to progradation (C1-C3), with the erosion indexes (E_1, Table 6) also assuming a minimum value of 1 in this case.

Considering then period of 2004–2016 (Figure 10), data show clear differences between the two test areas. The Petacciato area continues to exhibit a substantial stability, with all profiles being characterized by an erosion index of 1 (E_2, Table 6). The Campomarino area (C1–C5), however, is characterized by erosion indexes ranging from 1 to the maximum value of 4 (E_2, Table 6), highlighting its general destabilization and a significant trend toward erosion in the southern sector of B (C4–C5). These data are in agreement with those previously illustrated (Section 4.1, Figure 3) with regard to the stability conditions of segment S3 and the negative shoreline trend of the northernmost portion of segment S9 during the period of 2004–2016.

Table 7. NSM and LLR values calculated for profiles P1–P7 and C1–C5 for the periods 2016–2019, 2016–2020, and 2016–2021; related erosion indexes E_{2019}, E_{2020}, and E_{2021}; absolute shoreline variations (Net Shoreline Measurement) for the period of 2019–2021 (NSM 2019–2021).

Beach Profiles		2016–2019			2016–2020			2016–2021			2019–2021
		NSM (m)	LRR (m/y)	E_{2019}	NSM (m)	LRR (m/y)	E_{2020}	NSM (m)	LRR (m/y)	E_{2021}	NSM (m)
Petacciato beach	P1	22.8	6.5	1	18.9	4.8	1	18.5	3.7	1	−4.3
	P2	19.2	5.5	1	12.5	3.5	1	11.6	2.4	1	−7.6
	P3	16.9	4.8	1	14.3	3.6	1	9.7	2.3	1	−7.2
	P4	14.0	4.0	1	9.7	2.6	1	10.1	2.0	1	−3.9
	P5	17.7	5.0	1	12.7	3.4	1	14.2	2.7	1	−3.5
	P6	−1.2	−0.4	1	0.6	0.1	1	−1.7	−0.2	1	−0.5
	P7	2.3	0.7	1	3.4	0.7	1	0.7	0.3	1	−1.6
Campomarino beach	C1	−4.2	−1.2	3	1.4	−0.1	1	−7.3	−0.8	2	−3.1
	C2	−16.9	−4.8	4	−10.4	−3.0	4	−18.8	−3.1	4	−1.9
	C3	−18.8	−5.3	4	−18.2	−4.4	4	−22.6	−4.1	4	−3.8
	C4	−7.2	−2.0	4	−6.6	−1.6	3	−6.8	−1.3	3	0,4
	C5	−14.7	−4.2	4	−8.9	−2.5	4	−13.7	−2.4	4	1.0

Regarding the period of 2016–2021 (Table 7), our data for test area A confirm conditions of substantial stability as well as a slight trend toward shoreline progradation; for test area B, however, there is a slight to moderate trend toward shoreline retreat (Figure 10 and Table 7).

Considering the periods 2016–2019, 2016–2020, and 2019–2021 in detail, data obtained for test area A (Table 7) show that 2016–2019 was the most positive period, while the last three years (period 2019–2021) were characterized by a slight negative shoreline trend, with values between −0.5 m (P6) and −7.6 (P2) (NSM, Table 7). This negative shoreline trend, however, did not have any influence on the erosion index levels for profiles P1–P7, which remained equal to 1.

Conversely, the data calculated for test area B for the period of 2016–2019 (Table 7) show that this period was the most negative one, as it was characterized overall by the worst values for annual shoreline retreat rates (LRR up to −5.3 m/y, see C3) and by prevalent erosion indexes of 3 and 4 (E_{2019}, Table 7). Some amelioration in time is evidenced by profiles C1 and C4, whose erosion indexes decreased from 3 and 4 to 2 and 3, respectively, when considering the general period of 2016–2021 (Table 7).

Overall, these data highlight opposite trends for the two test areas. A slight trend toward destabilization for A, but without any effect on erosion indexes E_{2021}; a slight trend toward recovery for B, with some positive influence on erosion indexes E_{2021}. These trends appear to be clearly related to the increase of erosion along A and the decrease of erosion along B (see NSM values, Table 7) during the most recent period of 2019–2021.

4.4. Coastal Vulnerability Assessment

For the Coastal Vulnerability Assessment (CVA) of the test areas, the following three indexes were considered: the wave run-up height index I_{RU}, the short-term erosion index I_R (Tables 6 and 7), and the beach erosion index E, which were calculated using Formulas (2), (3), and (4), respectively.

4.4.1. The Wave Run-Up Height Index I_{RU}

$I_{RU2\%}$ levels were obtained by calculating the run-up values $R_{u2\%}$ and the related parameters $X_{Ru2\%}$ and $X_{Ru2\%}/L$ (width of the beach affected by run-up and relative percentage) for the beach profiles P1–P7 and C1–C5 for years 2019, 2020, and 2021 (Tables 8 and 9). Both normal wave (H_s = 0.7 m) and average extreme wave conditions (H_t = 3.5 m) were considered.

Table 8. Estimated run-up index values $I_{Ru2\%}$ and related parameters $X_{Ru2\%}$ and $X_{Ru2\%}/L$ under normal wave (Hs = 0.7 m) and average extreme wave conditions (H_t = 3.5 m).

		Petacciato Beach							Campomarino Beach				
		P1	P2	P3	P4	P5	P6	P7	C1	C2	C3	C4	C5
2019 H_s = 0.7 m	$X_{Ru2\%}$ (m)	4.1	2.9	3	3.1	3.6	3.1	3.2	3.5	3.3	3.4	3.1	5.3
	$X_{Ru2\%}/L$ (%)	12.0	10.2	10.7	11.0	15.0	13.6	16.9	33.5	35.9	14.0	45.9	109.1
	$\mathbf{I_{Ru2\%}}$	1	1	1	1	1	1	1	1	2	1	2	4
2019 H_t = 3.5 m	$X_{Ru2\%}$ (m)	17.3	12.4	12.6	13.3	15.2	12.9	13.4	14.8	13.9	14.2	13.2	22.5
	$X_{Ru2\%}/L$ (%)	50.7	43.0	45.0	46.5	63.4	57.2	71.1	141.1	151.4	59.0	193.5	459.9
	$\mathbf{I_{Ru2\%}}$	2	2	2	2	3	2	3	4	4	2	4	4
2020 H_s = 0.7 m	$X_{Ru2\%}$ (m)	3.2	3.1	3.2	3	3.4	3	3	3.2	3.1	3.1	2.9	3.3
	$X_{Ru2\%}/L$ (%)	18.0	15.9	13.2	12.7	19.1	19.8	19.6	27.1	31.6	12	48.4	36.2
	$\mathbf{I_{Ru2\%}}$	1	1	1	1	1	1	1	1	1	1	2	1
2020 H_t = 3.5 m	$X_{Ru2\%}$ (m)	13.3	13.1	13.4	12.5	14.2	12.5	12.8	13.6	13.2	12.9	12.3	13.8
	$X_{Ru2\%}/L$ (%)	42.8	66.3	55.6	53.5	80.4	83.5	82.8	114.4	133.3	50.8	204.1	152.8
	$\mathbf{I_{Ru2\%}}$	3	3	2	2	4	4	4	4	4	2	4	4
2021 H_s = 0.7 m	$X_{Ru2\%}$ (m)	4	3.5	3.3	3.4	3.9	3.5	3.8	3	3.4	4.2	4.2	3.4
	$X_{Ru2\%}/L$ (%)	13.1	16.2	16.2	13.8	19.0	16.5	23.1	36.2	44.9	21.2	69.5	56.7
	$\mathbf{I_{Ru2\%}}$	1	1	1	1	1	1	1	1	2	1	3	2
2021 H_t = 3.5 m	$X_{Ru2\%}$ (m)	16.9	14.9	14	14.2	16.5	14.8	16.1	12.8	14.2	17.9	17.9	14.1
	$X_{Ru2\%}/L$ (%)	55.0	68.1	68.4	53.8	80.2	69.7	97.3	152.5	189.1	89.4	293.2	239.3
	$\mathbf{I_{Ru2\%}}$	2	3	3	2	4	3	4	4	4	4	4	4

Table 9. Evaluation of short-term erosion indexes I_R for profiles P1–P7 and C1–C5, based on R values calculated for normal wave (H_s = 0.7 m) and average extreme wave conditions (H_t = 3.5 m), respectively.

		Petacciato Beach							Campomarino Beach				
		P1	P2	P3	P4	P5	P6	P7	C1	C2	C3	C4	C5
2019 H_s = 0.7 m	R (m)	2.9	0.1	0.2	1.0	0.1	1.9	0.26	2.0	3.3	0.1	3.1	3.7
	R/L (%)	8.6	0.5	0.8	3.4	0.1	8.4	1.4	19.4	29.6	0.6	45.4	74.7
	$\mathbf{I_R}$	1	1	1	1	1	1	1	2	3	1	3	4
2019 H_t = 3.5 m	R (m)	5.2	6.3	6.7	9.3	7.4	11.0	8.8	12.4	13.7	7.0	14.1	15.8
	R/L (%)	15.1	22.0	24.0	32.5	30.9	48.9	44.1	118.2	148.6	29.1	207.7	321.5
	$\mathbf{I_R}$	1	3	2	3	3	3	3	4	4	2	4	4
2020 H_s = 0.7 m	R (m)	0.7	0.6	1.2	1.9	2.4	2.2	1.5	2.8	1.1	0.1	3.3	3.7
	R/L (%)	2.3	2.8	5.9	7.4	10.0	12.2	8.8	23.8	11.0	0.5	54.7	74.7
	$\mathbf{I_R}$	1	1	1	1	1	1	1	2	1	1	4	4
2020 H_t = 3.5 m	R (m)	8.9	8.9	10.7	11.5	12.3	11.2	10.1	12.9	9.7	8.2	13.6	14.8
	R/L (%)	28.7	44.8	48.9	44.6	50.6	61.6	59.0	108.6	98.2	32.6	227.0	164.6
	$\mathbf{I_R}$	2	3	3	3	4	4	4	4	4	3	4	4
2021 H_s = 0.7 m	R (m)	15.9	5.5	0.2	0.1	0.1	0.1	0.2	8.9	5.3	0.2	3.4	1.3
	R/L (%)	51.7	24.9	1.0	0.1	0.1	0.5	1.0	106.1	70.2	0.9	57.0	22.8
	$\mathbf{I_R}$	4	2	1	1	1	1	1	4	4	1	4	2
2021 H_t = 3.5 m	R (m)	1.0	3.5	5.7	6.9	8.5	8.6	9.3	20.3	16.7	6.4	14.6	11.4
	R/L (%)	3.2	15.8	27.8	28.1	41.7	40.6	56.4	241.9	222.2	31.9	239.4	192.4
	$\mathbf{I_R}$	1	2	2	2	3	3	4	4	4	3	4	4

Regarding test area A, the results highlight the stable run-up height conditions under normal wave conditions (H_s = 0.7 m) from 2019 to 2021, with indexes $\mathbf{I_{Ru2\%}}$ maintaining a minimum value of 1 for all profiles.

However, with regard to mean extreme wave conditions (H_t = 3.5 m), $I_{RU2\%}$ levels reached values between 2 and 3 in 2019 (P5 and P7, Table 8). In 2020, several negative variations in the $I_{RU2\%}$ levels were registered in the northern (P1–P2) and especially in the southern part (P5–P7) of the test area, where maximum values of 4 were calculated. Then, in 2021, both negative and positive variations were registered for parts of the profiles, resulting in final values between 2 and 4, and a condition substantially similar to that in 2020.

Regarding test area B, data highlight overall higher $I_{RU2\%}$ values under normal wave conditions (Table 8) for 2019 when compared with test area A, as well as several positive and negative variations in the $I_{RU2\%}$ values from 2019 to 2021, oscillating between 1 and 4. These variations, however, do not suggest any persisting positive or negative trend. Worthy of mention is the $I_{RU2\%}$ value of C3, which remained equal to 1 during all the three years.

When considering the average extreme wave conditions, all profiles were characterized by a maximum $I_{RU2\%}$ value of 4 in 2020 and 2021, except for C3 ($I_{RU2\%}$ value of 2, Table 8). This result highlights the generally worse run-up conditions of this beach during average extreme wave conditions, which can be generally related to the scarce width of the backshore. This is not the case for C3, i.e., the central portion of test area B, where the backshore is significantly wider and undergoes only a slight restriction from 2019 to 2021 (from 24.1 m to 20.0 m, Table 4). In this case, the increase of the run-up index value to 4 can be related to the decrease of the foreshore slope (from 10% to 5.3%, Table 5). In fact, as illustrated in Section 4.2, this portion of the beach has undergone important human interventions, resulting in enlargements, levelling, and the overall topographic lowering of the beach. Therefore, in our opinion, the worsening of the run-up conditions along C3 in 2021 is most likely related to these human interventions, showing that the latter did not contribute to the maintenance of the beach but, conversely, contributed to its major vulnerability to erosion.

4.4.2. The Short-Term Erosion Index I_R

Short-term erosion indexes I_{R2019}, I_{R2020}, and I_{R2021} were evaluated for both test areas (Petacciato beach, profiles P1–P7, and Campomarino beach, profiles C1–C5, Table 9) on the basis of R and R/L values obtained for both normal wave conditions (H_s = 0.7 m) and average extreme wave conditions (H_t = 3.5 m).

With regard to normal wave conditions, the results obtained for test area A highlight the persistence of low short-term erosion indexes (I_R values of 1, Table 9), with the only exception of the I_R2021 values for profiles P1 and P2, which increased to 4 and 2, respectively. However, regarding average extreme wave conditions, the profiles showed a notable variation in I_R values that oscillated between 1 and 4, but without showing any clear trend from 2019 to 2021.

The results obtained for test area B for normal wave conditions indicate a major variability of the short-term erosion indexes, which oscillated between values of 1 and 4, showing an overall worsening of erosion conditions from 2019 to 2021. The only exception is profile C3, which maintained an I_R value of 1, thus confirming the important role played by human interventions that have annually affected this portion of the beach. Concerning average extreme wave conditions, the I_R values reached maximum values of 4 for all profiles, with the only exception of C3, which is characterized by I_R values of 2 in 2019 and 3 in the following years, 2020 and 2021 (Table 9).

4.4.3. Variations in the CVA Parameters

The variations in the E, I_{RU}, and I_R indexes from 2019 to 2021 (Tables 7–9, for I_{RU} and I_R values, average extreme wave conditions were considered) led to significant variations in the related CVA levels, which were calculated according to Formula 1 (see Section 3.3).

In the case of test area A (Figure 11), with regard to data for 2019 and 2021, some profiles show the shift from low to medium CVA levels (P1, P3) or vice versa (P4); however, the other profiles show an overall slight increase of CVA values while remaining within the medium level. These data give evidence of an overall slight increase of erosion susceptibility along the Petacciato beach. Furthermore, CVA values obtained for profiles P5, P6, and P7 for 2020 are even higher than those of 2021, indicating a temporary major worsening of CVA conditions for the southern portion of study area A from 2019 to 2020, which then partially recovered in 2021.

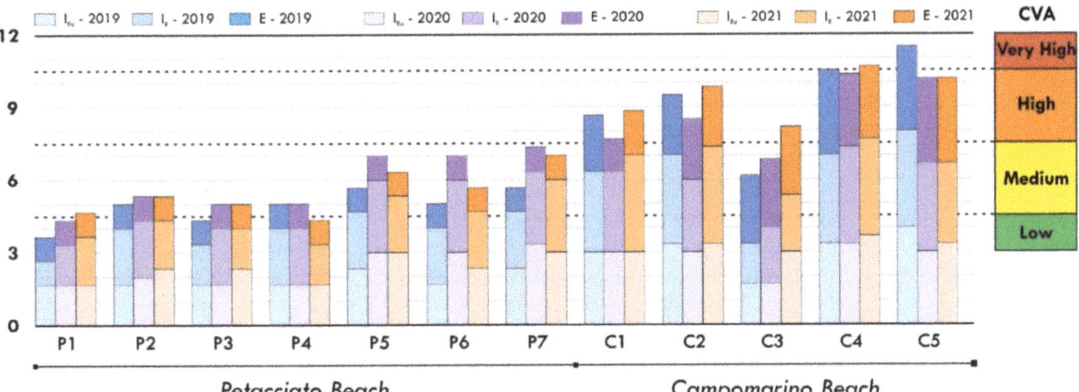

Figure 11. Comparison of I_{RU}, I_R, and E indexes and related CVA levels obtained respectively for 2019, 2020, and 2021 for the beach profiles surveyed in test areas A and B (Petacciato and Campomarino beaches).

Test area B, however, started 2019 with CVA values at the medium up to the very high CVA level, decidedly higher with respect to those of test area A. A comparison of 2019 and 2021 highlights very modest changes in the CVA values for C1, C2, and C4, the shift of C3 from a medium to a high CVA level, and that of C5 from a very high to a high CVA level. The CVA levels of 2020 are lower with respect to those of 2021, with the only exception of C5, where no change occurs from 2020 to 2021. Overall, only profile C3 shows a significant change in CVA values from 2019 to 2021, which can be mainly attributed to the negative variations, first in I_R and then in I_{RU} values (see also Tables 8 and 9). This profile has undergone, among all, the most significant negative changes in CVA values and the consequent increase of coastal vulnerability, which must be related, in our opinion, to the impact of human interventions in the central portion of test area B, which have been already reported and discussed before.

5. Discussion and Conclusions

The present study confirms that the Molise coast has been affected by severe shoreline erosion in the long-term period (1954–2016), which mainly affected the coastal segments that include the Trigno and Biferno river mouths (S1 and S7, Table 4).

A comparison of the most up-to-date previous shoreline change data for the period of 1954–2014 [21] with those obtained in this study for 1954–2016 shows that annual retreat rates in coastal segments S1 and S7 have slightly decreased from 2.7 m/y to 2.2 m/y and from 2.9 m/y to 2.7 m/y, respectively.

Concerning the mid-term shoreline evolution of the Molise coast (for the periods 2004–2014 [21] and 2004–2016 in this study), the comparisons highlight that erosion has increasingly affected S2 (−0.7 m/y in [21], −1.2 m/y in this study) and has also begun to affect S9 (0.5 m/y in [21], −0.5 m/y in this study), thus evidencing a slight strengthening of the erosion trend.

Regarding specifically the mid-term shoreline changes in test areas A and B (for the period of 2004–2016, Figure 10 and Table 6), the data highlight their stability and slight trend toward erosion, which are in line with the mid-term shoreline evolution of the coastal segments S3 and S9 (Table 4), which include them.

Finally, concerning the period of 2016–2021 (Table 7), the acquired data show that test area A prograded overall. However, morpho-topographical features and shoreline positions measured along profiles P1–P7 give evidence of a slight trend toward shoreline retreat from 2019 to 2021, although without any effect on the overall positive balance. Test area B, however, has already been affected by shoreline retreat and dune erosion at least since 2004 (Table 6), and it continued to be affected by erosion until 2021 (Table 7), resulting in a progressive degradation of the beach–dune system.

Changing beach and shoreline conditions in the test areas from 2019 to 2021 has induced several changes in the values of the parameters used in the CVA assessment, and therefore changes in the resulting CVA values/levels. The CVA values obtained for 2019 and 2021 highlight a general although modest trend towards increasing coastal vulnerability for both test areas. In detail, the most significant negative changes in CVA values concern profile C3, pointing out the negative effect of human interventions on the coastal vulnerability of the central portion of test area B.

The obtained results highlight, according to other studies realized in similar coastal contexts [44–46], that beach–dune systems can undergo significant changes in the short-term period and even from one year to the next, thus becoming part of a persisting trend or simply presenting as evidence of reversible fluctuations. Consequently, short-term/annual monitoring of shoreline dynamics and morphometric changes in the beach–dune system appears to be essential in order to detect in time and supervise erosion trends.

Moreover, the integration of traditional investigation methods—mainly those based on available photogrammetric and/or satellite imagery and GPS measurements—with the more innovative UAV survey technology allows for an increase of the scale of observation and monitoring as well as for the detection and updating of the most recent beach–dune and shoreline changes in an efficient, cheaper, and rapid way.

The demonstration that shoreline and beach morphology changes from 2019 to 2021 have caused variations in the indexes that enter in the CVA assessment highlights the need for and the opportunity to update such indexes in a rapid and efficient manner by using the proposed UAV approach, especially in critical erosion hotspot areas under monitoring.

In conclusion, the relatively simple use of UAV technology, along with the possibility to acquire DEMs and georeferenced images with high spatio-temporal resolution, allow this technology to excellently lend itself to the integration of existing coastal change and shoreline migration mapping methodologies and databases. Moreover, the integrated use of UAV and GIS approaches has proven to be an effective instrument, not only for a quick spatial data analysis, but also in order to offer an objective approach with consistent measurement and calculation processes.

Author Contributions: Conceptualization. G.D.P., A.M.A., G.R. and C.M.R.; methodology. G.D.P. and A.M.A.; software. G.D.P. and A.M.A.; validation. G.D.P., A.M.A., G.R. and C.M.R.; formal analysis. G.D.P., G.D., A.M.A., G.R. and C.M.R.; investigation. G.D.P., G.D., A.M.A., G.R. and C.M.R.; data curation. G.D.P. and A.M.A.; writing—original draft preparation. G.D.P., A.M.A., G.R. and C.M.R.; writing—review and editing. G.D.P. and C.M.R.; supervision. C.M.R. All authors have read and agreed to the published version of the manuscript.

Funding: This research received no external funding.

Institutional Review Board Statement: Not applicable.

Informed Consent Statement: Not applicable.

Data Availability Statement: Not applicable.

Conflicts of Interest: The authors declare no conflict of interest.

References

1. Spencer, T.; French, J.R. Coastal processes and landforms. *Geol. Soc. Lond. Mem.* **2022**, *58*, M58-2021-2034. [CrossRef]
2. Masselink, G.; Hughes, M.; Knight, J. *Introduction to Coastal Processes & Geomorphology*, 2nd ed.; Routledge: Oxfordshire, UK, 2011.
3. Short, A.; Jackson, D. *Beach Morphodynamics*, 2nd ed.; Elsevier: Amsterdam, The Netherlands, 2020.

4. Nicholls, R.; Wong, P.; Burket, V.; Codignotto, J.; Hay, J.; McLean, R.; Ragoonaden, S.; Woodroffe, C. Coastal systems and low-lying areas. In *Climate Change 2007: Impacts, Adaptation and Vulnerability, Contribution of Working Group II to the Fourth Assessment Report of the Intergovernmental Panel on Climate Change*; Parry, M.L., Canziani, O.F., Palutikof, J.P., Van der Linden, P.J., Hanson, C.E., Eds.; Cambridge University Press: Cambridge, UK, 2007; pp. 315–356.
5. Woodroffe, C. *The Natural Resilience of Coastal Systems: Primary Concepts*; University of Wollongong: Wollongong, NSW, Australia, 2007.
6. He, Y.; Wu, Y.; Lu, C.; Wu, M.; Chen, Y.; Yang, Y. Morphological change of the mouth bar in relation to natural and anthropogenic interferences. *Cont. Shelf Res.* **2019**, *175*, 42–52. [CrossRef]
7. Zimmermann, M.; Erikson, L.H.; Gibbs, A.E.; Prescott, M.M.; Escarzaga, S.M.; Tweedie, C.E.; Kasper, J.L.; Duvoy, P.X. Nearshore bathymetric changes along the Alaska Beaufort Sea coast and possible physical drivers. *Cont. Shelf Res.* **2022**, *242*, 104745. [CrossRef]
8. Valentine, K.; Mariotti, G. Wind-driven water level fluctuations drive marsh edge erosion variability in microtidal coastal bays. *Cont. Shelf Res.* **2019**, *176*, 76–89. [CrossRef]
9. Pirazzoli, P.A.; Tomasin, A. Recent near-surface wind changes in the central Mediterranean and Adriatic areas. *Int. J. Climatol.* **2003**, *23*, 963–973. [CrossRef]
10. Buccino, M.; Paola, G.D.; Ciccaglione, M.C.; Giudice, G.D.; Rosskopf, C.M. A medium-term study of molise coast evolution based on the one-line equation and "equivalent wave" concept. *Water* **2020**, *12*, 2831. [CrossRef]
11. UN. The Ocean Conference. *Factsheet: People and Oceans. New York 2017.* Available online: https://www.un.org/sustainabledevelopment/wp-content/uploads/2017/05/Ocean-fact-sheet-package.pdf (accessed on 1 February 2022).
12. Nicholls, R.J.; Marinova, N.; Lowe, J.A.; Brown, S.; Vellinga, P.; De Gusmão, D.; Hinkel, J.; Tol, R.S.J. Sea-level rise and its possible impacts given a 'beyond 4 °C world' in the twenty-first century. *Philos. Trans. R. Soc. A* **2011**, *369*, 161–181. [CrossRef]
13. MATTM. National Macro Data on the Coast Line Changes from 1960 to 2012. Available online: http://www.pcn.minambiente.it/mattm/en/project-coasts/ (accessed on 1 February 2022).
14. Minervino Amodio, A.; Di Paola, G.; Rosskopf, C.M. Monitoring Coastal Vulnerability by Using DEMs Based on UAV Spatial Data. *ISPRS Int. J. Geo-Inf.* **2022**, *11*, 155. [CrossRef]
15. Gioia, D.; Amodio, A.M.; Maggio, A.; Sabia, C.A. Impact of land use changes on the erosion processes of a degraded rural landscape: An analysis based on high-resolution DEMs, historical images, and soil erosion models. *Land* **2021**, *10*, 673. [CrossRef]
16. Flores-de-Santiago, F.; Valderrama-Landeros, L.; Rodríguez-Sobreyra, R.; Flores-Verdugo, F. Assessing the effect of flight altitude and overlap on orthoimage generation for UAV estimates of coastal wetlands. *J. Coast. Conserv.* **2020**, *24*, 1–11. [CrossRef]
17. Colomina, I.; Molina, P. Unmanned aerial systems for photogrammetry and remote sensing: A review. *ISPRS J. Photogramm. Remote Sens.* **2014**, *92*, 79–97. [CrossRef]
18. Nield, J.M.; Wiggs, G.F.; Squirrell, R.S. Aeolian sand strip mobility and protodune development on a drying beach: Examining surface moisture and surface roughness patterns measured by terrestrial laser scanning. *Earth Surf. Processes Landf.* **2011**, *36*, 513–522. [CrossRef]
19. Coveney, S.; Stewart Fotheringham, A.; Charlton, M.; McCarthy, T. Dual-scale validation of a medium-resolution coastal DEM with terrestrial LiDAR DSM and GPS. *Comput. Geosci.* **2010**, *36*, 489–499. [CrossRef]
20. Aucelli, P.P.C.; Iannantuono, E.; Rosskopf, C.M. Recent evolution and erosion risk of the Molise coast (southern Italy). *Boll. Della Soc. Geol. Ital.* **2009**, *128*, 759–771. [CrossRef]
21. Rosskopf, C.M.; Di Paola, G.; Atkinson, D.E.; Rodríguez, G.; Walker, I.J. Recent shoreline evolution and beach erosion along the central Adriatic coast of Italy: The case of Molise region. *J. Coast. Conserv.* **2018**, *22*, 879–895. [CrossRef]
22. Aucelli, P.P.C.; Di Paola, G.; Rizzo, A.; Rosskopf, C.M. Present day and future scenarios of coastal erosion and flooding processes along the Italian Adriatic coast: The case of Molise region. *Environ. Earth Sci.* **2018**, *77*, 1–19. [CrossRef]
23. Bracone, V.; Amorosi, A.; Aucelli, P.P.C.; Rosskopf, C.M.; Scarciglia, F.; Di Donato, V.; Esposito, P. The Pleistocene tectono-sedimentary evolution of the Apenninic foreland basin between Trigno and Fortore rivers (Southern Italy) through a sequence-stratigraphic perspective. *Basin Res.* **2012**, *24*, 213–233. [CrossRef]
24. Berardo, F.; Carranza, M.L.; Frate, L.; Stanisci, A.; Loy, A. Seasonal habitat preference by the flagship species Testudo hermanni: Implications for the conservation of coastal dunes. *Comptes Rendus-Biol.* **2015**, *338*, 343–350. [CrossRef]
25. Stanisci, A.; Acosta, A.; Carranza, M.L.; Feola, S.; Giuliano, M. Habitats of European Community Interest along Molise coast and their naturalistic value based on flora. *Fitosociologia* **2007**, *44*, 171–175.
26. Cavaleri, L.; Bertotti, L.; Tescaro, N. The modelled wind climatology of the Adriatic Sea. *Theor. Appl. Climatol.* **1997**, *56*, 231–254. [CrossRef]
27. Signell, R.P.; Carniel, S.; Cavaleri, L.; Chiggiato, J.; Doyle, J.D.; Pullen, J.; Sclavo, M. Assessment of wind quality for oceanographic modelling in semi-enclosed basins. *J. Mar. Syst.* **2005**, *53*, 217–233. [CrossRef]
28. Bonaldo, D.; Bucchignani, E.; Ricchi, A.; Carniel, S. Wind storminess in the adriatic sea in a climate change scenario. *Acta Adriat.* **2017**, *58*, 195–208. [CrossRef]
29. Parea, G.C. Trasporto dei Sedimento ed Erosione Costiera Lungo il Litorale fra il Tronto ed il Fortore (Adriatico Centrale). 1978, pp. 361–367. Available online: https://discovery.sba.uniroma3.it/primo-explore/fulldisplay/39cab_almap2156010230002653/ (accessed on 15 July 2022).

30. Aucelli, P.P.C.; De Pippo, T.; Iannantuono, E.; Rosskopf, C.M. Caratterizzazione morfologico-dinamica e meteomarina della costa molisana nel settore compreso tra la foce del torrente Sinarca e Campomarino Lido (Italia meridionale). *Studi Costieri* **2007**, 75–92.
31. Lionello, P.; Barriopedro, D.; Ferrarin, C.; Nicholls, R.J.; Orlić, M.; Raicich, F.; Reale, M.; Umgiesser, G.; Vousdoukas, M.; Zanchettin, D. Extreme floods of Venice: Characteristics, dynamics, past and future evolution (review article). *Nat. Hazards Earth Syst. Sci.* **2021**, *21*, 2705–2731. [CrossRef]
32. ISPRA. Stazioni mariografiche—Ortona. 2022. Available online: https://www.mareografico.it/?session=0S3823604638750708 7KIA65&syslng=ita&sysmen=-1&sysind=-1&syssub=-1&sysfnt=0&code=STAZ&idst=17&idreq=1@1@2&set=date (accessed on 1 February 2022).
33. Martínez Del Pozo, J.Á.; Anfuso, G. Spatial approach to medium-term coastal evolution in south Sicily (Italy): Implications for coastal erosion management. *J. Coast. Res.* **2008**, *24*, 33–42. [CrossRef]
34. Himmelstoss, E.A.; Henderson, R.E.; Kratzmann, M.G.; Farris, A.S. *Digital Shoreline Analysis System (DSAS) Version 5.0 User Guide*; 2018-1179; US Geological Survey: Reston, VA, USA, 2018.
35. Eltner, A.; Baumgart, P.; Maas, H.G.; Faust, D. Multi-temporal UAV data for automatic measurement of rill and interrill erosion on loess soil. *Earth Surf. Processes Landf.* **2015**, *40*, 741–755. [CrossRef]
36. Snavely, N.; Seitz, S.M.; Szeliski, R. Modeling the world from Internet photo collections. *Int. J. Comput. Vis.* **2008**, *80*, 189–210. [CrossRef]
37. Di Paola, G.; Aucelli, P.P.C.; Benassai, G.; Rodríguez, G. Coastal vulnerability to wave storms of Sele littoral plain (southern Italy). *Nat. Hazards* **2014**, *71*, 1795–1819. [CrossRef]
38. Anfuso, G.; Postacchini, M.; Di Luccio, D.; Benassai, G. Coastal sensitivity/vulnerability characterization and adaptation strategies: A review. *J. Mar. Sci. Eng.* **2021**, *9*, 72. [CrossRef]
39. Di Paola, G.; Aucelli, P.P.C.; Benassai, G.; Iglesias, J.; Rodríguez, G.; Rosskopf, C.M. The assessment of the coastal vulnerability and exposure degree of Gran Canaria Island (Spain) with a focus on the coastal risk of Las Canteras Beach in Las Palmas de Gran Canaria. *J. Coast. Conserv.* **2018**, *22*, 1001–1015. [CrossRef]
40. Di Luccio, D.; Benassai, G.; Di Paola, G.; Rosskopf, C.M.; Mucerino, L.; Montella, R.; Contestabile, P. Monitoring and modelling coastal vulnerability and mitigation proposal for an archaeological site (Kaulonia, Southern Italy). *Sustainability* **2018**, *10*, 2017. [CrossRef]
41. Stockdon, H.F.; Holman, R.A.; Howd, P.A.; Sallenger, A.H., Jr. Empirical parameterization of setup, swash, and runup. *Coast. Eng.* **2006**, *53*, 573–588. [CrossRef]
42. Kriebel, D.L.; Dean, R.G. Convolution method for time-dependent beach-profile response. *J. Waterw. Port Coast. Ocean Eng.* **1993**, *119*, 204–226. [CrossRef]
43. Crowell, M.; Leatherman, S.P.; Buckley, M. Shoreline change rate analysis: Long term versus short term data. *Shore Beach* **1993**, *61*, 13–20.
44. Laporte-Fauret, Q.; Marieu, V.; Castelle, B.; Michalet, R.; Bujan, S.; Rosebery, D. Low-Cost UAV for high-resolution and large-scale coastal dune change monitoring using photogrammetry. *J. Mar. Sci. Eng.* **2019**, *7*, 63. [CrossRef]
45. Scarelli, F.M.; Sistilli, F.; Fabbri, S.; Cantelli, L.; Barboza, E.G.; Gabbianelli, G. Seasonal dune and beach monitoring using photogrammetry from UAV surveys to apply in the ICZM on the Ravenna coast (Emilia-Romagna, Italy). *Remote Sens. Appl. Soc. Environ.* **2017**, *7*, 27–39. [CrossRef]
46. Turner, I.L.; Harley, M.D.; Drummond, C.D. UAVs for coastal surveying. *Coast. Eng.* **2016**, *114*, 19–24. [CrossRef]

Review

Understanding the Dynamics of a Coastal Lagoon: Drivers, Exchanges, State of the Environment, Consequences and Responses

Samantha Chacón Abarca [1], Valeria Chávez [1], Rodolfo Silva [1], M. Luisa Martínez [2] and Giorgio Anfuso [3,*]

[1] Instituto de Ingeniería, Universidad Nacional Autónoma de Mexico, Mexico City 04510, Mexico; SChaconA@iingen.unam.mx (S.C.A.); vchavezc@iingen.unam.mx (V.C.); rsilvac@iingen.unam.mx (R.S.)
[2] Instituto de Ecología, A.C. (INECOL), Antigua Carretera a Coatepec no. 351, Xalapa, Veracruz 91073, Mexico; marisa.martinez@inecol.mx
[3] Faculty of Marine and Environmental Sciences, University of Cádiz, 11071 Cádiz, Spain
* Correspondence: giorgio.anfuso@uca.es

Abstract: At present, many coastal ecosystems worldwide are highly affected by anthropic activities. La Mancha lagoon, in the state of Veracruz, Mexico, is an important ecosystem due to the wide array of ecosystem services that it provides. In this paper, an analysis of the environmental balances of the lagoon is outlined, using the Drivers, Exchanges, State of the Environment, Consequences and Responses (DESCR) tool. The methodological framework considers the interrelationships between the natural systems and the forces of change that alter the performance of the natural environment, in order to provide an overview of actions that may reduce negative consequences. The study area has been impacted by anthropic development, such as changes in land use for agricultural and livestock activities, loss of mangroves due to logging and modifications, carried out by local fishermen, to the natural hydrodynamics of the lagoon that alter the salinity and affect the ecosystem dynamics. Following analysis of the area, using the DESCR tool, the responses proposed include long-term environmental impact evaluation, with the aim of preserving the local coastal ecosystems.

Keywords: La Mancha lagoon; coastal; dynamics; DESCR framework

Citation: Abarca, S.C.; Chávez, V.; Silva, R.; Martínez, M.L.; Anfuso, G. Understanding the Dynamics of a Coastal Lagoon: Drivers, Exchanges, State of the Environment, Consequences and Responses. *Geosciences* **2021**, *11*, 301. https://doi.org/10.3390/geosciences11080301

Academic Editors: Gianluigi Di Paola, Germán Rodríguez, Carmen M. Rosskopf and Jesus Martinez-Frias

Received: 21 May 2021
Accepted: 19 July 2021
Published: 21 July 2021

1. Introduction

Coastal zones are amongst the most dynamic areas of the planet, where processes occurring in the atmosphere, hydrosphere, lithosphere and biosphere are endlessly interacting in a continually changing balance. For human beings, these balances are highly relevant, as the coasts are home to socio-cultural and economic processes vitally important to humankind [1]. Benefits provided by these systems drive their exploitation, often leading to severe degradation [2]. Such negative changes stimulated the need for coastal study and monitoring in recent years [3].

The reduction in space available for the natural functioning of coastal ecosystems, due to phenomena such as sea level rise and the construction of infrastructure, produces what is known as "coastal squeeze", which may lead to the disappearance of some species, ecosystems and, consequently, ecosystem services. According to Silva et al. [4], this phenomenon includes local, regional or global anthropogenic processes and favors negative consequences at different time scales that induce alterations in the natural dynamic of ecosystems and inhibit the capacity of ecosystems to adapt to climate change. Human actions are therefore one of the most important pressures faced by ecosystems [1].

The Drivers–Pressure–State–Impact–Response (DPSIR) framework [5] has been widely used to assess coastal squeeze. However, in 2016, Elliot et al. [6] reported that, since 1999, 25 schemes for management and decision making across ecosystems have used derivations of the DPSIR conceptual framework. Elliot et al. [6] recognized that clearer, more comprehensive, nested conceptual models are needed to quantify the links between pressure–state change in marine and coastal ecosystems.

The Drivers, Exchanges, State of the Environment, Consequences and Responses (DESCR) framework is based on the Drivers–Pressure–State–Impact–Response (DPSIR) framework, which is considered a useful system for the organization and presentation of environmental sustainability factors [5]. DESCR is a variant of this methodology, in which, instead of analyzing the pressures, the bidirectional exchanges of fluxes of matter and energy with the environment are considered simultaneously, evaluating their natural dynamics and connectivity. These exchanges, in which the drivers modify the state of the environment and vice versa, are considered in this cycle, as suggested by [7]. This analysis of the exchanges reflects the intensity of the pressures that exist in the study area. The impacts are evaluated as consequences in the ecosystems and can be positive, negative or neutral [4]. This methodology was described in a study on coastal squeeze in Puerto Morelos, Mexico, as a tool to evaluate and manage the consequences of the phenomenon [4].

Over recent years, environmental management has developed considerably in Latin America, but problems associated with anthropic factors such as pollution, destruction and degradation of renewable natural resources and the environment are still all too common [8]. In the case of Mexico, the tourist boom of recent years [9] has left its mark. In the state of Veracruz, the coastline, ca. 745 km in length [10], is dotted with urban and touristic developments and is of great importance ecologically, socially and economically [9]. In particular, the coastal lagoon of La Mancha (Figure 1), in the state of Veracruz, belongs to the municipality of Actopan. The lagoon has been on the Ramsar List of Wetlands of International Importance since 2004 [11] and was also established as a Priority Mangrove Conservation Site by CONABIO, the National Commission for the Knowledge and Use of Biodiversity [12]. Tourism in the vicinity of La Mancha is managed by a sustainable ecotourism company, called *La Mancha en Movimiento S.S.S.*, where local fishermen, housewives, farmers and others have been trained to work as eco-guides [13].

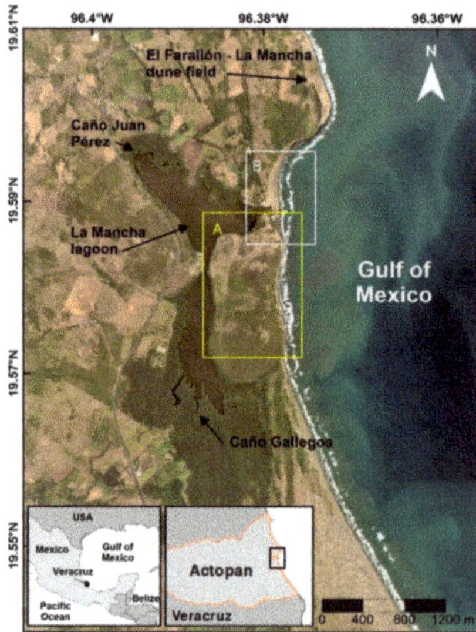

Figure 1. Location of the study area (satellite image source [14]). Area A is the site of a recent urban development project. Area B is the beach and lagoon inlet area.

It is important to identify, describe and analyze the anthropic and natural actions that regulate the lagoon ecosystem of La Mancha, in order to understand the conservation status of this environment and subsequently generate viable responses to manage it in a sustainable way. To explain the development and evolution of the coastal ecosystems of the area around La Mancha, a literature review was carried out, analyzing physical and environmental characteristics of the area, impacts of anthropic activities and consequences of the pressures on the lagoon, dune and mangrove ecosystems in the area, according to the DESCR framework. The characteristics of the lagoon and associated ecosystems were analyzed from a qualitative perspective. This approach could be used as a basis to outline policies for the sound management and use of this area.

2. Materials and Methods

2.1. Study Area

The study site is on the Gulf of Mexico in the state of Veracruz, between the coordinates 19°33′ and 19°36′ N and 96°22′ and 96°24′ W (Figure 1). Geologically, this area is a cumulative plain, formed by lacustrine, fluvial and biogenic sediments, occasionally combined with marine deposits [12]. The geomorphological evolution of the site is marked by long-term minimal marine sediment input and intense aeolian transport [15], which are both recorded during storm conditions related to northerly winds. Up until the 1970s, the aeolian transport had favored the formation of significant coastal dunes and active and fully mobile transverse dune fields. From the 1980s onwards, these dunes, particularly extensive on the north side of the rocky ledges (Figure 1), have stabilized naturally [16].

The lagoon of La Mancha is a brackish coastal lagoon, liable to siltation, with a surface of 126 ha, an average depth of 1.4 m and a maximum depth of 3 m [12,17]. The tides are mixed diurnal with a range of 0.69 m [18], and the lagoon is fed by two permanent river tributaries in the dry season (see Figure 1) and intermittently communicates with the sea [18]. The balance among sea water, rain and groundwater discharges regulates the hydrological variability of the lagoon [12]. The beach communicates with the lagoon in the lagoon inlet, and its natural opening and closing lead to modifications in the environment that alter the dynamics of the lagoon and cause changes in other surrounding ecosystems. The hydrodynamics of the inlet are very dynamic; in winter, a sandbar is formed, which naturally disappears again during the rainy season [9] (see Figure 2). The opening and closing of the lagoon inlet determine the migration of sediments, fish species and nutrients into the lagoon body [18]: when the inlet of the lagoon is closed, the water level of the lagoon rises and the salinity decreases; when it is open, the water level decreases and there is greater salinity and sediment accumulation [15] (see Figure 2). The lagoon inlet is also opened by local fishermen about three times per year in the dry season, in order to increase their catch by altering the normal flooding regime [12].

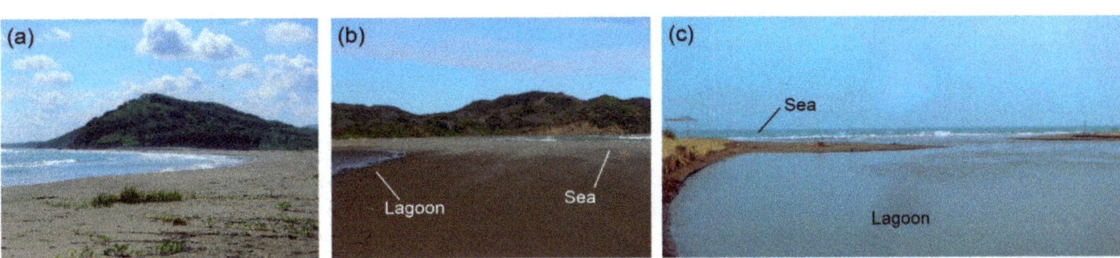

Figure 2. La Mancha: (**a**) beach; (**b**) the lagoon inlet closed by the sandbar in the dry season; (**c**) the open lagoon inlet in the wet season.

The climate of the study area has three pronounced seasons: the rainy season, from June to October; a period of "Nortes", winter storms that occur in the western Gulf of Mexico, from November to March; and a dry season, in April and May [12].

There is a wide variety of natural ecosystems in the study area, including mangroves around the lagoon (Figure 3a,b), coastal dune vegetation (Figure 3c), rainforest, tropical grassland, swamps, secondary vegetation, crop pasture and fruit trees. Mobile dune fields were reported to be mostly vegetated with *C. punctatus* and *Palafoxia lindenii* in 2008 [16], but recently, *Schizachyrium scoparium* became dominant and replaced those species (Figure 3d).

Figure 3. La Mancha: red mangrove in the (**a**) lagoon and (**b**) lagoon margins (*Rhizophora mangle*); (**c**) beach and coastal dunes; (**d**) vegetated dune field.

As the study area forms part of one of the world's largest migratory corridors for birds of prey, La Mancha receives many visiting birds of prey, shorebirds and waterfowl species, as well as marine and terrestrial fauna [11]. The coastal wetlands of the study area provide ecosystem services, such as water filtration, temperature regulation and storm protection, and goods such as fish, as well [19].

Other ecosystem services provided are shelter for animal species in their youth, scenic value for ecotourism and serving as a carbon dioxide sink [12]. The productivity of the mangrove ecosystem can be measured in terms of its primary components, i.e., litter production [20], which is a source of nutrients for the species inhabiting this ecosystem. Utrera-López et al. [20] stated that the mangrove in the area of La Mancha has medium-high litter production, with annual values of 6.92–13.50 t/ha/year.

2.2. DESCR Framework

The review of available information was conducted in December 2020, by searching the Scopus, Google Scholar and the *Instituto de Ecología, A. C.* databases, using the following keywords: lagoon, dunes, mangroves, coastal erosion and land use, in association with La Mancha, Veracruz and Mexico. Additionally, based on the authors' knowledge, the names of specific authors who have worked on coastal ecosystems and coastal evolution in this area were searched: M.L. Martínez, P. Moreno Casasola and J. López-Portillo. Masters and PhD theses were also consulted, as well as the gray literature and maritime climate databases. A list of relevant references is presented in Appendix A.

The information thus obtained was analyzed in line with the DESCR methodology [4], consisting of the following:

1. Drivers, as the social, economic or environmental forces that put pressure on the environment, which are both natural (storms, hurricanes, etc.) and human (urbanization, pollution, etc.);
2. Exchanges, which are the measure of how much the driving forces have produced changes in the system;
3. The State of the Environment, where the current environmental conditions are defined after analyzing the pressures;

4. Consequences, which are the effects of the processes on the environment, e.g., sediment transport may lead to erosion/accretion processes, and loss of ecosystem services;
5. Responses, which are the mitigation actions adopted by local authorities and stakeholders to solve environmental problems and/or improve the quality of the environment.

3. Results

3.1. Drivers in the DESCR Framework

3.1.1. Natural Drivers

Coastal ecosystems are affected by long- and short-term fluctuations in sea states and spatial changes on the coast [21], induced by the action of currents, winds and waves produced by natural phenomena such as hurricanes and storms [22]. The location of La Mancha, in the southwest of the Gulf of Mexico (Figure 1), means its dynamics are mainly the result of phenomena occurring within the gulf. The main natural drivers in La Mancha are as follows:

1. Winds;
2. Waves;
3. Storms, tropical cyclones and hurricanes.

Winds

Figure 4 shows the annual and seasonal wind roses for the coordinates 19.5° N 96° W, the location of which is approximately 40 km seaward of the La Mancha lagoon inlet. The data were obtained from the Era5 Climate Reanalysis produced by the European Centre for Medium-Range Weather Forecasts (ECMWF): hourly data on sea state parameters for 1979–2019, on a regular latitude–longitude grid at 0.5° × 0.5° resolution [23]. The annual Climate Reanalysis shows that the most frequent winds in La Mancha come from the NW, with speeds of 20–35 km/h. During winter storms and in the rainy season, winds reaching >50 km/h are observed.

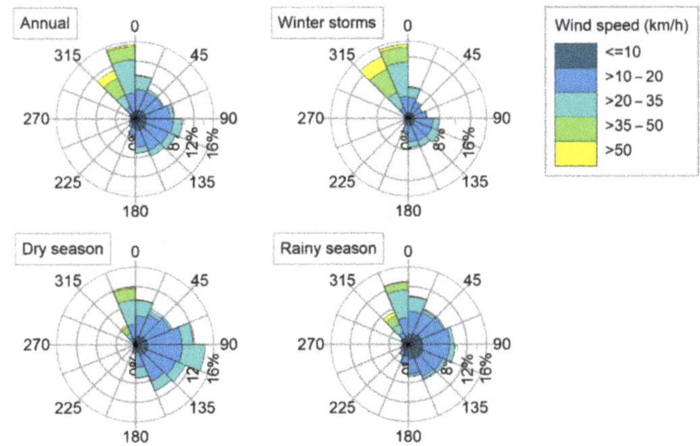

Figure 4. Annual and seasonal wind roses for La Mancha, data from Era5 [23].

Waves

Wave roses (Figure 5) for the 19.5° N 96° W coastal cell, obtained from the Era5 database [23], show that the most energetic waves come from the NE, with a 30% frequency and heights of 0.5–1 m. In winter, waves approach from the NE, with heights of 1–1.5 m, and the highest wave heights come from the NW (315° to 360°). Overall, the highest wave heights occur during winter storms, and the lowest values are in the dry and rainy seasons.

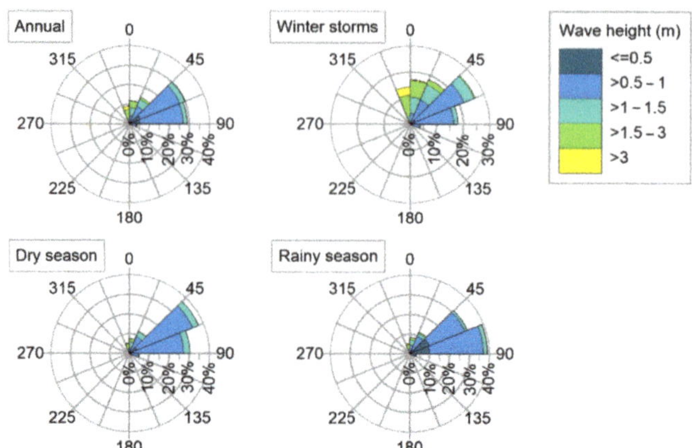

Figure 5. Annual and seasonal wave roses for La Mancha, data from Era5 [23].

Storms, Tropical Cyclones and Hurricanes

According to Martínez et al. [24], Veracruz has a sui generis climatic confluence, caused by winds and atmospheric disturbances, which can generate considerable flooding on the coast, since the rivers overflow, and the phreatic level rises. The prevailing winds in Veracruz are generated by high-pressure systems in the winter and low-pressure systems in the summer. These atmospheric systems usually approach from the north and the east, impacting the coast of Veracruz [25], with winds of up to 100 km/h in the summer, and up to 180 km/h in the winter [24]. In the last 50 years, six hurricanes of category 3–5 have hit the area [24]. Although hurricanes are not very frequent, the area is at risk from flooding and erosion due to the energetic waves and storm surge induced by these phenomena [24].

Figure 6 shows the annual wind and wave roses for extreme events for the area of La Mancha. The data correspond to the coordinates 19.50° N 96.25° W and were obtained from the Maritime Climate Atlas of *Instituto de Ingeniería UNAM* [26], calculated from 1948 to 2010 with the WAM-HURAC numerical model [27]. The wind rose includes hourly data with a wind velocity of >50 km/h (0.29% of the 1948–2010 time series). The wave rose includes hourly data with a wave height of >5 m (0.17% of the 1948–2010 time series). It can be seen that the more energetic winds and waves come from the north.

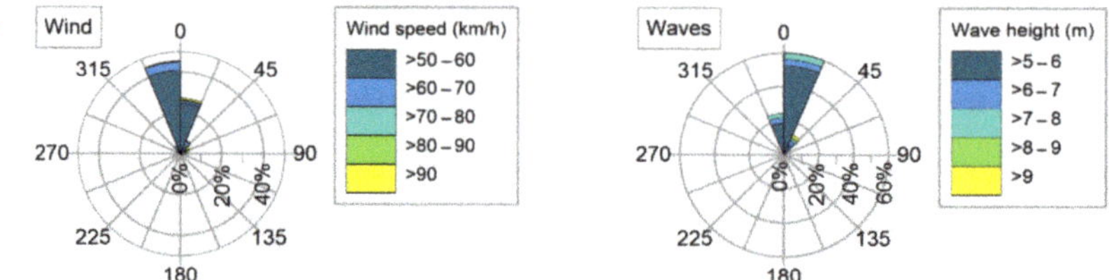

Figure 6. Annual wind and wave roses for extreme events for La Mancha, data from [26].

Tides

The measured tidal levels near the study area are reported by the Secretaría de Marina (SEMAR, Mexican Secretariat of the Navy) for two different locations: Tuxpan (97°20′48″ W 20°57′12″ N) [28], and Veracruz (96°07′51″ W 19°12′03″ N) [29]. These values are estimated with measured data from July 1999 to December 2017 and are presented in Table 1. The values for La Mancha in Table 1 were calculated with a linear interpolation between the data of Tuxpan and Veracruz.

Table 1. Tidal levels referring to the MLLW (mean lower low water), data from SEMAR [28,29].

Tidal Level	Tuxpan	La Mancha [1]	Veracruz
HAT (Highest Astronomical Tide) (m)	1.040	1.087	1.100
MHHW (Mean Higher High Water) (m)	0.496	0.500	0.501
MHW (Mean High Water) (m)	0.410	0.415	0.417
MSL (Mean Sea Level) (m)	0.284	0.286	0.287
MLW (Mean Low Water) (m)	0.220	0.230	0.233
MLLW (Mean Lower Low Water) (m)	0.000	0.000	0.000
LAT (Lowest Astronomical Tide) (m)	−0.470	−0.493	−0.500

[1] Interpolated values.

3.1.2. Anthropic Drivers

The basin of La Mancha is a rural area with little urban development, where incomes are based on livestock and agricultural activities. The contamination due to untreated wastewater from nearby settlements, and the small-scale felling of mangrove trees for agricultural land are the main problems in the study area, regarding the health and integrity of the ecosystems [18]. Notwithstanding, Ramírez Méndez [18] underlined the impacts of the opening of the lagoon inlet, two or three times a year, which modifies the natural hydrosedimentary regime in the lagoon. These openings are carried out by the *Sociedad Cooperativa de Producción Pesquera La Mancha de S.C.L.* (Fisheries Production Cooperative of La Mancha), in order to improve their fish catches [17]. Most of these fishermen prefer to fish inside the lagoon (72%), and almost half of them (42%) fish all year round. Consequently, they deliberately alter their fishing grounds [18].

The anthropic drivers considered are as follows:

1. Tourism;
2. Planed urban infrastructure.

Tourism in Coastal Areas and the Effects of Dune Trampling

The *Centro de Investigaciones costeras "La Mancha"* (CICOLMA) is a research center which aims to generate knowledge related to local ecological processes. This knowledge is transferred through activities such as ecotourism programs. In this case, people in the local communities who are interested in the conservation of the area are trained as eco-guides, and a sustainable business model has been developed, which includes ecotourism, training activities and fishing [30].

Hesp et al. [31] stated that in the dunes of La Mancha, pressure from tourists (see Figure 1) causes trampling, which can become a relevant disturbance factor. They pointed out that natural dunes are among the most sensitive environments to trampling. They also suggested that there is a relationship between the slope of the footpath and the disturbance to the dune vegetation and considered the height of the dune an important factor when quantifying the trampling damage to these ecosystems through trampling. As the dunes are trampled on, sediment dynamics may be altered, and this could indirectly affect the lagoon. Due to the feedback between dunes and the beach, modifications of dune dynamics may affect the sand budget of the beach which can eventually alter the opening and closure of the inlet. Psuty et al. [15] described the complex, indirect dynamics of the study site.

Planned Urban Infrastructure

A real estate project was begun in 2018 by *Innova Dintel Guanajuato*, with 800 "sustainable" houses planned for a 167 ha site (Figure 7), as well as a bridge, or connecting dock [32]. The land and houses in this area have been advertised for sale in 2021, and access roads to this area have also been laid, modifying the natural environment (Figure 7). Local people are concerned by the invasive nature of this project, citing the massive increase in the urban density and the environmental impacts produced by the construction work [33]. Once built, it is feared that these "sustainable" houses will use water extracted from the lagoon, which would cause an imbalance in the system. Furthermore, regulations are needed to ensure that wastewater is not discharged into the sea or the lagoon without proper treatment.

Figure 7. The area of the urban project in La Mancha (Area A in Figure 1): (**a**) in 2017, before construction (adapted from Google Earth); (**b**) in 2021, with new roads (adapted from Google Earth); (**c**,**d**) some of the new houses in the area (adapted from [34,35], respectively).

3.2. Exhanges in the DESCR Framework

The exchanges and the characteristics of the drivers will eventually be affected by the changes in the state of the environment [4]. The main exchanges in the study area are shown in Figure 8.

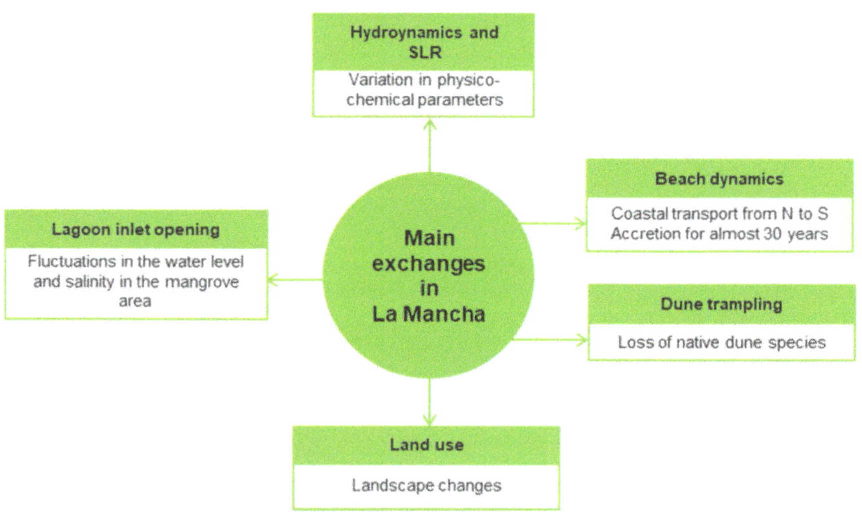

Figure 8. Main physical exchanges in La Mancha.

3.2.1. Hydrodynamics and SLR

In August (the rainy season), the values of pH, salinity and density are affected by the amount of freshwater available. In April (the dry season), these parameters depend only on the sea water, and this is also when the greatest spatial variability of the temperature is observed.

Under normal conditions, the salinity in the lagoon is stratified from south to north: the Caño Gallegos stream provides freshwater, with values of less than 0.5 PSU, while in the north of the lagoon, the salinity is higher, since it is connected to the sea [12]. As it was mentioned previously, the opening of the lagoon inlet by local fishermen also affects the salinity of the lagoon. From the results of the numerical model implemented by Rivera [36], it is seen that if the inlet is open continuously for more than three months, the residence time of the water in the lagoon increases. This increase is directly related to eutrophication in coastal waters [37].

Regarding coastal flooding, storm surge levels near the study area were reported by Silva et al. [38], where the values of the storm surge caused by the atmospheric pressure gradient, wind and waves were calculated for different return periods using climate data from 1948 to 2010. These values are shown in Table 2.

Table 2. Storm surge levels in La Mancha area, data from Silva et al. [38].

Storm Surge	Return Period (Years)									
	2	5	10	15	20	25	30	40	50	100
Atmospheric pressure gradient storm surge (m)	0.03	0.11	0.17	0.20	0.22	0.23	0.24	0.26	0.28	0.32
Wind storm surge (m)	0.06	0.16	0.26	0.33	0.38	0.42	0.45	0.50	0.54	0.68
Wave storm surge (m)	1.27	1.40	1.47	1.51	1.54	1.58	1.62	1.66	1.70	1.80
Pressure + wind + wave storm surge (m)	1.36	1.67	1.90	2.04	2.14	2.23	2.31	2.42	2.52	2.80

The relative rise in the sea level has a range of effects on the morphological changes in this area at different scales [19]. Sea level rise affects the flood levels, which are related to the geographic location, coastal orientation and beach slope [24]. According to Pérez et al. [19], in Veracruz, the trend of sea level rise is +1.89 mm/year. This is in line with global IPCC reports and is conditioned by the subsidence of the natural deltaic surface [39]. The geodynamics of the Gulf of Mexico are regulated by the movements of the North America, Circum-Pacific Kula, Farallón, Cocos and Caribbean tectonic plates. These tectonic plates cause continental and oceanic movements that intermittently induce the continental accumulations of deep marine sediments in the continental marginal accretional prism of the Gulf of Mexico [40].

3.2.2. Problems Affecting the Vegetation and Fauna

The structure and composition of the mangroves depend on a variety of factors, i.e., oceanographic, climatic, geomorphological, edaphic conditions, level and duration of flooding and sediment load [41]. According to several studies, the location of the mangrove forest affects the diameter and height of some specimens: they grow the most in areas protected from wave action. Moreno-Casasola et al. [42] also listed other variables relevant for mangroves in freshwater wetlands: salinity, water regime, conductivity and redox potential.

The dynamics of the vegetation around La Mancha are also clearly related to the opening of the inlet. When the inlet is closed and the water depth of the lagoon reaches <120 cm near the sandbar, the inlet opens naturally, discharging the supratidal accumulation through it. This ebb and flood exchange of water causes variations in parameters such as salinity and sediment distribution. The inlet dynamics are essential for the mangroves: if the inlet is permanently closed, there will be excessive flooding, and the mangroves will die due to a lack of oxygen in the stagnant water; if the inlet is permanently open, salinity will increase, also causing mortality [15].

Animal species are also affected by fluctuations in the water level and the salinity of the lagoon. For instance, oysters, and some other species, have high mortality rates when the water level increases and the oxygen in the water decreases, leading the fishermen to open the inlet, which causes imbalances in the lagoon dynamics [15]. Other environmental factors, such as evaporation and insolation, are <1% significant for mangrove development here [43].

Regarding the vegetation that develops on the coastal dunes surrounding the lagoon, there has been a decrease in the species which were previously most abundant. Previous studies indicated that although trampling on the dunes is infrequent, it is a relevant factor in the area, resulting in vegetation loss and compaction of the sand [31]. The seedlings are totally destroyed by crushing and flattening, leaving only bare soil in some cases [31]. The intensity of trampling is best seen on the steeper slopes, where the vegetation is detached and crushed [44]. The occurrence of this activity is not widespread, and trampling has a low impact; however, it may induce local extinction of some plant species [31].

3.3. *State of the Environment in the DESCR Framework*

Cortés [44] described the beach at La Mancha as having a dissipative profile with a bar, following the classification of Masselink and Short [45]. The sedimentary input comes from the northern dune field, with a net north-to-south longshore sediment transport. Ramírez Méndez [18] identified an overall trend of beach accretion, recording an advance of 58 m for 2005–2015, i.e., 0.109 km^2 of beach. Chávez [46] reported that the beach sediment is a fine sand (according to the ASTM D 2487 norm), with a mean diameter of 0.176–0.305 mm.

The main geomorphological components at the inlet are as follows: the tidal flood delta, and the sedimentary bar on the northern side of the inlet, which is a storm berm and is part of the beach (see Figure 9, the area corresponds to Area B in Figure 1). The berm is built up by sediment during storm surge conditions (winter storms and hurricanes). When there is a sediment deficiency, usually associated with erosive wave conditions, the berm breaks up [15]. The inlet is usually open from August to November and closed in December, with no definite pattern from January to September [18]. When the sedimentary bar is low and the inlet is open for a long period, intertidal exchanges sequester sediments in the north of the lagoon, the area of the tidal flood delta extends and this part of the lagoon becomes shallower [15].

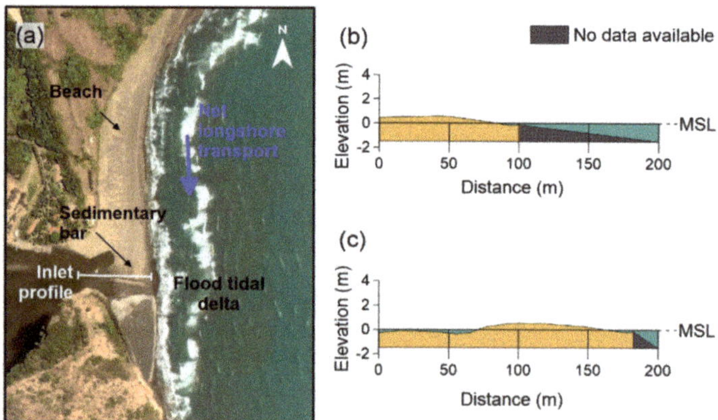

Figure 9. (**a**) Inlet area, and inlet profiles in La Mancha measured during winter storms in November (**b**) 2013 and (**c**) 2014 (data from [12]), see Area B in Figure 1.

3.3.1. Lagoon Vegetation

The physiognomy of the plant communities which surround the lagoon can be divided into tree-dominated and herbaceous species. Around the lagoon (Figure 1), there is a heterogeneous mangrove forest of about 190 ha, with a height range of 5–15 m and typical, associated plants [18]. Mangrove species found here include *Rhizophora mangle* (red mangrove), *Avicennia germinans* (black mangrove), *Laguncularia racemosa* (white mangrove) and *Conocarpus erectus* (button mangrove), with *Avicennia* and *Laguncularia* being the most abundant (88%) [15] (Figure 10). Between the 1980s and 2010, the area of the mangrove fell by 33.8%, although another 0.3% of mangrove area was converted from agricultural livestock to mangrove, between 2005 and 2010 [18].

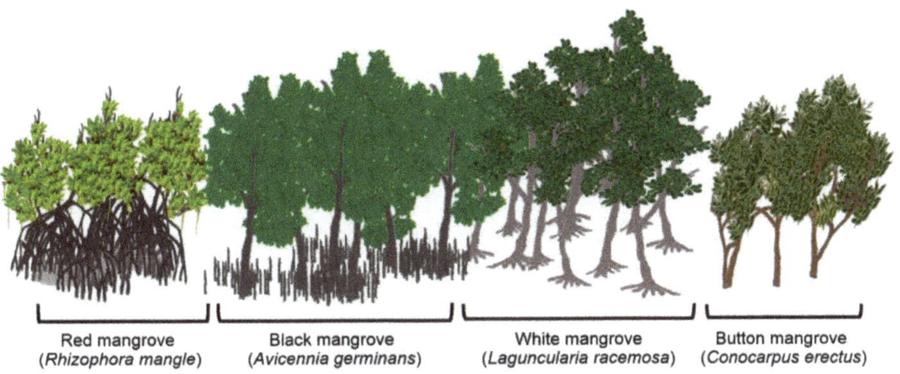

Figure 10. Distribution of mangrove species in La Mancha.

The hydrological and geomorphological conditions of the area determine the development of the species there [17], which grow the most in periods of flooding. The authors of [17] stated that the tallest mangroves are found in the area affected by floods, since this is where there are most nutrients. As for the health of the mangroves, this depends on the season, where more robust trees are seen in the rainy season.

Psuty et al. [15] explained the importance of flooding, as the seedlings of some species, such as *Avicennia*, are dispersed by water currents at the end of the rainy season, germinating when the sandbank closes the inlet, and the water level starts to rise. Litter production is an important factor to be taken into account in the evaluation of mangrove ecosystems, and the amount of fallen leaves in this ecosystem in La Mancha is 1025 g/m^2 per year [17].

3.3.2. Anthropic Impacts

While the state of Veracruz is very densely populated, that is not the case of the study area [44]. The land use and vegetation for 2015 [47] are shown in Figure 11 at a 1:250,000 scale. The main land uses are agriculture and pasture. According to Ramírez Méndez [18], the natural landscape of La Mancha is fragmented due to anthropogenic activities related to agriculture, livestock, fishing, tourism and human development, which have mainly affected the low deciduous forest.

3.4. *Consequences in the DESCR Framework*

3.4.1. Beach and Dunes

According to Doody [48], a beach that is extending offshore due to accretion needs sediment. As the amount of sediment increases and the sediment balance of the beach moves from neutral to positive, large dune areas will develop, causing regenerative trends in the terrestrial ecosystems. In the study area, the coastline and the vegetation bordering it have been altered. The aerial photographs in the analysis of Psuty et al. [15], in the 1980s, show mobile coastal dunes moving across the headland, supplying the beach with sand. However, by the mid-1990s, vegetation had covered these mobile dunes. This trend has continued, and, with less sand available, sediment transport to the south has decreased. In addition, as the use of 90 ha of the mobile sand dunes has changed, only 10% of the original area of mobile dunes is now in its previous, natural state.

Regarding the morphology of the inlet, there is a southward sediment transport. As sediment is now scarce, the berm is less developed, which means that the inlet stays open for longer. This causes the tidal exchanges to sequester sediment in the northern part of the lagoon, increasing the area of the delta, inducing subsidence and deepening the northern part of the lagoon [15].

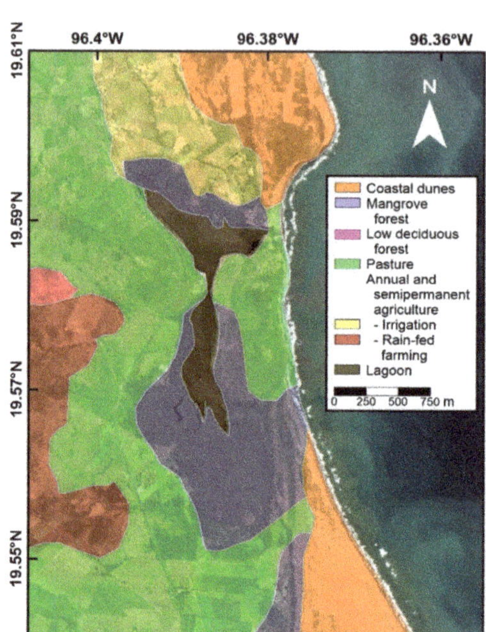

Figure 11. Land use and vegetation in La Mancha. Data from INEGI [47].

3.4.2. Lagoon and Mangroves

According to Psuty et al. [15], the variability of the water level and salinity has important effects on the species growing in the mangrove ecosystem. The opening and closing of the inlet are the determining factors in the behavior of this system; the closure of the bar, which prevents entry of seawater to the lagoon, produces a decrease in oxygen, salinity, pH and chlorophyll [18]. Some fish species in the lagoon change their balances depending on whether the inlet is open or closed, and the existence, or dominance, of many species varies depending on the quality of the lagoon water [15].

The mangrove species in the area provide shelter to native and endemic species, such as tilcampo, crocodiles, iguanas and turtles, which are vulnerable to ecosystem loss; crabs, ocelots and tigrillos, which are in danger of extinction; and also flounder, bass, yellowfin crappie and striped crappie, which are valuable commercial species for the local fishermen [18]. Mangrove is also an ecosystem that acts as an ecological membrane. Martínez et al. [49] analyzed the accumulation of heavy metals in La Mancha, noting that it was greater in mangrove leaves than in nearby sediments, noting the ability of this ecosystem to protect and regulate the balance and mitigate the effects of these substances in the environment, thus decreasing marine pollution.

Psuty et al. [15] pointed out that the germination of seedlings may eventually become more limited, depending on the opening and closing of the inlet, as some species, such as *Avicennia* seedlings, are dispersed by the water currents at the end of the rainy season (September–October), and they germinate when the sandbar closes the inlet.

3.5. Responses in the DESCR Framework

As the dynamic equilibrium of the lagoon system depends on changes in the physico-chemical parameters, the artificial opening of the inlet alters the natural dynamics. Since the opening of the inlet is beneficial to the fishermen's activities, the magnitude of the long-term ecological consequences of this practice should be quantified, taking into consideration large-scale processes such as SLR, in order to devise an adequate management plan.

Human activity has played an important role in the destabilization of the beach and dunes by modifying the hydrodynamics, the sediment availability and the sediment transport patterns. It is therefore essential to create conditions for a sand reservoir, and alternatives to preserve the vegetation of this ecosystem [48]. Trampling and other activities, such as the use of scooters, put the dune vegetation and their natural dynamics at risk. However, tourism benefits many people in the area. Therefore, ecotourism plans could be focused on the dunes, where the cultural ecosystem services can be taken advantage of while seeking the sustainability of the ecosystem. To restore the coastal dunes, it is necessary to restore the natural dynamics and recover habitats for dune-building species (e.g., previously, the mobile dunes were mostly covered by the grass *Schizachyrium scoparium*). Native species such as *Croton punctatus, Palafoxia lindenii* and *Chamaecrista chamaecristoides* are abundant on mobile dunes, and thus actions are needed to re-establish their environment so that they can increase their plant cover. Fencing the dunes, in order to restrict access to them, to reduce air flow and to increase sand deposition, is also an option to restore the equilibrium [50].

Regarding the recent urban development, this may have intense repercussions on the environment, as well as on the socio-economic development of the population of the area. Despite the fact that Mexico has legal instruments to conserve and protect natural ecosystems (General Law of Ecological Balance and Environmental Protection, General Law of Wildlife, General Law of Sustainable Forest Development, Fisheries Law), there is a lack of commitment to promoting sustainable development objectives by the state. The conflict is part of the conservation versus extraction debate. The "Comprehensive Management Plan for the La Mancha-El Llano Basin" (2006) was conceived as an instrument that links communities, academia and government authorities and allows discussion of the current environmental situation and conflicts, evaluating the participation of the actors in the area [51]. This document focuses on: (i) the biological and cultural conservation and restoration of the ecosystems and landscapes of the basin, (ii) community participation and (iii) comprehensive planning, taking into account the management of the area [52]. Despite the implementation of this plan, the resistance to change of the population is both noticeable and very intense, creating conflicts with government entities and within the population in general, based on economic interests [52]. This is currently occurring with the development of the "Diada" project in Veracruz which, according to Gómez Ramirez [53], will probably favor only wealthy people, and the local, native population will see hardly any economic benefits.

It is necessary to carry out an environmental impact assessment of the anthropogenic expansion in the area, since this can cause fragmentation of ecosystems and put some species at risk, such as migratory birds [53]. Mendoza et al. [54] analyzed the changes in land use and the valuation of ecosystem services on the coast of the Gulf of Mexico, noting the importance of maintaining a healthy balance in the area, related to the value and provision that this system provides to the community. It may be useful to carry out an assessment of ecosystem services in the area and determine how the urbanization projects will alter them, eventually affecting society. Environmental regulations should look for alternatives that promote conservation. Science-based information is an important pillar for motivation towards preservation, since if ecological, economic and social values are ranked, governments and citizens will have clearer technical elements for better decision making.

Given that anthropic pressures generate great changes in ecosystems, it is necessary to analyze the components of the planning and management of the coastal zone arising from the social environment [55]. The Decalogue for the Integrated Management of Coastal Areas, proposed by Barragán [56], is a tool that links public policy processes, based on ten structural elements of the legal-administrative subsystem of a zone, allowing decision makers to assess and compare the spaces in relation to their objectives. The ten elements of this decalogue are as follows: policy (explicit government policies directed to the management of coastal areas), participation (institutional and social support for public policies), normative (specific laws for coastal management), institutions (part of the public administration concerned with coastal marine spaces or resources), managers (government and stakeholders), information (physical, social-economic and administrative knowledge of the area), resources (economic tools that help apply and develop a management model), education (initiatives or proposals that promote education towards sustainability), strategies (for coastal management), instruments (tools such as zoning, concessions and authorizations for the use of coastal resources). In La Mancha, this decalogue includes the following elements:

- Policy: National Policy for Seas and Coasts of Mexico-Proposal of the Intersecretarial Commission for Sustainable Management of Seas and Coasts-2018 [57]; National Development Plan, 2019–2024 [58]; National Forestry Program 2020–2024 [59].
- Participation: representatives of the local, state and federal government; federation of fishermen; ecotourism organizations; academic institutions; grouping of fishing cooperatives with political interests; civil society [60].
- Normative: General Law of Ecological Balance and Environmental Protection (LEGEEPA); National Water Law; Federal Law of the Sea; Sustainable Rural Development Law; Sustainable Forestry Development Law; General Wildlife Law; Fishing Law; Federal Law of Rights; General Law of Human Settlements; Mexican Official Standard NOM-059-SEMARNAT-200 (Environmental Protection-Mexican native species of wild flora and fauna); NOM-126-SEMARNAT-2000; Official Mexican Standard NOM-022-SEMARNAT-2003; Official Mexican Standard NOM-001-SEMARNAT-1996; Mexican Official Standard NOM-075-SEMARNAT-1994 [61].
- Institutions: Secretary of the Environment and Natural Resources (SEMARNAT); Secretariat of Agriculture, Livestock, Rural Development, Fisheries and Food (CONAPES CA) [62]; marine secretary; Ministry of Communications and Transportation; Secretary of Tourism; Secretariat of Health; Office of the Attorney General of the Republic; Secretariat of Management for Environmental Protection; General Directorate of the Federal Maritime Terrestrial Zone and Coastal Environments; National Water Commission; National Commission for Protected Natural Areas; Federal Attorney for Environmental Protection; National Institute of Ecology [61].
- Managers: administration of the Institute of the Ecology Sector of the Institute of Ecology; Secretariat of the Environment and Natural Resources (SEMARNAT); ZOFEMATAC Federal Maritime Terrestrial Zone, dependency of SEMARNAT; Municipality of Actopan; State Coordination of the Environment (SEDERE) [60]; *La Mancha en Movimiento S.S.S.* [63]; PROFEPA Social Participation Committee; state nuclear power plant *Nucleoeléctrica Laguna Verde* (Federal Electricity Commission); Directorate of Fisheries State Agency; National Water Commission [61].
- Information: Community and Sustainable Development Program and Management Plan for the Protection and Conservation of the La Mancha-El Llano Ramsar Site [64]; La Mancha-El Llano Community Management Plan. In Search of a Sustainable Coastal Development [51]; Strategy for Comprehensive Coastal Management-The Municipal Approach [60]; Environments of Veracruz. The coast of La Mancha [62]; government; general bibliography.
- Resources: no specific data.
- Education: CICOLMA (La Mancha Coastal Research Center); Group La Mancha [61]; INECOL [51]; *La Mancha en Movimiento S.S.S.* [63]; Institute of Anthropology; Universidad Veracruzana [61].
- Strategies: conservation and management program for the blue crab; productive enclosures in rivers, lagoons and ponds; sowing and propagation of native species in nurseries [64]; ecotourism [13]; Strategy for Comprehensive Coastal Management-The Municipal Approach [64].

- Instruments: La Mancha-El Llano Community Management Plan [51].

3.6. Summary of the DESCR Framework

Being a coastal area, marine physical parameters such as wind, waves, tides and storm surge are the main processes regulating the lagoon systems. These natural drivers, as well as being pressures, are inherent to the system. Human activities in La Mancha are anthropic drivers: the urban project of recent years is a potential risk due to the alterations to the natural conditions of the ecosystems and changes in land use, among others. Exchanges refer to the interaction between the drivers and the system, where in La Mancha, the lagoon's own biophysical processes regulate hydrodynamics, sea level rise and the opening and closing of the inlet of the lagoon. Energetic periods of winds and waves cause the lagoon inlet to close, due to intense sediment transport, and the water level in the lagoon increases, favoring the development of mangrove species. The level of flooding and the fluctuations in salinity are factors that condition the growth of some species of fish.

The state of the environment in the lagoon area has been modified by anthropic drivers: the opening of the lagoon inlet by the fisherman alters its natural hydrological regime, and the urban project produces an imbalance in the natural dynamics of the system, which could cause conflict in the future if adequate management of the project is not afforded.

Responses should include strengthening local education programs in the area. Such actions will help to temper the lack of socio-environmental awareness and respect for laws. Continuous environmental monitoring will produce reliable information that should be made public, reduce uncertainties and ensure that appropriate action is taken to adapt measures where necessary. Efforts should be made to reduce socio-economic inequality. By encouraging other economic activities, pressure on coastal ecosystems will decrease, and economic vulnerability will be reduced due to the emergent urbanization projects. Measures should be put in place for wastewater treatment in the basin to control its consequences. Mitigation and management plans should be implemented to address pressure in emerging problems.

The following diagram (Figure 12) summarizes the DESCR framework for La Mancha.

From the elements listed in Figure 12, the interaction of some natural and anthropic drivers, through energy and matter exchanges in La Mancha, may induce coastal squeeze by the following:

- Alteration in the natural patterns of the physicochemical parameters (e.g., salinity and pH) due to the forced opening of the inlet;
- Sea level rise;
- Sediment deficit;
- Sediment availability reduction;
- Changes in native species coverage;
- Land use changes, including development of urbanization.

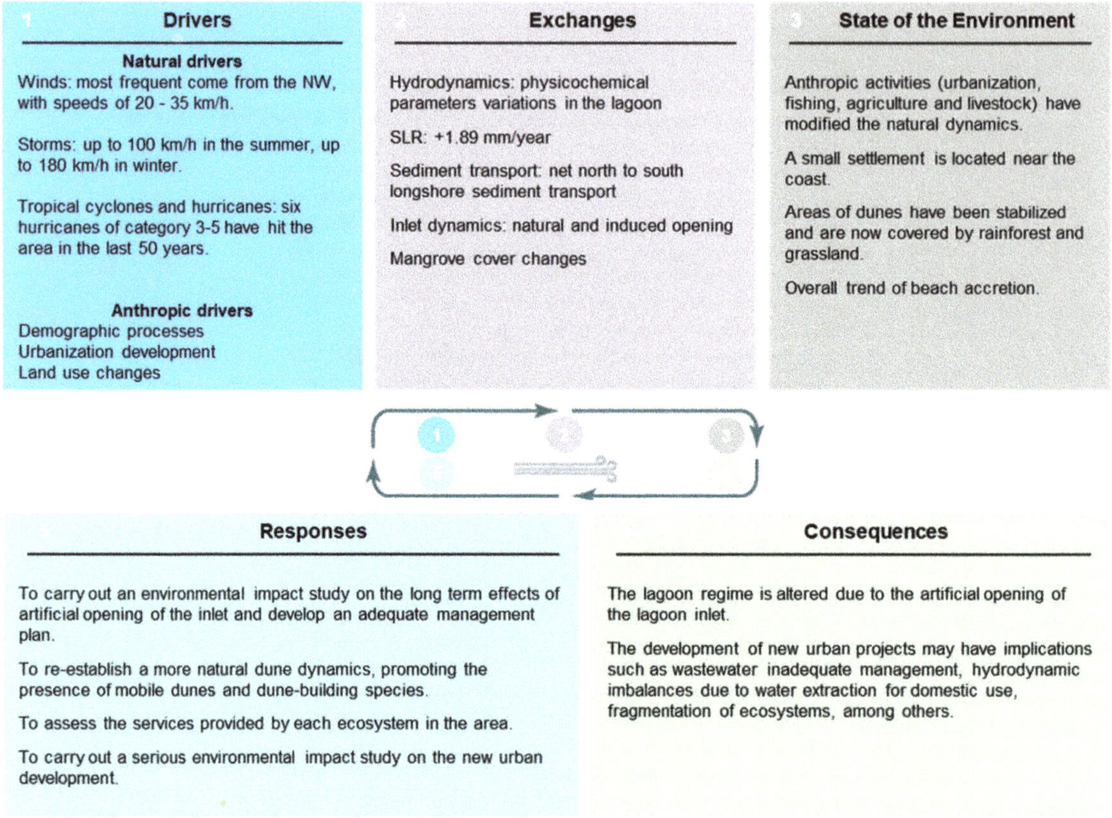

Drivers

Natural drivers

Winds: most frequent come from the NW, with speeds of 20 - 35 km/h.

Storms: up to 100 km/h in the summer, up to 180 km/h in winter.

Tropical cyclones and hurricanes: six hurricanes of category 3-5 have hit the area in the last 50 years.

Anthropic drivers

Demographic processes
Urbanization development
Land use changes

Exchanges

Hydrodynamics: physicochemical parameters variations in the lagoon

SLR: +1.89 mm/year

Sediment transport: net north to south longshore sediment transport

Inlet dynamics: natural and induced opening

Mangrove cover changes

State of the Environment

Anthropic activities (urbanization, fishing, agriculture and livestock) have modified the natural dynamics.

A small settlement is located near the coast.

Areas of dunes have been stabilized and are now covered by rainforest and grassland.

Overall trend of beach accretion.

Responses

To carry out an environmental impact study on the long term effects of artificial opening of the inlet and develop an adequate management plan.

To re-establish a more natural dune dynamics, promoting the presence of mobile dunes and dune-building species.

To assess the services provided by each ecosystem in the area.

To carry out a serious environmental impact study on the new urban development.

Consequences

The lagoon regime is altered due to the artificial opening of the lagoon inlet.

The development of new urban projects may have implications such as wastewater inadequate management, hydrodynamic imbalances due to water extraction for domestic use, fragmentation of ecosystems, among others.

Figure 12. DESCR for La Mancha, Veracruz.

4. Conclusions

The DESCR framework is a useful tool to evaluate processes in this study area, taking into account the causes that trigger variations in the state of the environment, the physical environment with its natural dynamism and the social actors who will be responsible for formulating solutions to mitigate the problems detected.

In La Mancha, the natural dynamics of the ecosystems have been altered, by natural and anthropic factors, such as the artificial opening of the inlet. It is necessary to determine the long-term consequences of these actions in order to provide adequate elements for decision making. On the other hand, the urban development plans put the environmental integrity of La Mancha at risk. A lack of commitment to protecting the ecological resources and environmental services provided by the ecosystems of La Mancha is evident.

Although further scientific-based quantitative analysis of the interactions and responses is needed, the conceptual implementation of the DESCR framework allows a diagnosis to be made, and the identification of key interactions that occur in the system. The results presented here should serve as a basis for implementing better coastal management strategies. Projects with evaluation tools such as the DESCR framework can help link the local community with the scientific community and the authorities and provide means for developing future sustainable projects in coastal areas. Innovative forms of linking stakeholders, such as environmental education towards sustainability, seeking a high level of scope for these projects and monitoring over time will yield tangible results for a better use of the environmental resources of the La Mancha lagoon region.

Author Contributions: Conceptualization, R.S.; methodology, R.S., V.C.; formal analysis, S.C.A.; investigation, S.C.A.; resources, R.S.; writing—original draft preparation, S.C.A., V.C.; writing—review and editing, S.C.A., V.C., R.S., M.L.M., G.A.; supervision, V.C., R.S., M.L.M., G.A.; project administration, R.S.; funding acquisition, R.S., G.A. All authors have read and agreed to the published version of the manuscript.

Funding: CONACYT-SENER-Sustentabilidad Energética project: FSE-2014-06-249795 Centro Mexicano de Innovación en Energía del Océano (CEMIE-Océano).

Acknowledgments: This work is a contribution to the PAI Research Group RNM-328 (Andalusia, Spain).

Conflicts of Interest: The authors declare no conflict of interest.

Appendix A

Numerous research works have been carried out in the area of La Mancha since the 1970s. Some of these studies analyzed its ecological and biological characteristics, determined by the presence of phytoplankton, vegetation, primary productivity, zooplankton, benthos, nekton, birds and pollution, as well as studies on fishing activity and aquaculture (e.g., [65–70]). The hydro-morphological dynamics and coastal management in the area have also been studied. Lists of relevant comprehensive studies of the local dynamics (articles and theses) are presented in Tables A1 and A2, respectively.

Table A1. List of relevant articles of La Mancha.

Author, Year	Title	Main Contributions
Moreno-Casasola et al., 2006 [51]	Plan de manejo comunitario La Mancha-El Llano. En busca de un desarrollo costero sustentable. Estrategias para el manejo integral de la zona costera: un enfoque municipal	A management strategy based on the environmental protection and the experiences of the Latin American Forum of Environmental Sciences.
Moreno-Casasola, 2006 [62]	Entornos Veracruzanos. La costa de La Mancha	This book describes the main characteristics, physical, ecological, social and cultural, in great detail.
Moreno-Casasola et al., 2007 [52]	Los conflictos de la conservación: el caso de La Mancha. Hacia una cultura de conservación de la diversidad biológica	A view of the experiences of the social sectors of La Mancha and the conflicts of interests between them and the government regarding the introduction of the La Mancha-El Llano Community Management Plan.
Moreno-Casasola and Salinas Pulido, 2007 [64]	Programa de desarrollo comunitario sustentable y plan de manejo para la protección y conservación del Sitio Ramsar La Mancha-El Llano	An environmental conservation and community project for La Mancha, Veracruz, based on the sustainable management models of three groups of people from the local community.
Hesp and Martínez, 2008 [16]	Transverse dune trailing ridges and vegetation succession	A description of features, such as evolution and vegetation, of the Farallon dunes located in La Mancha, Veracruz.
Utrera-López and Moreno-Casasola, 2008 [20]	Mangrove litter dynamics in La Mancha Lagoon, Veracruz, Mexico	A description of litter dynamics among mangrove types in La Mancha, to help understand functional heterogeneity within this coastal ecosystem.
Psuty et al., 2009 [15]	Interaction of alongshore sediment transport and habitat conditions at Laguna La Mancha, Veracruz, Mexico	An overview of the hydrological regime, the dynamic of sediments and the ecological features in La Mancha.
Mata et al., 2011 [71]	Floristic composition and soil characteristics of tropical freshwater forested wetlands of Veracruz on the coastal plain of the Gulf of Mexico	Analysis of the geomorphological setting, influence and soil properties on the structure of vegetation of five coastal lagoons in Veracruz, one being La Mancha.

Table A1. *Cont.*

Author, Year	Title	Main Contributions
Martínez et al., 2014 [9]	Land use changes and sea level rise may induce a "coastal squeeze" on the coasts of Veracruz, Mexico	Analysis of the coastal line geodynamics and geodynamic trends to model niches under SLR scenarios.
Ruiz and López-Portillo, 2014 [72]	Variación espacio-temporal de la comunidad de macroinvertebrados epibiontes en las raíces del mangle rojo Rhizophora mangle (Rhizophoraceae) en la laguna costera de La Mancha, Veracruz, México	An analysis of spatiotemporal variations of epibiont macroinvertebrates in red mangrove roots (Rhizophoraceae), based on the hydrological dynamics of the system.
Ramírez Méndez et al., 2015 [73]	Estudio de la dinámica y calidad de agua en la laguna de La Mancha, Veracruz	A study of the dynamics of hydrodynamic, morphodynamic and hydrological conditions and physicochemical parameters of La Mancha and its littoral cell.
Chávez et al., 2017 [12]	Impact of Inlet Management on the Resilience of a Coastal Lagoon: La Mancha, Veracruz, Mexico	A study on features of La Mancha lagoon, such as ecosystem vulnerability, physical processes, such as erosion and accretion of the beach, inlet dynamics, and hydrodynamics of circulation patterns in the lagoon.
Rivera et al., 2019 [36]	Modelling the effects of the artificial opening of an inlet: Salinity distribution in a coastal lagoon	Numerical results are used to analyze and describe changes in the salinity distribution in La Mancha lagoon, showing results which would be useful in developing an adequate management plan.
Gómez Ramírez, 2020 [53]	El estudio de los ciclones tropicales se minimizó, en la manifestación del impacto ambiental para el Proyecto Diada La Mancha, en la costa del municipio de Actopan, Estado de Veracruz	A review of the problems caused by the lack of a serious environmental impact study prior to the development of the La Mancha-Diada project.

Table A2. List of relevant theses of La Mancha.

Author, Year	Title	Main Contributions
Cortés López, 2017 [44]	Desarrollo de un Índice de Riesgo sobre la ocurrencia de Opresión Costera en el centro-norte del Estado de Veracruz Master Thesis	Development of a risk index, based on the occurrence of coastal oppression, considering the evolution of the coastline and of the sediments in 14 beaches, one being La Mancha.
Chávez, 2018 [46]	Balance Hidrodinámico en Humedales Costeros y su valor como elemento de protección litoral PhD Thesis	Analysis and characterization of the physical protection against floods provided by wetlands, based on the monitoring of the main physical processes and the determination of their balances.
Ramírez Méndez, 2018 [18]	Rumbo a un Plan de Manejo Integral de la laguna de La Mancha, Veracruz Master Thesis	Study of physical, ecological and social parameters of La Mancha, Veracruz, in order to extend a management plan for this zone.

References

1. Silva, R.; Oumeraci, H.; Martínez, M.L.; Chávez, V.; Lithgow, D.; van Tussenbroek, B.I.; van Rijswick, H.F.; Bouma, T.J. Ten commandments for sustainable, safe, and w/healthy sandy coasts facing global change. *Front. Mar. Sci.* **2021**, *8*, 126. [CrossRef]
2. Martínez, M.L.; Landgrave, R.; Silva, R.; Hesp, P. Shoreline Dynamics and Coastal Dune Stabilization in Response to Changes in Infrastructure and Climate. *J. Coast. Res.* **2019**, *92*, 6–12. [CrossRef]
3. Chávez, V.; Lithgow, D.; Losada, M.; Silva-Casarín, R. Coastal green infrastructure to mitigate coastal squeeze. *J. Infrastruct. Preserv. Resil.* **2021**, *2*, 7. [CrossRef]
4. Silva, R.; Martínez, M.L.; van Tussenbroek, B.I.; Guzmán-Rodríguez, L.O.; Mendoza, E.; López-Portillo, J. A Framework to Manage Coastal Squeeze. *Sustainability* **2020**, *12*, 10610. [CrossRef]

5. Smeets, E.; Weterings, R. *Environmental Indicators: Typology and Overview*; European Environment Agency: Copenhagen, Denmark, 1999; pp. 6–15.
6. Patrício, J.; Elliott, M.; Mazik, K.; Papadopoulou, K.-N.; Smith, C.J. DPSIR—Two decades of trying to develop a unifying framework for marine environmental management? *Front. Mar. Sci.* **2016**, *3*, 177. [CrossRef]
7. Silva, R.; Chávez, V.; Bouma, T.J.; van Tussenbroek, B.I.; Arkema, K.K.; Martínez, M.L.; Oumeraci, H.; Heymans, J.J.; Osorio, A.F.; Mendoza, E. The incorporation of biophysical and social components in coastal management. *Estuaries Coasts* **2019**, *42*, 1695–1708. [CrossRef]
8. Silva, R.; Martínez, M.L.; Hesp, P.A.; Catalán, P.; Osorio, A.F.; Martell, R.; Fossati, M.; Miot da Silva, G.; Mariño-Tapia, I.; Pereira, P. Present and future challenges of coastal erosion in Latin America. *J. Coast. Res.* **2014**, 1–16. [CrossRef]
9. Martínez, M.L.; Mendoza-González, G.; Silva-Casarín, R.; Mendoza-Baldwin, E. Land use changes and sea level rise may induce a "coastal squeeze" on the coasts of Veracruz, Mexico. *Glob. Environ. Chang.* **2014**, *29*, 180–188. [CrossRef]
10. Moreno Casasola, P.; Galaviz Rojas, J.L.; Lomelí Zárate, D.; Pérez Ortiz, M.A.; Domínguez Lara, A.L.; Vázquez Saavedra, T. Diagnóstico de los manglares de Veracruz: Distribución, vínculo con los recursos pesqueros y su problemática. *Madera Bosques* **2002**, *8*, 61–88. [CrossRef]
11. Ramsar Sites Information Service-La Mancha y El Llano. Available online: https://rsis.ramsar.org/ris/1336 (accessed on 11 May 2021).
12. Chávez, V.; Mendoza, E.; Ramírez, E.; Silva, R. Impact of inlet management on the resilience of a coastal lagoon: La Mancha, Veracruz, Mexico. *J. Coast. Res.* **2017**, 51–61. [CrossRef]
13. La Mancha, E.M. Ecoturismo La Mancha en Movimiento. Available online: https://www.ecoturismolamancha.com/ (accessed on 10 March 2021).
14. Digital Globe. Available online: https://www.digitalglobe.com (accessed on 12 June 2012).
15. Psuty, N.P.; Martínez, M.L.; López-Portillo, J.; Silveira, T.M.; García-Franco, J.G.; Rodríguez, N.A. Interaction of alongshore sediment transport and habitat conditions at Laguna La Mancha, Veracruz, Mexico. *J. Coast. Conserv.* **2009**, *13*, 77–87. [CrossRef]
16. Hesp, P.A.; Martínez, M.L.M. Transverse dune trailing ridges and vegetation succession. *Geomorphology* **2008**, *99*, 205–213. [CrossRef]
17. Hernández-Trejo, H.; Priego-Santander, A.; López-Portillo, J.; Isunza-Vera, E. Los paisajes Físico-Geográficos de los manglares de la laguna de La Mancha, Veracruz, México. *Interciencia* **2006**, *31*, 211–219.
18. Ramírez Méndez, E. Rumbo a un Plan de Manejo Integral de la Laguna de La Mancha, Veracruz. Master Thesis, Univesidad Nacional Autónoma de Mexico, Ciudad de México, Mexico, 2018.
19. Pérez Landa, I.; Galaviz-Villa, I.; Garay Marín, J. Incremento en el nivel de mar en la zona costera de Veracruz. In *Temas Selectos de Vulnerabilidad Costera en el Estado de Veracruz*; Lango-Reynoso, F.B., Alfonso, V., Castañeda-Chávez, M.R., Eds.; Universidad Autónoma de Campeche, Instituto de Ecología, Pesquerías y Oceanografía del Golfo de México (EPOMEX): Campeche, México, 2019; pp. 21–32.
20. Utrera-López, M.E.; Moreno-Casasola, P. Mangrove Litter Dynamics in la Mancha Lagoon, Veracruz, Mexico. *Wetl. Ecol. Manag.* **2008**, *16*, 11–22. [CrossRef]
21. Nordstrom, K.F.; Jackson, N.L. Removing shore protection structures to facilitate migration of landforms and habitats on the bayside of a barrier spit. *Geomorphology* **2013**, *199*, 179–191. [CrossRef]
22. Small, C.; Nicholls, R.J. A global analysis of human settlement in coastal zones. *J. Coast. Res.* **2003**, *19*, 584–599.
23. Reanalysis, C. Available online: https://climate.copernicus.eu/climate-reanalysis (accessed on 5 June 2020).
24. Martínez, M.L.; Silva, R.; Lithgow, D.; Mendoza, E.; Flores, P.; Martínez, R.; Cruz, C. Human impact on coastal resilience along the coast of Veracruz, Mexico. *J. Coast. Res.* **2017**, 143–153. [CrossRef]
25. Jáuregui, E.; Zitácuaro, I. El impacto de los ciclones tropicales del Golfo de México, en el estado de Veracruz. *Cienc. Hombre* **1995**, *623*, 75–119.
26. Silva, R.; Ruiz, G.; Posada, G.; Pérez, D.; Rivillas, G.; Espinal, J.; Mendoza, E. *Atlas de Clima Marítimo de la Vertiente Atlántica Mexicana*; Universidad Nacional Autónoma de México: Mexico City, Mexico, 2008.
27. Ruiz-Martínez, G.; Silva-Casarín, R.; Pérez-Romero, D.M.; Posada-Vanegas, G.; Bautista-Godínez, E.G. Modelo híbrido para la caracterización del oleaje. *Ing. Hidráulica México* **2009**, *24*, 5–22.
28. Estación Mareográfica de Tuxpan, Veracruz. Available online: https://oceanografia.semar.gob.mx/Templates/grafnum_tuxpan.html (accessed on 14 July 2021).
29. Estación Mareográfica de Veracruz, Veracruz. Available online: https://oceanografia.semar.gob.mx/Templates/grafnum_veracruz.html (accessed on 14 July 2021).
30. Estaciones de Campo INECOL. Available online: https://www.inecol.mx/inecol/index.php/es/ct-menu-item-1/estaciones-de-campo (accessed on 11 May 2021).
31. Hesp, P.; Schmutz, P.; Martínez, M.M.; Driskell, L.; Orgera, R.; Renken, K.; Revelo, N.A.R.; Orocio, O.A.J. The effect on coastal vegetation of trampling on a parabolic dune. *Aeolian Res.* **2010**, *2*, 105–111. [CrossRef]
32. E-consulta. Inicia Construcción de Fraccionamiento de 800 Casas en la Mancha. Available online: http://www.e-veracruz.mx/nota/2018-01-19/ecologia/inicia-construccion-de-fraccionamiento-de-800-casas-en-la-mancha (accessed on 12 March 2021).
33. Veracruz, L.J. Destrucción en Sitio Ramsar de La Mancha. Available online: http://www.jornadaveracruz.com.mx/Post.aspx?id=180120_115217_446 (accessed on 12 March 2021).

34. Diada-Ecommunity. Diada-Desarrollos- La Mancha-5. Available online: https://i2.wp.com/www.diada.mx/wp-content/uploads/2020/03/DIADA-Desarrollos-La-Mancha-5.jpg (accessed on 29 April 2021).
35. Diada-Ecommunity. Diada-Desarrollos- La Mancha-8. Available online: https://i0.wp.com/www.diada.mx/wp-content/uploads/2020/03/DIADA-Desarrollos-La-Mancha-8.jpg (accessed on 29 April 2021).
36. Rivera, J.; Chávez, V.; Silva, R.; Mendoza, E. Modelling the effects of the artificial opening of an inlet: Salinity distribution in a coastal lagoon. *J. Coast. Res.* **2019**, *92*, 128–135. [CrossRef]
37. Camacho-Ibar, V.F.; Rivera-Monroy, V.H. Coastal lagoons and estuaries in Mexico: Processes and vulnerability. *Estuaries Coasts* **2014**, *37*, 1313–1318. [CrossRef]
38. Silva, R.; Moreno, P.; Martínez, M.L.; Mendoza, E.; López-Portillo, J.; Lithgow, D.; Vázquez, G.; Martínez, R.; Ibarra, R. *La Zona Costera del Estado de Veracruz: Clima Marítimo, Medio Físico y Medio Biótico*; Instituto de Ecología, A.C.: Xalapa, Mexico, 2018.
39. De la Lanza Espino, G.; Pérez, M.A.O.; Pérez, J.L.C. Diferenciación hidrogeomorfológica de los ambientes costeros del Pacífico, del Golfo de México y del Mar Caribe. *Investig. Geográficas Boletín Inst. Geogr.* **2013**, *2013*, 33–50. [CrossRef]
40. Aguayo-Camargo, J. Neotectónica y facies sedimentarias cuaternarias en el suroeste del Golfo de México, dentro del marco tectono-estratigráfico regional evolutivo del Sur de México. *Ing. Investig. Tecnol.* **2005**, *6*, 19–45. [CrossRef]
41. Mukherjee, N.; Sutherland, W.J.; Khan, M.N.I.; Berger, U.; Schmitz, N.; Dahdouh-Guebas, F.; Koedam, N. Using expert knowledge and modeling to define mangrove composition, functioning, and threats and estimate time frame for recovery. *Ecol. Evol.* **2014**, *4*, 2247–2262. [CrossRef] [PubMed]
42. Moreno-Casasola, P.; Cejudo-Espinosa, E.; Capistrán-Barradas, A.; Infante-Mata, D.; López-Rosas, H.; Castillo-Campos, G.; Pale-Pale, J.; Campos-Cascaredo, A. Composición florística, diversidad y ecología de humedales herbáceos emergentes en la planicie costera central de Veracruz, México. *Bol. Soc. Bot. Méx* **2010**, *87*, 29–50. [CrossRef]
43. Portillo, J.L.; Ezcurra, E. Los manglares de México: Una revisión. *Madera Bosques* **2002**, *8*, 27–51. [CrossRef]
44. Cortés López, L. Desarrollo de un Índice de Riesgo Sobre Opresión Costera en el Centro-Norte del Estado de Veracruz. Master's Thesis, Universidad Nacional Autónoma de México, Ciudad de México, Mexico, 2017.
45. Masselink, G.; Short, A.D. The effect of tide range on beach morphodynamics and morphology: A conceptual beach model. *J. Coast. Res.* **1993**, *9*, 785–800.
46. Chávez, V. *Balance Hidrodinámico en Humedales Costeros y su Valor Como Elemento de Protección Litoral*; Universidad Nacional Autónoma de México: Mexico City, Mexico, 2018.
47. Conjunto de Datos Vectoriales de Uso del Suelo y Vegetación E14-3 (Veracruz) escala 1:250 000 serie V (Conjunto Nacional). Ver. Available online: https://datos.gob.mx/busca/dataset/mapas-de-uso-del-suelo-y-vegetacion-escala-1-250-000-serie-v-veracruz-de-ignacio-de-la-llave/resource/6ecb85a1-3f86-46fd-b37b-dd19f2cb5429?inner_span=True (accessed on 29 April 2021).
48. Doody, J.P. *Sand Dune Conservation, Management and Restoration*; Springer Science & Business Media: Berlin, Germany; Department of Geosciences, Florida Atlantic University: Boca Raton, FL, USA, 2012; Volume 4.
49. Martínez, M.L.; Silva, R.; López-Portillo, J.; Feagin, R.A.; Martínez, E. Coastal Ecosystems as an Ecological Membrane. *J. Coast. Res.* **2020**, *95*, 97–101. [CrossRef]
50. Martínez, M.L.; Vázquez, G.; Sánchez Colón, S. Spatial and temporal variability during primary succession on tropical coastal sand dunes. *J. Veg. Sci.* **2001**, *12*, 361–372. [CrossRef]
51. Moreno-Casasola, P.; Salinas, M.; Amador, L.E.; Cruz, H.; Juárez, A.; Ruelas, L. Plan de manejo comunitario La Mancha-El Llano. En busca de un desarrollo costero sustentable. In *Estrategias Para el Manejo INTEGRAl de la Zona Costera: Un Enfoque Municipal*; Moreno-Casasola, P., Peresbarbosa, E., Travieso-Bello, A.C., Eds.; Instituto de Ecología A.C.—Comisión Nacional de Áreas Naturales Protegidas (SEMARNAT)—Gobierno del Estado de Veracruz: Veracruz, Mexico, 2008; Volume 1, pp. 121–149.
52. Moreno-Casasola, P.; Paradowska, K.; Sada, S.G.; Salinas, G. Los conflictos de la conservación: El caso de La Mancha. In *Hacia una Cultura de Conservación de la Diversidad Biológica*; Halffter, G., Guevara, S., Melci, A., Eds.; Monografías Tercer Mileno: Zaragoza, España, 2007; Volume 6, pp. 225–236.
53. Gómez Ramírez, M. El estudio de los ciclones tropicales se minimizó, en la manifestación del impacto ambiental para el proyecto Diada La Mancha, en la costa del municipio de Actopan, Estado de Veracruz. *DELOS* **2020**, *12*, 26.
54. Mendoza-González, G.; Martínez, M.L.; Lithgow, D.; Pérez-Maqueo, O.; Simonin, P. Land use change and its effects on the value of ecosystem services along the coast of the Gulf of Mexico. *Ecol. Econ.* **2012**, *82*, 23–32. [CrossRef]
55. Caviedes, V.; Arenas-Granados, P.; Barragán-Muñoz, J.M. Regional public policy for Integrated Coastal Zone Management in Central America. *Ocean. Coast. Manag.* **2020**, *186*, 105114. [CrossRef]
56. Barragán Muñoz, J.M. *Manejo Costero Integrado en Iberoamérica: Diagnóstico y Propuestas Para una Nueva Política Pública*; Red IBEMAR (CYTED): Cádiz, España, 2012. [CrossRef]
57. Acuerdo Mediante el Cual se Expide la Política Nacional de Mares y Costas de México. Available online: https://www.dof.gob.mx/nota_detalle.php?codigo=5545511&fecha=30/11/2018 (accessed on 29 June 2021).
58. Diputados, H.C.D. Plan Nacional de Desarrollo 2019–2024. Available online: http://www.diputados.gob.mx/LeyesBiblio/compila/pnd.htm (accessed on 29 June 2021).
59. Programa Nacional Forestal 2020–2024. Available online: https://www.dof.gob.mx/nota_detalle.php?codigo=5609275&fecha=31/12/2020 (accessed on 29 June 2021).
60. Moreno-Casasola, P.; Rojas, E.P.; Travieso-Bello, A.C. *Estrategia Para el Manejo Costero Integral: Secc. IV–V*; Instituto de Ecología: Xalapa, Mexico, 2006; Volume 2.

61. Rivera Arriaga, E.; Villalobos, G.; Azuz Adeath, I.; Rosado May, F. (Eds.) *El Manejo Costero en México*; Universidad Autónoma de Campeche, SEMARNAT, CETYS-Universidad, Universidad de Quintana Roo: Campeche, Mexico, 2004.
62. Moreno Casasola, P. *Entornos Veracruzanos. La costa de La Mancha*; Instituto de Ecología: Xalapa, Mexico, 2006; ISBN 9707090677.
63. La Mancha en Movimiento: Capacitación. Available online: https://www.ecoturismolamancha.com/capacitaci%C3%B3n (accessed on 12 March 2021).
64. Moreno-Casasola, P.; Salinas, G. Programa de desarrollo comunitario sustentable y plan de manejo para la protección y conservación del Sitio Ramsar La Mancha-El Llano. In *Hacia una Cultura de Conservación de la Diversidad Biológica*; Halffter, G., Guevara, S., Melci, A., Eds.; Monografías Tercer Mileno: Zaragoza, España, 2007; Volume 6, pp. 173–185.
65. Moreno-Casasola, P.; Maarel, S.E.; Castillo, M.L.H.; Pisanty, I. Ecología de la vegetación de dunas costeras: Estructura y composición en el Morro de La Mancha. *Biotica* **1982**, *7*, 491–526.
66. Barreiro-Güemes, T.; Balderas-Cortés, J. Evaluación de algunas comunidades de productores primarios de la Laguna de la Mancha, Veracruz. *An. Inst. Cienc. del Mar y Limnol. Univ. Nal. Autón. México* **1991**, *2*, 229–245.
67. Barreiro-Güemes, M.; Matus, J. Diagnóstico ecológico y de uso de recursos de la laguna de La Mancha, Veracruz: Propuesta para su manejo. In Proceedings of the V Congreso Latinoamericano de Ciencias del Mar, La Paz, Mexico, 27 September–1 October 1993; p. 186.
68. Villalobos, F.; Ortiz-Pulido, R.; Moreno, C.; Pavón-Hernández, N.; Hernández-Trejo, H.; Bello, J.; Montiel, S. Patrones de la macrofauna edáfica en un cultivo de Zea maiz durante la fase postcosecha en" La Mancha", Veracruz, México. *Acta Zool. Mex.* **2000**, *80*, 167–183.
69. Cázares Hernández, E. *Monitoreo de Poblaciones de Tortugas Dulceacuícolas Como Parte del Proceso de Restauración de un Humedal del Sitio Ramsar la Mancha y el Llano*; Universidad Veracruzana, Facultad de Ingeniería Química, Región Xalapa: Xalapa, México, 2015.
70. Botello, A. Polycyclic aromatic hydrocarbons in sediments from coastal lagoons of Veracruz State, Gulf of Mexico. *Bull. Environ. Contam. Toxicol.* **2001**, *67*, 889–897. [CrossRef]
71. Mata, D.I., Moreno-Casasola, P.; Madero-Vega, C.; Castillo Campos, G.; Warner, B.G. Floristic composition and soil characteristics of tropical freshwater forested wetlands of Veracruz on the coastal plain of the Gulf of Mexico. *For. Ecol. Manag.* **2011**, *262*, 1514–1531. [CrossRef]
72. Ruiz, M.; López-Portillo, J. Variación espacio-temporal de la comunidad de macroinvertebrados epibiontes en las raíces del mangle rojo Rhizophora mangle (Rhizophoraceae) en la laguna costera de La Mancha, Veracruz, México. *Rev. Biol. Trop.* **2014**, *62*, 1309–1330. [CrossRef] [PubMed]
73. Ramírez-Méndez, E.; Mendoza, E.; Silva-Casarín, R. Estudio de la dinámica y calidad del agua en Laguna la Mancha Veracruz. In Proceedings of the IX Congreso "Los Puertos Mexicanos y su Conectividad, Veracruz, Mexico, 15–17 April 2015.

MDPI
St. Alban-Anlage 66
4052 Basel
Switzerland
Tel. +41 61 683 77 34
Fax +41 61 302 89 18
www.mdpi.com

Geosciences Editorial Office
E-mail: geosciences@mdpi.com
www.mdpi.com/journal/geosciences

www.ingramcontent.com/pod-product-compliance
Lightning Source LLC
LaVergne TN
LVHW070127170726
843515LV00007B/1763

* 9 7 8 3 0 3 6 5 7 4 3 8 7 *